W9-DGV-948

Means Residential Repair & Remodeling Costs

Contractor's Pricing Guide 2007

Senior Editor
Robert W. Mewis, CCC

Contributing Editors
Christopher Babbit
Ted Baker
Barbara Balboni
Robert A. Bastoni
John H. Chiang, PE
Gary W. Christensen
Cheryl Elsmore
Robert J. Kuchta
Robert C. McNichols
Melville J. Mossman, PE
Jeannene D. Murphy
Stephen C. Plotner
Eugene R. Spencer
Marshall J. Stetson
Phillip R. Waier, PE

*Senior Engineering
Operations Manager*
John H. Ferguson, PE

*Senior Vice President
& General Manager*
John Ware

*Vice President
of Direct Response*
John M. Shea

*Director of
Product Development*
Thomas J. Dion

Production Manager
Michael Kokernak

Production Coordinator
Wayne D. Anderson

Technical Support
Jonathan Forgit
Mary Lou Geary
Jill Goodman
Gary L. Hoitt
Genevieve Medeiros
Paula Reale-Camelio
Kathryn S. Rodriguez
Sheryl A. Rose

Book & Cover Design
Norman R. Forgit

RSMeans

Means
Residential
Repair & Remodeling
Costs

Contractor's
Pricing
Guide 2007

- Quick Costs to
  Help You Estimate

- New, Easy-to-Use
  Format . . .
  Organized the
  Way You Build

- Hundreds of
  Work Items with
  Material, Labor &
  Equipment

$39.95 per copy (in United States).
Price subject to change without prior notice.

Copyright © 2006

RSMeans
Construction Publishers & Consultants
63 Smiths Lane
Kingston, MA 02364-0800
(781) 422-5000

Printed in the United States of America.

10 9 8 7 6 5 4 3 2 1

ISSN 1542-9156

ISBN 0-87629-873-0

Foreword

Reed Construction Data's portfolio of project information products and services includes national, regional, and local construction data, project leads, and project plans, specifications, and addenda available online or in print. Reed Bulletin (www.reedbulletin.com) and Reed CONNECT™ (www.reedconnect.com) deliver the most comprehensive, timely, and reliable project information to support contractors, distributers, and building product manufacturers in identifying, bidding, and tracking projects. The Reed First Source (www.reedfirstsource.com) suite of products used by design professionals for the search, selection, and specification of nationally available building projects consists of the First Source™ annual, SPEC-DATA™, MANU-SPEC™, First Source CAD, and manufacturer catalogs. Reed Design Registry, a database of more than 30,000 U.S. architectural firms, is also published by Reed Construction Data. RSMeans (www.rsmeans.com) provides construction cost data, training, and consulting services in print, CD-ROM, and online. Associated Construction Publications (www.reedpubs.com) reports on heavy, highway, and non-residential construction through a network of 14 regional construction magazines. Reed Construction Data (www.reedconstructiondata.com), headquartered in Atlanta, is a subsidiary of Reed Business Information (www.reedbusinessinformation.com), North America's largest business-to-business information provider. With more than 80 market-leading publications and 55 websites, Reed Business Information's wide range of services also includes research, business development, direct marketing lists, training and development programs, and technology solutions. Reed Business Information is a member of the Reed Elsevier plc group (NYSE: RUK and ENL)—a leading provider of global information-driven services and solutions in the science and medical, legal, education, and business-to-business industry sectors.

Our Mission

Since 1942, RSMeans has been actively engaged in construction cost publishing and consulting throughout North America.

Today, over 60 years after RSMeans began, our primary objective remains the same: to provide you, the construction and facilities professional, with the most current and comprehensive construction cost data possible.

Whether you are a contractor, an owner, an architect, an engineer, a facilities manager, or anyone else who needs a reliable construction cost estimate, you'll find this publication to be a highly useful and necessary tool.

Today, with the constant flow of new construction methods and materials, it's difficult to find the time to look at and evaluate all the different construction cost possibilities. In addition, because labor and material costs keep changing, last year's cost information is not a reliable basis for today's estimate or budget.

That's why so many construction professionals turn to RSMeans. We keep track of the costs for you, along with a wide range of other key information, from city cost indexes . . . to productivity rates . . . to crew composition . . . to contractor's overhead and profit rates.

RSMeans performs these functions by collecting data from all facets of the industry and organizing it in a format that is instantly accessible to you. From the preliminary budget to the detailed unit price estimate, you'll find the data in this book useful for all phases of construction cost determination.

The Staff, the Organization, and Our Services

When you purchase one of RSMeans' publications, you are, in effect, hiring the services of a full-time staff of construction and engineering professionals.

Our thoroughly experienced and highly qualified staff works daily at collecting, analyzing, and disseminating comprehensive cost information for your needs. These staff members have years of practical construction experience and engineering training prior to joining the firm. As a result, you can count on them not only for the cost figures, but also for additional background reference information that will help you create a realistic estimate.

The RSMeans organization is always prepared to help you solve construction problems

through its five major divisions: Construction and Cost Data Publishing, Electronic Products and Services, Consulting and Research Services, Insurance Services, and Professional Development Services.

Besides a full array of construction cost estimating books, RSMeans also publishes a number of other reference works for the construction industry. Subjects include construction estimating and project and business management; special topics such as HVAC, roofing, plumbing, and hazardous waste remediation; and a library of facility management references.

In addition, you can access all of our construction cost data electronically using *Means CostWorks*® CD or on the Web.

What's more, you can increase your knowledge and improve your construction estimating and management performance with an RSMeans Construction Seminar or In-House Training Program. These two-day seminar programs offer unparalleled opportunities for everyone in your organization to get updated on a wide variety of construction-related issues.

RSMeans is also a worldwide provider of construction cost management and analysis services for commercial and government owners, and of claims and valuation services for insurers.

In short, RSMeans can provide you with the tools and expertise for constructing accurate and dependable construction estimates and budgets in a variety of ways.

Robert Snow Means Established a Tradition of Quality That Continues Today

Robert Snow Means spent years building RSMeans, making certain he always delivered a quality product.

Today, at RSMeans, we do more than talk about the quality of our data and the usefulness of our books. We stand behind all of our data, from historical cost indexes to construction materials and techniques to current costs.

If you have any questions about our products or services, please call us toll-free at 1-800-334-3509. Our customer service representatives will be happy to assist you. You can also visit our Web site at www.rsmeans.com.

Table of Contents

How the Book is Built: An Overview

A Powerful Construction Tool

You have in your hands one of the most powerful construction tools available today. A successful project is built on the foundation of an accurate and dependable estimate. This book will enable you to construct just such an estimate.

For the casual user the book is designed to be:

- quickly and easily understood so you can get right to your estimate.
- filled with valuable information so you can understand the necessary factors that go into the cost estimate.

For the regular user, the book is designed to be:

- a handy desk reference that can be quickly referred to for key costs.
- a comprehensive, fully reliable source of current construction costs, so you'll be prepared to estimate any project.
- a source book for project cost, product selections, and alternate materials and methods.

To meet all of these requirements we have organized the book into the following clearly defined sections.

Unit Price Section

The cost data has been divided into 23 sections representing the order of construction. Within each section is a listing of components and variations to those components applicable to that section. Costs are shown for the various actions that must be done to that component.

Reference Section

This section includes information on Location Factors and a listing of Abbreviations.

Location Factors: You can adjust total project costs to over 900 locations throughout the U.S. and Canada by using the data in this section.

Abbreviations: A listing of the abbreviations used throughout this book, along with the terms they represent, is included.

Index

A comprehensive listing of all terms and subjects in this book will help you quickly find what you need when you are not sure where it falls in the order of construction.

The Scope of This Book

This book is designed to be as comprehensive and as easy to use as possible. To that end we have made certain assumptions and limited its scope in two key ways:

1. We have established material prices based on a national average.
2. We have computed labor costs based on a seven major-region average of open shop wage rates.

Project Size

This book is intended for use by those involved primarily in Residential Repair and Remodeling construction costing between $10,000-$100,000.

With reasonable exercise of judgment the figures can be used for any building work. For other types of projects, such as new home construction or commercial buildings, consult the appropriate RSMeans publication for more information.

How to Use the Book: The Details

What's Behind the Numbers? The Development of Cost Data

The staff at RSMeans continuously monitors developments in the construction industry in order to ensure reliable, thorough, and up-to-date cost information.

While *overall* construction costs may vary relative to general economic conditions, price fluctuations within the industry are dependent upon many factors. Individual price variations may, in fact, be opposite to overall economic trends. Therefore, costs are continually monitored and complete updates are published yearly. Also, new items are frequently added in response to changes in materials and methods.

Costs—$ (U.S.)

All costs represent U.S. national averages and are given in U.S. dollars. The RSMeans Location Factors can be used to adjust costs to a particular location. The Location Factors for Canada can be used to adjust U.S. national averages to local costs in Canadian dollars. No exchange rate conversion is necessary.

Material Costs

The RSMeans staff contacts manufacturers, dealers, distributors, and contractors all across the U.S. and Canada to determine national average material costs. If you have access to current material costs for your specific location, you may wish to make adjustments to reflect differences from the national average.

Included within material costs are fasteners for a normal installation. RSMeans engineers use manufacturers' recommendations, written specifications, and/or standard construction practice for size and spacing of fasteners. Adjustments to material costs may be required for your specific application or location. Material costs do not include sales tax.

Labor Costs

Labor costs are based on the average of residential wages from across the U.S. for the current year. Rates, along with overhead and profit markups, are listed on the inside back cover of this book.

• If wage rates in your area vary from those used in this book, or if rate increases are expected within a given year, labor costs should be adjusted accordingly.

Labor costs reflect productivity based on actual working conditions. These figures include time spent during a normal workday on tasks other than actual installation, such as material receiving and handling, mobilization at site, site movement, breaks, and cleanup.

Productivity data is developed over an extended period so as not to be influenced by abnormal variations and reflects a typical average.

Equipment Costs

Equipment costs include not only rental costs, but also operating costs for equipment under normal use. Equipment and rental rates are obtained from industry sources throughout North America—contractors, suppliers, dealers, manufacturers, and distributors.

Factors Affecting Costs

Costs can vary depending upon a number of variables. Here's how we have handled the main factors affecting costs.

Quality—The prices for materials and the workmanship upon which productivity is based represent sound construction work. They are also in line with U.S. government specifications.

Overtime—We have made no allowance for overtime. If you anticipate premium time or work beyond normal working hours, be sure to make an appropriate adjustment to your labor costs.

Productivity—The labor costs for each line item are based on working an eight-hour day in daylight hours in moderate temperatures. For work that extends beyond normal work hours or is performed under adverse conditions, productivity may decrease.

Size of Project—The size, scope of work, and type of construction project will have a significant impact on cost. Economies of scale can reduce costs for large projects. Unit costs can often run higher for small projects. Costs in this book are intended for the size and type of project as previously described in "How the Book Is Built: An Overview." Costs for projects of a significantly different size or type should be adjusted accordingly.

Location—Material prices in this book are for metropolitan areas. However, in dense urban areas, traffic and site storage limitations may increase costs. Beyond a 20-mile radius of large cities, extra trucking or transportation charges may also increase the material costs slightly. On the other hand, lower wage rates may be in effect. Be sure to consider both of these factors when preparing an estimate, particularly if the job site is located in a central city or remote rural location.

In addition, highly specialized subcontract items may require travel and per diem expenses for mechanics.

Other Factors—

• season of year
• contractor management
• weather conditions
• building code requirements
• availability of:
  · adequate energy
  · skilled labor
  · building materials
• owner's special requirements/restrictions
• safety requirements
• environmental considerations

General Conditions—The extreme right-hand column of each chart gives the "Totals." These figures contain the installing contractor's O&P. Therefore, it is necessary for a general contractor to add a percentage of all subcontracted items. For a detailed breakdown of O&P see the inside back cover of this book.

Unpredictable Factors—General business conditions influence "in-place" costs of all items. Substitute materials and construction methods may have to be employed. These may affect the installed cost and/or life cycle costs. Such factors may be difficult to evaluate and cannot necessarily be predicted on the basis of the job's location in a particular section of the country. Thus, where these factors apply, you may find significant but unavoidable cost variations for which you will have to apply a measure of judgment to your estimate.

Final Checklist

Estimating can be a straightforward process provided you remember the basics. Here's a checklist of some of the steps you should remember to complete before finalizing your estimate.

Did you remember to . . .

- factor in the Location Factor for your locale?
- take into consideration which items have been marked up and by how much?
- mark up the entire estimate sufficiently for your purposes?
- include all components of your project in the final estimate?
- double check your figures to be sure of your accuracy?
- call RSMeans if you have any questions about your estimate or the data you've found in our publications?

Remember, RSMeans stands behind its publications. If you have any questions about your estimate . . . about the costs you've used from our books . . . or even about the technical aspects of the job that may affect your estimate, feel free to call the RSMeans editors at 1-800-334-3509.

Unit Price Section

Table of Contents

Unit Price Section

Table of Contents (cont.)

How to Use the Unit Price Pages

The prices in this publication include all overhead and profit mark-ups for the installing contractor. If a general contractor is involved, an additional 10% should be added to the cost shown in the "Total" column.

The costs for material, labor and any equipment are shown here. Material costs include delivery and a 10% mark-up. No sales taxes are included. Labor costs include all mark-ups as shown on the inside back cover of this book. Equipment costs also include a 10% mark-up.

The cost in the total column should be multiplied by the quantity of each item to determine the total cost for that item. The sum of all item's total costs becomes the project's total cost.

The book is arranged into 23 categories that follow the order of construction for residential projects. The category is shown as white letters on a black background at the top of each page.

A brief specification of the cost item is shown here.

Graphics representing some of the cost items are found at the beginning of each category.

The cost data is arranged in a tree structure format. Major cost sections are left justified and are shown as bold headings within a grey background. Cost items and sub-cost items are found under the bold headings. The complete description should read as "bold heading, cost item, sub-cost item."
For example, the identified item's description is "Beam/Girder, Solid Wood Beam, 2″ x 8″."

Each action performed to the cost item is conveniently listed to the right of the description. See the explanation below for the scope of work included as part of the action.

The unit of measure upon which all the costs are based is shown here. An abbreviations list is included within the reference section found at the back of this book.

Rough Frame / Structure

Rough Framing

Beam / Girder	Unit	Material	Labor	Equip.	Total	Specification
Solid Wood Beam						
2″ x 8″						
Demolish	L.F.		.68		.68	Cost includes material and labor to
Install	L.F.	.92	1.07		1.99	install 2″ x 8″ beam.
Demolish and Install	L.F.	.92	1.75		2.67	
Clean	L.F.	.10	.23		.33	
Paint	L.F.	.12	1.65		1.77	
Minimum Charge	Job		174		174	
2″ x 10″						
Demolish	L.F.		.85		.85	Cost includes material and labor to
Install	L.F.	1.31	1.16		2.47	install 2″ x 10″ beam.
Demolish and Install	L.F.	1.31	2.01		3.32	
Clean	L.F.	.12	.27		.39	
Paint	L.F.	.19	1.02		1.21	
Minimum Charge	Job		174		174	
2″ x 12″						
Demolish	L.F.		1.03		1.03	Cost includes material and labor to
Install	L.F.	1.57	1.27		2.84	install 2″ x 12″ beam.
Demolish and Install	L.F.	1.57	2.30		3.87	
Clean	L.F.	.14	.32		.46	
Paint	L.F.	.18	2.44		2.62	
Minimum Charge	Job		174		174	
4″ x 8″						
Demolish	L.F.		2.12		2.12	Cost includes material and labor to
Install	L.F.	4.11	1.57	.80	6.48	install 4″ x 8″ beam.
Demolish and Install	L.F.	4.11	3.69	.80	8.60	
Clean	L.F.	.12	.27		.39	
Paint	L.F.	.19	1.02		1.21	
Minimum Charge	Job		174		174	
4″ x 10″						
Demolish	L.F.		2.66		2.66	Cost includes material and labor to
Install	L.F.	6.05	1.65	.84	8.54	install 4″ x 10″ beam.
Demolish and Install	L.F.	6.05	4.31	.84	11.20	
Clean	L.F.	.14	.32		.46	
Paint	L.F.	.21	1.19		1.40	
Minimum Charge	Job		174		174	
4″ x 12″						
Demolish	L.F.		3.19		3.19	Cost includes material and labor to
Install	L.F.	5.60	1.74	.88	8.22	install 4″ x 12″ beam.
Demolish and Install	L.F.	5.60	4.93	.88	11.41	
Clean	L.F.	.17	.36		.53	
Paint	L.F.	.18	2.44		2.62	
Minimum Charge	Job		174		174	

Action	Scope
Demolish	Includes removal of the item and hauling the debris to a truck, dumpster, or storage area.
Install	Includes the installation of the item with normal placement and spacing of fasteners.
Demolish and Install	Includes the demolish action and the install action.
Reinstall	Includes the reinstallation of a removed item.
Clean	Includes cleaning of the item.
Paint	Includes up to two coats of finish on the item (unless otherwise indicated).
Minimum Charge	Includes the minimum labor or the minimum labor and equipment cost. Use this charge for small quantities of material.

Scaffolding	Temporary Electrical Service	Air Compressor

Construction Fees		Unit	Material	Labor	Equip.	Total	Specification
Asbestos Test	Minimum Charge	Job		210		210	Minimum charge for testing.
Architectural	Minimum Charge	Day	730			730	Minimum labor charge for architect.
Engineering	Minimum Charge	Day	256			256	Minimum labor charge for field engineer.

Construction Permits		Unit	Material	Labor	Equip.	Total	Specification
Job (less than 5K)	Minimum Charge	Job	110			110	Permit fees, job less than 5K.
Job (5K to 10K)	Minimum Charge	Job	220			220	Permit fees, job more than 5K, but less than 10K.
Job (10K to 25K)	Minimum Charge	Job	550			550	Permit fees, job more than 10K, but less than 25K.
Job (25K to 50K)	Minimum Charge	Job	1100			1100	Permit fees, job more than 25K, but less than 50K.
Job (50K to 100K)	Minimum Charge	Job	2200			2200	Permit fees, job more than 50K, but less than 100K.
Job (100K to 250K)	Minimum Charge	Job	5500			5500	Permit fees, job more than 100K, but less than 250K.
Job (250K to 500K)	Minimum Charge	Job	11000			11000	Permit fees, job more than 250K, but less than 500K.
Job (500K to 1M)	Minimum Charge	Job	22000			22000	Permit fees, job more than 500K, but less than 1 million.

Job Costs

Trade Labor		Unit	Material	Labor	Equip.	Total	Specification
Carpenter							
Daily (8 hours)							
	Install	Day		350		350	Daily labor rate for one carpenter.
	Minimum Charge	Job		174		174	
Weekly (40 hours)							
	Install	Week		1750		1750	Weekly labor rate for one carpenter.
	Minimum Charge	Job		174		174	
Drywaller							
Daily (8 hours)							
	Install	Day		350		350	Daily labor rate for one carpenter.
	Minimum Charge	Job		174		174	
Weekly (40 hours)							
	Install	Week		1750		1750	Weekly labor rate for one carpenter.
	Minimum Charge	Job		174		174	
Roofer							
Daily (8 hours)							
	Install	Day		320		320	Daily labor rate for one skilled roofer.
	Minimum Charge	Job		161		161	
Weekly (40 hours)							
	Install	Week		1600		1600	Weekly labor rate for one skilled
	Minimum Charge	Job		161		161	roofer.
Painter							
Daily (8 hours)							
	Install	Day		305		305	Daily labor rate for one painter.
	Minimum Charge	Job		153		153	
Weekly (40 hours)							
	Install	Week		1525		1525	Weekly labor rate for one painter.
	Minimum Charge	Job		153		153	
Mason							
Daily (8 hours)							
	Install	Day		355		355	Daily labor rate for one skilled mason.
	Minimum Charge	Job		177		177	
Weekly (40 hours)							
	Install	Week		1775		1775	Weekly labor rate for one skilled
	Minimum Charge	Job		177		177	mason.
Electrician							
Daily (8 hours)							
	Install	Day		385		385	Daily labor rate for an electrician.
	Minimum Charge	Job		192		192	
Weekly (40 hours)							
	Install	Week		1925		1925	Weekly labor rate for an electrician.
	Minimum Charge	Job		192		192	
Plumber							
Daily (8 hours)							
	Install	Day		390		390	Daily labor rate for a plumber.
	Minimum Charge	Job		194		194	
Weekly (40 hours)							
	Install	Week		1950		1950	Weekly labor rate for a plumber.
	Minimum Charge	Job		194		194	
Common Laborer							
Daily (8 hours)							
	Install	Day		254		254	Daily labor rate for a common
	Minimum Charge	Job		127		127	laborer.

Job Costs

Trade Labor

Trade Labor		Unit	Material	Labor	Equip.	Total	Specification
Weekly (40 hours)							
	Install	Week		1275		1275	Weekly labor rate for a common
	Minimum Charge	Job		127		127	laborer.

Temporary Utilities

Temporary Utilities		Unit	Material	Labor	Equip.	Total	Specification
Power Pole							
Job Site Electricity							
	Install	Month	40.50			40.50	Includes average cost of electricity per
	Minimum Charge	Job		192		192	month.
Water Service							
Job Site Water							
	Install	Month	68			68	Includes average costs for water for
	Minimum Charge	Month	68			68	one month.
Telephone							
	Install	Month	231			231	Includes average cost for local and
	Minimum Charge	Month	231			231	long distance telephone service for
							one month.
Chemical Toilet							
	Install	Month			262	262	Monthly rental of portable toilet.
	Minimum Charge	Day			13.10	13.10	
Office Trailer							
Place and Remove							
	Install	Job	298			298	Includes cost to a contractor to deliver,
	Minimum Charge	Job		141		141	place and level an office trailer and
							remove at end of job.
Storage Container							
	Install	Job	193			193	Includes cost to a contractor to deliver
	Minimum Charge	Job		141		141	and place a storage container and
							remove at end of job.
Fencing							
	Install	L.F.	2.92			2.92	Includes material costs for temporary
	Minimum Charge	Job	275			275	fencing.
Railing							
	Install	L.F.	13.35	10	.72	24.07	Includes material and labor to install
	Minimum Charge	Job	275			275	primed steel pipe.

Equipment Rental

Equipment Rental		Unit	Material	Labor	Equip.	Total	Specification
Flatbed Truck							
Daily Rental							
	Install	Day			156	156	Daily rental of flatbed truck.
	Minimum Charge	Job			156	156	
Weekly Rental							
	Install	Week			780	780	Weekly rental of flatbed truck.
	Minimum Charge	Job			156	156	
Monthly Rental							
	Install	Month			3100	3100	Monthly rental of flatbed truck.
	Minimum Charge	Job			156	156	

Job Costs

Equipment Rental		Unit	Material	Labor	Equip.	Total	Specification
Dump Truck							
Daily Rental							
	Install	Day			405	405	Daily rental of dump truck.
	Minimum Charge	Day			405	405	
Weekly Rental							
	Install	Week			2025	2025	Weekly rental of dump truck.
	Minimum Charge	Day			405	405	
Monthly Rental							
	Install	Month			8100	8100	Monthly rental of dump truck.
	Minimum Charge	Day			405	405	
Forklift							
Daily Rental							
	Install	Day			295	295	Daily rental of forklift.
	Minimum Charge	Day			295	295	
Weekly Rental							
	Install	Week			1475	1475	Weekly rental of forklift.
	Minimum Charge	Day			295	295	
Monthly Rental							
	Install	Month			5900	5900	Monthly rental of forklift.
	Minimum Charge	Day			295	295	
Bobcat							
Daily Rental							
	Install	Day			134	134	Daily rental of wheeled, skid steer
	Minimum Charge	Day			134	134	loader.
Weekly Rental							
	Install	Week			670	670	Weekly rental of wheeled, skid steer
	Minimum Charge	Day			134	134	loader.
Monthly Rental							
	Install	Month			2675	2675	Monthly rental of wheeled, skid steer
	Minimum Charge	Day			134	134	loader.
Air Compressor							
Daily Rental							
	Install	Day			120	120	Daily rental of air compressor.
	Minimum Charge	Day			120	120	
Weekly Rental							
	Install	Week			600	600	Weekly rental of air compressor.
	Minimum Charge	Day			120	120	
Monthly Rental							
	Install	Month			2400	2400	Monthly rental of air compressor.
	Minimum Charge	Day			120	120	
Jackhammer							
Daily Rental							
	Install	Day			9.30	9.30	Daily rental of air tool, jackhammer.
	Minimum Charge	Day			9.30	9.30	
Weekly Rental							
	Install	Week			46.50	46.50	Weekly rental of air tool, jackhammer.
	Minimum Charge	Day			9.30	9.30	
Monthly Rental							
	Install	Month			186	186	Monthly rental of air tool,
	Minimum Charge	Day			9.30	9.30	jackhammer.
Concrete Mixer							
Daily Rental							
	Install	Day			125	125	Daily rental of concrete mixer.
	Minimum Charge	Day			125	125	

Job Costs

Equipment Rental		Unit	Material	Labor	Equip.	Total	Specification
Weekly Rental							
	Install	Week			625	625	Weekly rental of concrete mixer.
	Minimum Charge	Day			125	125	
Monthly Rental							
	Install	Month			2500	2500	Monthly rental of concrete mixer.
	Minimum Charge	Day			125	125	
Concrete Bucket							
Daily Rental							
	Install	Day			16.70	16.70	Daily rental of concrete bucket.
	Minimum Charge	Day			16.70	16.70	
Weekly Rental							
	Install	Week			83.50	83.50	Weekly rental of concrete bucket.
	Minimum Charge	Day			16.70	16.70	
Monthly Rental							
	Install	Month			335	335	Monthly rental of concrete bucket.
	Minimum Charge	Day			16.70	16.70	
Concrete Pump							
Daily Rental							
	Install	Day			780	780	Daily rental of concrete pump.
	Minimum Charge	Day			780	780	
Weekly Rental							
	Install	Week			3900	3900	Weekly rental of concrete pump.
	Minimum Charge	Day			780	780	
Monthly Rental							
	Install	Month			15600	15600	Monthly rental of concrete pump.
	Minimum Charge	Day			780	780	
Generator							
Daily Rental							
	Install	Day			29.50	29.50	Daily rental of generator.
	Minimum Charge	Day			29.50	29.50	
Weekly Rental							
	Install	Week			148	148	Weekly rental of generator.
	Minimum Charge	Day			29.50	29.50	
Monthly Rental							
	Install	Month			590	590	Monthly rental of generator.
	Minimum Charge	Day			29.50	29.50	
Welder							
Daily Rental							
	Install	Day			50.50	50.50	Daily rental of welder.
	Minimum Charge	Day			50.50	50.50	
Weekly Rental							
	Install	Week			253	253	Weekly rental of welder.
	Minimum Charge	Day			50.50	50.50	
Monthly Rental							
	Install	Month			1000	1000	Monthly rental of welder.
	Minimum Charge	Day			50.50	50.50	
Sandblaster							
Daily Rental							
	Install	Day			17.20	17.20	Daily rental of sandblaster.
	Minimum Charge	Day			17.20	17.20	
Weekly Rental							
	Install	Week			86	86	Weekly rental of sandblaster.
	Minimum Charge	Day			17.20	17.20	

Job Costs

Equipment Rental

Equipment Rental		Unit	Material	Labor	Equip.	Total	Specification
Monthly Rental							
	Install	Month			345	345	Monthly rental of sandblaster.
	Minimum Charge	Day			17.20	17.20	
Space Heater							
Daily Rental							
	Install	Day			19.45	19.45	Daily rental of space heater.
	Minimum Charge	Day			18.55	18.55	
Weekly Rental							
	Install	Day			97	97	Weekly rental of space heater.
	Minimum Charge	Day			18.55	18.55	
Monthly Rental							
	Install	Day			390	390	Monthly rental of space heater.
	Minimum Charge	Day			18.55	18.55	

Move / Reset Contents

Move / Reset Contents		Unit	Material	Labor	Equip.	Total	Specification
Small Room							
	Install	Room		31		31	Includes cost to remove and reset the
	Minimum Charge	Job		127		127	contents of a small room.
Average Room							
	Install	Room		37.50		37.50	Includes cost to remove and reset the
	Minimum Charge	Job		127		127	contents of an average room.
Large Room							
	Install	Room		51.50		51.50	Includes cost to remove and reset the
	Minimum Charge	Job		127		127	contents of a large room.
Extra Large Room							
	Install	Room		81.50		81.50	Includes cost to remove and reset the
	Minimum Charge	Job		127		127	contents of a extra large room.
Store Contents							
Weekly							
	Install	Week	171			171	Storage fee, per week.
Monthly							
	Install	Month	510			510	Storage fee, per month.

Cover and Protect

Cover and Protect		Unit	Material	Labor	Equip.	Total	Specification
Cover / Protect Walls							
	Install	SF Wall		.10		.10	Includes labor costs to cover and
	Minimum Charge	Job		127		127	protect walls.
Cover / Protect Floors							
	Install	SF Flr.		.10		.10	Includes labor costs to cover and
	Minimum Charge	Job		127		127	protect floors.

Temporary Bracing

Temporary Bracing		Unit	Material	Labor	Equip.	Total	Specification
Shim Support Piers							
	Install	Ea.	3.76	101		104.76	Includes labor and material to install
	Minimum Charge	Job		300		300	shims support for piers.

Job Costs

Jack and Relevel

		Unit	Material	Labor	Equip.	Total	Specification
Single Story							
	Install	S.F.	1.43	1.21		2.64	Includes labor and equipment to jack
	Minimum Charge	Job		300		300	and relevel single-story building.
Two-story							
	Install	S.F.	1.43	2.01		3.44	Includes labor and equipment to jack
	Minimum Charge	Job		300		300	and relevel two-story building.

Dumpster Rental

		Unit	Material	Labor	Equip.	Total	Specification
Small Load							
	Install	Week			475	475	Dumpster, weekly rental, 2 dumps, 5
	Minimum Charge	Week			475	475	yard container.
Medium Load							
	Install	Week			570	570	Dumpster, weekly rental, 2 dumps, 10
	Minimum Charge	Week			570	570	yard container.
Large Load							
	Install	Week			985	985	Dumpster, weekly rental, 2 dumps, 30
	Minimum Charge	Week			985	985	yard container.

Debris Hauling

		Unit	Material	Labor	Equip.	Total	Specification
Per Ton							
	Install	Ton		79.50	43	122.50	Includes hauling of debris by trailer or
	Minimum Charge	Ton		595	325	920	dump truck.
Per Cubic Yard							
	Install	C.Y.		26.50	14.40	40.90	Includes hauling of debris by trailer or
	Minimum Charge	Job		595	325	920	dump truck.
Per Pick-up Truck Load							
	Install	Ea.		141		141	Includes hauling of debris by pick-up
	Minimum Charge	Ea.		141		141	truck.

Dump Fees

		Unit	Material	Labor	Equip.	Total	Specification
Per Cubic Yard							
	Install	C.Y.	7.45			7.45	Dump charges, building materials.
	Minimum Charge	C.Y.	7.45			7.45	
Per Ton							
	Install	Ton	105			105	Dump charges, building materials.
	Minimum Charge	Ton	105			105	

Scaffolding

		Unit	Material	Labor	Equip.	Total	Specification
4' to 6' High							
	Install	Month	67			67	Includes monthly rental of scaffolding,
	Minimum Charge	Month	67			67	steel tubular, 30" wide, 7' long, 5' high.
7' to 11' High							
	Install	Month	80.50			80.50	Includes monthly rental of scaffolding,
	Minimum Charge	Month	80.50			80.50	steel tubular, 30" wide, 7' long, 10' high.

Scaffolding		Unit	Material	Labor	Equip.	Total	Specification
12' to 16' High							
	Install	Month	93.50			93.50	Includes monthly rental of scaffolding, steel tubular, 30" wide, 7' long, 15' high.
	Minimum Charge	Month	93.50			93.50	
17' to 21' High							
	Install	Month	122			122	Includes monthly rental of scaffolding, steel tubular, 30" wide, 7' long, 20' high.
	Minimum Charge	Month	122			122	
22' to 26' High							
	Install	Month	135			135	Includes monthly rental of scaffolding, steel tubular, 30" wide, 7' long, 25' high.
	Minimum Charge	Month	135			135	
27' to 30' High							
	Install	Month	149			149	Includes monthly rental of scaffolding, steel tubular, 30" wide, 7' long, 30' high.
	Minimum Charge	Month	149			149	

Construction Clean-up		Unit	Material	Labor	Equip.	Total	Specification
Initial							
	Clean	S.F.		.32		.32	Includes labor for general clean-up.
	Minimum Charge	Job		127		127	
Progressive							
	Clean	S.F.		.16		.16	Includes labor for progressive site/job clean-up on a weekly basis.
	Minimum Charge	Job		127		127	

Building Demolition	Unit	Material	Labor	Equip.	Total	Specification
Wood-framed Building						
1 Story Demolition						
Demolish	S.F.		1.20	1.46	2.66	Includes demolition of 1 floor only
Minimum Charge	Job		1900	2325	4225	light wood-framed building, including hauling of debris. Dump fees not included.
2nd Story Demolition						
Demolish	S.F.		1.22	1.49	2.71	Includes demolition of 2nd floor only
Minimum Charge	Job		1900	2325	4225	light wood-framed building, including hauling of debris. Dump fees not included.
3rd Story Demolition						
Demolish	S.F.		1.59	1.94	3.53	Includes demolition of 3rd floor only
Minimum Charge	Job		1900	2325	4225	light wood-framed building, including hauling of debris. Dump fees not included.
Masonry Building						
Demolish	S.F.		1.70	2.08	3.78	Includes demolition of one story
Minimum Charge	Job		1900	2325	4225	masonry building, including hauling of debris. Dump fees not included.
Concrete Building						
Unreinforced						
Demolish	S.F.		3.49	1.11	4.60	Includes demolition of a one story
Minimum Charge	Job		1900	2325	4225	non-reinforced concrete building, including hauling of debris. Dump fees not included.
Reinforced						
Demolish	S.F.		4	1.27	5.27	Includes demolition of a one story
Minimum Charge	Job		1900	2325	4225	reinforced concrete building, including hauling of debris. Dump fees not included.
Steel Building						
Demolish	S.F.		2.27	2.78	5.05	Includes demolition of a metal
Minimum Charge	Job		1900	2325	4225	building, including hauling of debris. Dump fees not included. No salvage value assumed for scrap metal.

Job Preparation

General Clean-up

General Clean-up		Unit	Material	Labor	Equip.	Total	Specification
Remove Debris							
	Clean	S.F.		.78	.16	.94	Includes removal of debris caused by the loss including contents items and loose debris.
	Minimum Charge	Job		390	78	468	
Muck-out							
	Clean	S.F.		.49		.49	Includes labor to remove 2″ of muck and mud after a flood loss.
	Minimum Charge	Job		127		127	
Deodorize / Disinfect							
	Clean	SF Flr.		.17		.17	Includes deodorizing building with application of mildicide and disinfectant to floors and walls - will vary depending on the extent of deodorization required.
	Minimum Charge	Job		127		127	
Clean Walls							
Heavy							
	Clean	S.F.		.12		.12	Includes labor and material for heavy cleaning (multiple applications) with detergent and solvent.
	Minimum Charge	Job		127		127	
Clean Ceiling							
Light							
	Clean	S.F.		.20		.20	Includes labor and material to clean light to moderate smoke and soot.
	Minimum Charge	Job		127		127	
Heavy							
	Clean	S.F.		.31		.31	Includes labor and material to clean heavy smoke and soot.
	Minimum Charge	Job		127		127	

Flood Clean-up

Flood Clean-up		Unit	Material	Labor	Equip.	Total	Specification
Emergency Service Call							
After Hours / Weekend							
	Minimum Charge	Job		169		169	Includes minimum charges for deflooding clean-up after hours or on weekends.
Grey Water / Sewage							
	Minimum Charge	Job		127		127	Includes minimum charges for deflooding clean-up.
Water Extraction							
Clean (non-grey)							
	Clean	SF Flr.	.02	.51		.53	Includes labor to remove clean uncontaminated water from the loss.
	Minimum Charge	Job		127		127	
Grey (non-solids)							
	Clean	SF Flr.	.02	1.02		1.04	Includes labor to remove non-solid grey water from the loss.
	Minimum Charge	Job		127		127	
Sewage / Solids							
	Clean	SF Flr.	.04	1.27		1.31	Includes labor to remove water from the loss that may contain sewage or solids.
	Minimum Charge	Job		127		127	

Job Preparation

Flood Clean-up		Unit	Material	Labor	Equip.	Total	Specification
Mildicide Walls							
Topographical							
	Clean	SF Wall		.23		.23	Clean-up of deflooding including
	Minimum Charge	Job		127		127	topographical mildicide application.
Injection Treatments							
	Clean	SF Flr.	20.50	3.91		24.41	Clean-up of deflooding including
	Minimum Charge	Job		127		127	mildicide injection.
Carpet Cleaning							
Uncontaminated							
	Clean	SF Flr.		.16		.16	Includes labor to clean carpet.
	Minimum Charge	Job		127		127	
Contaminated							
	Clean	SF Flr.		.25		.25	Includes labor to clean contaminated
	Minimum Charge	Job		127		127	carpet.
Stairway							
	Clean	Ea.	.10	2.99		3.09	Includes labor to clean contaminated
	Minimum Charge	Job		127		127	carpet on a stairway.
Carpet Treatment							
Disinfect / Deodorize							
	Clean	SF Flr.		.16		.16	Includes disinfecting and deodorizing
	Minimum Charge	Job		127		127	of carpet.
Disinfect For Sewage							
	Clean	SF Flr.		.41		.41	Includes labor to disinfect and
	Minimum Charge	Job		127		127	deodorize carpet exposed to sewage.
Mildicide							
	Clean	SF Flr.	.01	.16		.17	Includes labor and materials to apply
	Minimum Charge	Job		127		127	mildicide to carpet.
Thermal Fog Area							
	Clean	SF Flr.	.03	.39		.42	Includes labor and materials to
	Minimum Charge	Job		127		127	perform thermal fogging during
							deflooding clean-up.
Wet Fog / ULV Area							
	Clean	SF Flr.	.02	.22		.24	Includes labor and materials to
	Minimum Charge	Job		127		127	perform wet fogging / ULV during
							deflooding clean-up.
Steam Clean Fixtures							
	Clean	Ea.	.01	25.50		25.51	Includes labor and materials to steam
	Minimum Charge	Job		127		127	clean fixtures during deflooding
							clean-up.
Pressure Wash Sewage							
	Clean	SF Flr.		2.31		2.31	Includes labor to pressure wash
	Minimum Charge	Job		127		127	sewage contamination.
Airmover							
	Minimum Charge	Day			18.55	18.55	Daily rental of air mover.

Job Preparation

Flood Clean-up

Flood Clean-up		Unit	Material	Labor	Equip.	Total	Specification
Dehumidifier							
Medium Size							
	Minimum Charge	Day			38.50	38.50	Daily rental of dehumidifier, medium.
Large Size							
	Minimum Charge	Day			76.50	76.50	Daily rental of dehumidifier, large.
Air Purifier							
HEPA Filter-Bacteria							
	Minimum Charge	Ea.	252			252	HEPA filter for air purifier.

Asbestos Removal

Asbestos Removal		Unit	Material	Labor	Equip.	Total	Specification
Asbestos Analysis							
	Minimum Charge	Job		210		210	Minimum charge for testing.
Asbestos Pipe Insulation							
1/2" to 3/4" Diameter							
	Demolish	L.F.	.32	5.40		5.72	Minimum charge to remove asbestos. Includes full Tyvek suits for workers changed four times per eight-hour shift with respirators, air monitoring and supervision by qualified professionals.
	Minimum Charge	Job		3350		3350	
1" to 3" Diameter							
	Demolish	L.F.	.98	16.80		17.78	Minimum charge to remove asbestos. Includes full Tyvek suits for workers changed four times per eight-hour shift with respirators, air monitoring and supervision by qualified professionals.
	Minimum Charge	Job		3350		3350	
Asbestos-based Siding							
	Demolish	S.F.	2.18	37.50		39.68	Minimum charge to remove asbestos. Includes full Tyvek suits for workers changed four times per eight-hour shift with respirators, air monitoring and supervision by qualified professionals.
	Minimum Charge	Job		3350		3350	
Asbestos-based Plaster							
	Demolish	S.F.	1.67	28.50		30.17	Minimum charge to remove asbestos. Includes full Tyvek suits for workers changed four times per eight-hour shift with respirators, air monitoring and supervision by qualified professionals.
	Minimum Charge	Job		3350		3350	
Encapsulate Asbestos Ceiling							
	Install	S.F.	.30	.16		.46	Includes encapsulation with penetrating sealant sprayed onto ceiling and walls with airless sprayer.
	Minimum Charge	Job		3350		3350	
Scrape Acoustical Ceiling							
	Demolish	S.F.	.07	.95		1.02	Minimum charge to remove asbestos. Includes full Tyvek suits for workers changed four times per eight-hour shift with respirators, air monitoring and supervision by qualified professionals.
	Minimum Charge	Job		3350		3350	

Job Preparation

Asbestos Removal

		Unit	Material	Labor	Equip.	Total	Specification
Seal Ceiling Asbestos							
	Install	S.F.	1.12	3.16		4.28	Includes attaching plastic cover to ceiling area and sealing where asbestos exists, taping all seams, caulk or foam insulation in cracks and negative air system.
	Minimum Charge	Job		3350		3350	
Seal Wall Asbestos							
	Install	S.F.	1.12	2.81		3.93	Includes attaching plastic cover to wall area and sealing where asbestos exists, taping all seams, caulk or foam insulation in cracks and negative air system.
	Minimum Charge	Job		3350		3350	
Seal Floor Asbestos							
	Install	S.F.	1.79	5.60		7.39	Includes attaching plastic cover to floor area and sealing where asbestos exists, taping all seams, caulk or foam insulation in cracks and negative air system.
	Minimum Charge	Job		3350		3350	
Negative Air Vent System							
	Install	Day			44	44	Daily rental of negative air vent system.
	Minimum Charge	Week			219	219	
HEPA Vacuum Cleaner							
	Install	Day			22.50	22.50	Daily rental of HEPA vacuum, 16 gallon.
	Minimum Charge	Week			113	113	
Airless Sprayer							
	Install	Day			49.50	49.50	Daily rental of airless sprayer.
	Minimum Charge	Week			248	248	
Decontamination Unit							
	Install	Day			133	133	Daily rental of decontamination unit.
	Minimum Charge	Week			665	665	
Light Stand							
	Install	Day			10.65	10.65	Daily rental of floodlight w/tripod.
	Minimum Charge	Week			53	53	

Room Demolition

		Unit	Material	Labor	Equip.	Total	Specification
Strip Typical Room							
	Demolish	S.F.		2.54		2.54	Includes 4 labor hours for general demolition.
	Minimum Charge	Job		141		141	
Strip Typical Bathroom							
	Demolish	S.F.		3.39		3.39	Includes 4 labor hours for general demolition.
	Minimum Charge	Job		141		141	
Strip Typical Kitchen							
	Demolish	S.F.		3.26		3.26	Includes 4 labor hours for general demolition.
	Minimum Charge	Job		141		141	
Strip Typical Laundry							
	Demolish	S.F.		2.90		2.90	Includes 4 labor hours for general demolition.
	Minimum Charge	Job		141		141	

Selective Demolition		Unit	Material	Labor	Equip.	Total	Specification
Concrete Slab							
4″ Unreinforced							
	Demolish	S.F.		1.70	.26	1.96	Includes labor and equipment to
	Minimum Charge	Job		720	110	830	remove unreinforced concrete slab
							and haul debris to truck or dumpster.
6″ Unreinforced							
	Demolish	S.F.		2.40	.37	2.77	Includes labor and equipment to
	Minimum Charge	Job		720	110	830	remove unreinforced concrete slab
							and haul debris to truck or dumpster.
8″ Unreinforced							
	Demolish	S.F.		3.95	.60	4.55	Includes labor and equipment to
	Minimum Charge	Job		720	110	830	remove unreinforced concrete slab
							and haul debris to truck or dumpster.
4″ Reinforced							
	Demolish	S.F.		2.14	.33	2.47	Includes labor and equipment to
	Minimum Charge	Job		720	110	830	remove reinforced concrete slab and
							haul debris to truck or dumpster.
6″ Reinforced							
	Demolish	S.F.		3.20	.49	3.69	Includes labor and equipment to
	Minimum Charge	Job		720	110	830	remove reinforced concrete slab and
							haul debris to truck or dumpster.
8″ Reinforced							
	Demolish	S.F.		5.25	.80	6.05	Includes labor and equipment to
	Minimum Charge	Job		720	110	830	remove reinforced concrete slab and
							haul debris to truck or dumpster.
Brick Wall							
4″ Thick							
	Demolish	S.F.		2.36	.36	2.72	Includes labor and equipment to
	Minimum Charge	Job		720	110	830	remove brick wall and haul debris to
							a central location for disposal.
8″ Thick							
	Demolish	S.F.		3.24	.49	3.73	Includes labor and equipment to
	Minimum Charge	Job		720	110	830	remove brick wall and haul debris to
							a central location for disposal.
12″ Thick							
	Demolish	S.F.		4.32	.66	4.98	Includes labor and equipment to
	Minimum Charge	Job		720	110	830	remove brick wall and haul debris to
							a central location for disposal.
Concrete Masonry Wall							
4″ Thick							
	Demolish	S.F.		2.62	.40	3.02	Includes minimum labor and
	Minimum Charge	Job		720	110	830	equipment to remove 4″ block wall
							and haul debris to a central location
							for disposal.
6″ Thick							
	Demolish	S.F.		3.24	.49	3.73	Includes minimum labor and
	Minimum Charge	Job		720	110	830	equipment to remove 6″ block wall
							and haul debris to a central location
							for disposal.

Job Preparation

Selective Demolition		Unit	Material	Labor	Equip.	Total	Specification
8″ Thick							
	Demolish	S.F.		3.99	.61	4.60	Includes minimum labor and
	Minimum Charge	Job		720	110	830	equipment to remove 8″ block wall
							and haul debris to a central location
							for disposal.
12″ Thick							
	Demolish	S.F.		5.75	.88	6.63	Includes minimum labor and
	Minimum Charge	Job		720	110	830	equipment to remove 12″ block wall
							and haul debris to a central location
							for disposal.
Tree Removal							
Small, 4″ - 8″							
	Demolish	Ea.		91.50	66.50	158	Minimum charge, tree removal.
	Minimum Charge	Job		127		127	
Medium, 9″ - 12″							
	Demolish	Ea.		137	99.50	236.50	Minimum charge, tree removal.
	Minimum Charge	Job		127		127	
Large, 13″ - 18″							
	Demolish	Ea.		165	120	285	Minimum charge, tree removal.
	Minimum Charge	Job		420	133	553	
Very Large, 19″ - 24″							
	Demolish	Ea.		206	149	355	Minimum charge, tree removal.
	Minimum Charge	Job		420	133	553	
Stump Removal							
	Demolish	Ea.		57	126	183	Includes minimum charges for a site
	Minimum Charge	Job		234	74.50	308.50	demolition crew and backhoe /
							loader.

Excavation

Backhoe/Loader - Wheel Type	Tractor Loader - Wheel Type	Backhoe - Crawler Type

Excavation		Unit	Material	Labor	Equip.	Total	Specification
Footings							
Hand Trenching							
	Install	L.F.		1.59		1.59	Includes hand excavation for footing
	Minimum Charge	Job		63.50		63.50	or trench up to 10″ deep and 12″ wide.
Machine Trenching							
	Install	L.F.		.60	.27	.87	Includes machine excavation for
	Minimum Charge	Job		300	134	434	footing or trench up to 10″ deep and 12″ wide.
Trenching w / Backfill							
Hand							
	Install	L.F.		2.54		2.54	Includes hand excavation, backfill and
	Minimum Charge	Job		63.50		63.50	compaction.
Machine							
	Install	L.F.		3.02	8.85	11.87	Includes machine excavation, backfill
	Minimum Charge	Job		127	28	155	and compaction.

Backfill		Unit	Material	Labor	Equip.	Total	Specification
Hand (no compaction)							
	Install	L.C.Y.		21		21	Includes backfill by hand in average
	Minimum Charge	Job		127		127	soil without compaction.
Hand (w / compaction)							
	Install	E.C.Y.		30		30	Includes backfill by hand in average
	Minimum Charge	Job		127		127	soil with compaction.
Machine (no compaction)							
	Install	L.C.Y.		3.02	1.34	4.36	Includes backfill, by 55 HP wheel
	Minimum Charge	Job		300	134	434	loader, of trenches from loose material piled adjacent without compaction.
Machine (w / compaction)							
	Install	L.C.Y.		8.05	3.57	11.62	Includes backfill, by 55 HP wheel
	Minimum Charge	Job		300	134	434	loader, of trenches from loose material piled adjacent with compaction by vibrating tampers.

Concrete Footing and Formwork   **Wood Pile**

Piles		Unit	Material	Labor	Equip.	Total	Specification
Treated Wood, 12″ Butt							
To 30′							
	Demolish	L.F.		4.07	1.29	5.36	Includes material, labor and
	Install	V.L.F.	11.15	3.89	2.90	17.94	equipment to install up to 30′ long
	Demolish and Install	V.L.F.	11.15	7.96	4.19	23.30	treated wood piling.
	Paint	L.F.	.41	1.74		2.15	
	Minimum Charge	Job		2425	1825	4250	
30′ to 40′							
	Demolish	L.F.		4.07	1.29	5.36	Includes material, labor and
	Install	V.L.F.	10.95	3.47	2.59	17.01	equipment to install 30′ to 40′ long
	Demolish and Install	V.L.F.	10.95	7.54	3.88	22.37	treated wood piling.
	Paint	L.F.	.41	1.74		2.15	
	Minimum Charge	Job		2425	1825	4250	
40′ to 50′							
	Demolish	L.F.		4.07	1.29	5.36	Includes material, labor and
	Install	V.L.F.	11.70	3.38	2.52	17.60	equipment to install 40′ to 50′ long
	Demolish and Install	V.L.F.	11.70	7.45	3.81	22.96	treated wood piling.
	Paint	L.F.	.41	1.74		2.15	
	Minimum Charge	Job		2425	1825	4250	
50′ to 60′							
	Demolish	L.F.		4.07	1.29	5.36	Includes material, labor and
	Install	V.L.F.	12.95	3.04	2.27	18.26	equipment to install 50′ to 60′ long
	Demolish and Install	V.L.F.	12.95	7.11	3.56	23.62	treated wood piling.
	Paint	L.F.	.41	1.74		2.15	
	Minimum Charge	Job		2425	1825	4250	
60′ to 80′							
	Demolish	L.F.		4.07	1.29	5.36	Includes material, labor and
	Install	V.L.F.	22	3.39	2.89	28.28	equipment to install 60′ to 80′ long
	Demolish and Install	V.L.F.	22	7.46	4.18	33.64	treated wood piling.
	Paint	L.F.	.41	1.74		2.15	
	Minimum Charge	Job		2425	1825	4250	
Untreated Wood, 12″ Butt							
To 30′							
	Demolish	L.F.		4.07	1.29	5.36	Includes material, labor and
	Install	V.L.F.	7.60	3.89	2.90	14.39	equipment to install up to 30′ long
	Demolish and Install	V.L.F.	7.60	7.96	4.19	19.75	untreated wood piling.
	Paint	L.F.	.41	1.74		2.15	
	Minimum Charge	Job		2425	1825	4250	

Foundation

Piles

	Unit	Material	Labor	Equip.	Total	Specification
30' to 40'						
Demolish	L.F.		4.07	1.29	5.36	Includes material, labor and
Install	V.L.F.	7.60	3.47	2.59	13.66	equipment to install 30' to 40' long
Demolish and Install	V.L.F.	7.60	7.54	3.88	19.02	untreated wood piling.
Paint	L.F.	.41	1.74		2.15	
Minimum Charge	Job		2425	1825	4250	
40' to 50'						
Demolish	L.F.		4.07	1.29	5.36	Includes material, labor and
Install	V.L.F.	7.60	3.38	2.52	13.50	equipment to install 40' to 50' long
Demolish and Install	V.L.F.	7.60	7.45	3.81	18.86	untreated wood piling.
Paint	L.F.	.41	1.74		2.15	
Minimum Charge	Job		2425	1825	4250	
50' to 60'						
Demolish	L.F.		4.07	1.29	5.36	Includes material, labor and
Install	V.L.F.	7.65	3.04	2.27	12.96	equipment to install 50' to 60' long
Demolish and Install	V.L.F.	7.65	7.11	3.56	18.32	untreated wood piling.
Paint	L.F.	.41	1.74		2.15	
Minimum Charge	Job		2425	1825	4250	
60' to 80'						
Demolish	L.F.		4.07	1.29	5.36	Includes material, labor and
Install	V.L.F.	9.55	2.90	2.16	14.61	equipment to install 60' to 80' long
Demolish and Install	V.L.F.	9.55	6.97	3.45	19.97	untreated wood piling.
Paint	L.F.	.41	1.74		2.15	
Minimum Charge	Job		2425	1825	4250	

Precast Concrete

	Unit	Material	Labor	Equip.	Total	Specification
10" Square						
Demolish	L.F.		4.16	1.32	5.48	Includes material, labor and
Install	V.L.F.	11.60	3.47	2.59	17.66	equipment to install a precast,
Demolish and Install	V.L.F.	11.60	7.63	3.91	23.14	prestressed, 40' long, 10" square
Paint	L.F.	.41	1.74		2.15	concrete pile.
Minimum Charge	Job		2425	1825	4250	
12" Square						
Demolish	L.F.		4.46	1.42	5.88	Includes material, labor and
Install	V.L.F.	14.60	3.58	2.67	20.85	equipment to install a precast,
Demolish and Install	V.L.F.	14.60	8.04	4.09	26.73	prestressed, 40' long, 12" square
Paint	L.F.	.41	1.74		2.15	concrete pile.
Minimum Charge	Job		2425	1825	4250	
14" Square						
Demolish	L.F.		4.57	1.45	6.02	Includes material, labor and
Install	V.L.F.	17.25	4.05	3.02	24.32	equipment to install a precast,
Demolish and Install	V.L.F.	17.25	8.62	4.47	30.34	prestressed, 40' long, 14" square
Paint	L.F.	.41	1.74		2.15	concrete pile.
Minimum Charge	Job		2425	1825	4250	
16" Square						
Demolish	L.F.		4.80	1.53	6.33	Includes material, labor and
Install	V.L.F.	27.50	4.34	3.24	35.08	equipment to install a precast,
Demolish and Install	V.L.F.	27.50	9.14	4.77	41.41	prestressed, 40' long, 16" square
Paint	L.F.	.41	1.74		2.15	concrete pile.
Minimum Charge	Job		2425	1825	4250	
18" Square						
Demolish	L.F.		5.20	1.65	6.85	Includes material, labor and
Install	V.L.F.	33	5.50	4.67	43.17	equipment to install a precast,
Demolish and Install	V.L.F.	33	10.70	6.32	50.02	prestressed, 40' long, 18" square
Paint	L.F.	.41	1.74		2.15	concrete pile.
Minimum Charge	Job		2425	1825	4250	

Foundation

Piles	Unit	Material	Labor	Equip.	Total	Specification
Cross Bracing						
Treated 2″ x 6″						
Demolish	L.F.		.83		.83	Cost includes material and labor to
Install	L.F.	.91	4.86		5.77	install 2″ x 6″ lumber for cross-bracing
Demolish and Install	L.F.	.91	5.69		6.60	of foundation piling or posts.
Reinstall	L.F.		3.89		3.89	
Clean	S.F.		.32		.32	
Paint	L.F.	.18	1.17		1.35	
Minimum Charge	Job		605		605	
Treated 2″ x 8″						
Demolish	L.F.		.83		.83	Cost includes material and labor to
Install	L.F.	1.27	5.05		6.32	install 2″ x 8″ lumber for cross-bracing
Demolish and Install	L.F.	1.27	5.88		7.15	of foundation piling or posts.
Reinstall	L.F.		4.02		4.02	
Clean	S.F.		.32		.32	
Paint	L.F.	.23	1.22		1.45	
Minimum Charge	Job		605		605	
Treated 2″ x 10″						
Demolish	L.F.		.86		.86	Cost includes material and labor to
Install	L.F.	1.72	5.20		6.92	install 2″ x 10″ lumber for
Demolish and Install	L.F.	1.72	6.06		7.78	cross-bracing of foundation piling or
Reinstall	L.F.		4.16		4.16	posts.
Clean	S.F.		.32		.32	
Paint	L.F.	.28	1.25		1.53	
Minimum Charge	Job		605		605	
Treated 2″ x 12″						
Demolish	L.F.		1.03		1.03	Cost includes material and labor to
Install	L.F.	2.04	5.40		7.44	install 2″ x 12″ lumber for
Demolish and Install	L.F.	2.04	6.43		8.47	cross-bracing of foundation piling or
Reinstall	L.F.		4.31		4.31	posts.
Clean	S.F.		.32		.32	
Paint	L.F.	.32	1.27		1.59	
Minimum Charge	Job		605		605	
Metal Pipe						
6″ Concrete Filled						
Demolish	L.F.		5.80	1.84	7.64	Includes material, labor and
Install	V.L.F.	19.20	5.05	3.78	28.03	equipment to install concrete filled
Demolish and Install	V.L.F.	19.20	10.85	5.62	35.67	steel pipe pile.
Paint	L.F.	.31	1.74		2.05	
Minimum Charge	Job		2425	1825	4250	
6″ Unfilled						
Demolish	L.F.		5.80	1.84	7.64	Includes material, labor and
Install	V.L.F.	16.35	4.63	3.46	24.44	equipment to install unfilled steel pipe
Demolish and Install	V.L.F.	16.35	10.43	5.30	32.08	pile.
Paint	L.F.	.31	1.74		2.05	
Minimum Charge	Job		2425	1825	4250	
8″ Concrete Filled						
Demolish	L.F.		5.80	1.84	7.64	Includes material, labor and
Install	V.L.F.	17.55	5.30	3.95	26.80	equipment to install concrete filled
Demolish and Install	V.L.F.	17.55	11.10	5.79	34.44	steel pipe pile.
Paint	L.F.	.31	1.74		2.05	
Minimum Charge	Job		2425	1825	4250	
8″ Unfilled						
Demolish	L.F.		5.80	1.84	7.64	Includes material, labor and
Install	V.L.F.	16.45	4.86	3.63	24.94	equipment to install unfilled steel pipe
Demolish and Install	V.L.F.	16.45	10.66	5.47	32.58	pile.
Paint	L.F.	.31	1.74		2.05	
Minimum Charge	Job		2425	1825	4250	

Foundation

Piles	Unit	Material	Labor	Equip.	Total	Specification
10″ Concrete Filled						
Demolish	L.F.		5.95	1.90	7.85	Includes material, labor and
Install	V.L.F.	22.50	5.40	4.03	31.93	equipment to install concrete filled
Demolish and Install	V.L.F.	22.50	11.35	5.93	39.78	steel pipe pile.
Paint	L.F.	.31	1.74		2.05	
Minimum Charge	Job		2425	1825	4250	
10″ Unfilled						
Demolish	L.F.		5.95	1.90	7.85	Includes material, labor and
Install	V.L.F.	20.50	4.86	3.63	28.99	equipment to install unfilled steel pipe
Demolish and Install	V.L.F.	20.50	10.81	5.53	36.84	pile.
Paint	L.F.	.31	1.74		2.05	
Minimum Charge	Job		2425	1825	4250	
12″ Concrete Filled						
Demolish	L.F.		5.95	1.90	7.85	Includes material, labor and
Install	V.L.F.	27	5.85	4.37	37.22	equipment to install concrete filled
Demolish and Install	V.L.F.	27	11.80	6.27	45.07	steel pipe pile.
Paint	L.F.	.31	1.74		2.05	
Minimum Charge	Job		2425	1825	4250	
12″ Unfilled						
Demolish	L.F.		5.95	1.90	7.85	Includes material, labor and
Install	V.L.F.	25.50	5.10	3.82	34.42	equipment to install unfilled steel pipe
Demolish and Install	V.L.F.	25.50	11.05	5.72	42.27	pile.
Paint	L.F.	.31	1.74		2.05	
Minimum Charge	Job		2425	1825	4250	
Steel H Section						
H Section 8 x 8 x 36#						
Demolish	L.F.		4.04	1.28	5.32	Includes material, labor and
Install	V.L.F.	15.25	3.80	2.84	21.89	equipment to install steel H section
Demolish and Install	V.L.F.	15.25	7.84	4.12	27.21	pile.
Paint	L.F.	.31	1.74		2.05	
Minimum Charge	Job		2425	1825	4250	
H Section 10 x 10 x 57#						
Demolish	L.F.		4.42	1.40	5.82	Includes material, labor and
Install	V.L.F.	24	3.99	2.98	30.97	equipment to install steel H section
Demolish and Install	V.L.F.	24	8.41	4.38	36.79	pile.
Paint	L.F.	.31	1.74		2.05	
Minimum Charge	Job		2425	1825	4250	
H Section 12 x 12 x 74#						
Demolish	L.F.		4.57	1.45	6.02	Includes material, labor and
Install	V.L.F.	32	4.83	4.12	40.95	equipment to install steel H section
Demolish and Install	V.L.F.	32	9.40	5.57	46.97	pile.
Paint	L.F.	.31	1.74		2.05	
Minimum Charge	Job		2425	1825	4250	
H Section 14 x 14 x 89#						
Demolish	L.F.		4.80	1.53	6.33	Includes material, labor and
Install	V.L.F.	38.50	5.25	4.50	48.25	equipment to install steel H section
Demolish and Install	V.L.F.	38.50	10.05	6.03	54.58	pile.
Paint	L.F.	.31	1.74		2.05	
Minimum Charge	Job		2425	1825	4250	

Foundation

Foundation Post		Unit	Material	Labor	Equip.	Total	Specification
Wood Foundation Post							
4″ x 4″							
	Demolish	L.F.		.66	.21	.87	Cost includes material and labor to
	Install	L.F.	1.05	4.23		5.28	install 4″ x 4″ treated wood post by
	Demolish and Install	L.F.	1.05	4.89	.21	6.15	hand in normal soil conditions.
	Reinstall	L.F.		3.39		3.39	
	Minimum Charge	Job		605		605	
4″ x 6″							
	Demolish	L.F.		.66	.21	.87	Cost includes material and labor to
	Install	L.F.	1.56	4.23		5.79	install 4″ x 6″ treated wood post by
	Demolish and Install	L.F.	1.56	4.89	.21	6.66	hand in normal soil conditions.
	Reinstall	L.F.		3.39		3.39	
	Minimum Charge	Job		605		605	
6″ x 6″							
	Demolish	L.F.		.66	.21	.87	Cost includes material and labor to
	Install	L.F.	2.32	4.23		6.55	install 6″ x 6″ treated wood post by
	Demolish and Install	L.F.	2.32	4.89	.21	7.42	hand in normal soil conditions.
	Reinstall	L.F.		3.39		3.39	
	Minimum Charge	Job		605		605	
Water Jet							
	Demolish	Ea.		75.50		75.50	Includes material, labor and
	Install	Ea.	61.50	580		641.50	equipment to water jet a 6″ x 6″ wood
	Demolish and Install	Ea.	61.50	655.50		717	post under an existing building. Cost
	Reinstall	Ea.		580.57		580.57	may vary depending upon location
	Minimum Charge	Job		510		510	and access.
Shim							
	Install	Ea.	3.41	60.50		63.91	Includes labor and material to trim or
	Minimum Charge	Job		605		605	readjust existing foundation posts
							under an existing building. Access of
							2 - 4 foot assumed.
Cross-bracing							
Treated 2″ x 6″							
	Demolish	L.F.		.83		.83	Cost includes material and labor to
	Install	L.F.	.91	4.86		5.77	install 2″ x 6″ lumber for cross-bracing
	Demolish and Install	L.F.	.91	5.69		6.60	of foundation piling or posts.
	Reinstall	L.F.		3.89		3.89	
	Clean	S.F.		.32		.32	
	Paint	L.F.	.18	1.17		1.35	
	Minimum Charge	Job		605		605	
Treated 2″ x 8″							
	Demolish	L.F.		.83		.83	Cost includes material and labor to
	Install	L.F.	1.27	5.05		6.32	install 2″ x 8″ lumber for cross-bracing
	Demolish and Install	L.F.	1.27	5.88		7.15	of foundation piling or posts.
	Reinstall	L.F.		4.02		4.02	
	Clean	S.F.		.32		.32	
	Paint	L.F.	.23	1.22		1.45	
	Minimum Charge	Job		605		605	
Treated 2″ x 10″							
	Demolish	L.F.		.86		.86	Cost includes material and labor to
	Install	L.F.	1.72	5.20		6.92	install 2″ x 10″ lumber for
	Demolish and Install	L.F.	1.72	6.06		7.78	cross-bracing of foundation piling or
	Reinstall	L.F.		4.16		4.16	posts.
	Clean	S.F.		.32		.32	
	Paint	L.F.	.28	1.25		1.53	
	Minimum Charge	Job		605		605	

Foundation

Foundation Post

	Unit	Material	Labor	Equip.	Total	Specification
Treated 2″ x 12″						
Demolish	L.F.		1.03		1.03	Cost includes material and labor to
Install	L.F.	2.04	5.40		7.44	install 2″ x 12″ lumber for
Demolish and Install	L.F.	2.04	6.43		8.47	cross-bracing of foundation piling or
Reinstall	L.F.		4.31		4.31	posts.
Clean	S.F.		.32		.32	
Paint	L.F.	.32	1.27		1.59	
Minimum Charge	Job		605		605	

Concrete Footing

	Unit	Material	Labor	Equip.	Total	Specification
Continuous						
12″ w x 6″ d						
Demolish	L.F.		1.24	.99	2.23	Includes material and labor to install
Install	L.F.	4.24	2.87	.01	7.12	continuous reinforced concrete footing
Demolish and Install	L.F.	4.24	4.11	1	9.35	poured by chute cast against earth
Minimum Charge	Job		700		700	including reinforcing, forming and
						finishing.
12″ w x 12″ d						
Demolish	L.F.		2.38	1.91	4.29	Includes material and labor to install
Install	L.F.	9.10	5.75	.03	14.88	continuous reinforced concrete footing
Demolish and Install	L.F.	9.10	8.13	1.94	19.17	poured by chute cast against earth
Minimum Charge	Job		700		700	including reinforcing, forming and
						finishing.
18″ w x 10″ d						
Demolish	L.F.		3.09	2.48	5.57	Includes material and labor to install
Install	L.F.	8.45	7.05	.04	15.54	continuous reinforced concrete footing
Demolish and Install	L.F.	8.45	10.14	2.52	21.11	poured by chute cast against earth
Minimum Charge	Job		700		700	including reinforcing, forming and
						finishing.
24″ w x 24″ d						
Demolish	L.F.		5.15	4.13	9.28	Includes material and labor to install
Install	L.F.	26	23	.12	49.12	continuous reinforced concrete footing
Demolish and Install	L.F.	26	28.15	4.25	58.40	poured by chute cast against earth
Minimum Charge	Job		700		700	including reinforcing, forming and
						finishing.
Stem Wall						
Single Story						
Demolish	L.F.		11.45	9.20	20.65	Includes material and labor to install
Install	L.F.	22	60.50	48.50	131	stem wall 24″ above and 18″ below
Demolish and Install	L.F.	22	71.95	57.70	151.65	grade for a single story structure.
Minimum Charge	Job		700		700	Includes 12″ wide and 8″ deep
						footing.
Two Story						
Demolish	L.F.		11.45	9.20	20.65	Includes material and labor to install
Install	L.F.	27	87	69.50	183.50	stem wall 24″ above and 18″ below
Demolish and Install	L.F.	27	98.45	78.70	204.15	grade typical for a two story structure,
Minimum Charge	Job		700		700	includes 18″ wide and 10″ deep
						footing.

Concrete Slab		Unit	Material	Labor	Equip.	Total	Specification
Reinforced							
4″							
	Demolish	S.F.		2.14	.33	2.47	Cost includes material and labor to
	Install	S.F.	2.22	.33		2.55	install 4″ reinforced concrete
	Demolish and Install	S.F.	2.22	2.47	.33	5.02	slab-on-grade poured by chute
	Clean	S.F.	.04	.28		.32	including forms, vapor barrier, wire
	Paint	S.F.	.11	.66		.77	mesh, 3000 PSI concrete, float finish,
	Minimum Charge	Job		160		160	and curing.
6″							
	Demolish	S.F.		3.20	.49	3.69	Cost includes material and labor to
	Install	S.F.	3.05	.49		3.54	install 6″ reinforced concrete
	Demolish and Install	S.F.	3.05	3.69	.49	7.23	slab-on-grade poured by chute
	Clean	S.F.	.04	.28		.32	including forms, vapor barrier, wire
	Paint	S.F.	.11	.66		.77	mesh, 3000 PSI concrete, float finish,
	Minimum Charge	Job		160		160	and curing.
8″							
	Demolish	S.F.		5.25	.80	6.05	Cost includes material and labor to
	Install	S.F.	3.93	.65	.01	4.59	install 8″ reinforced concrete
	Demolish and Install	S.F.	3.93	5.90	.81	10.64	slab-on-grade poured by chute
	Clean	S.F.	.04	.28		.32	including forms, vapor barrier, wire
	Paint	S.F.	.11	.66		.77	mesh, 3000 PSI concrete, float finish,
	Minimum Charge	Job		160		160	and curing.
Unreinforced							
4″							
	Demolish	S.F.		1.70	.26	1.96	Cost includes material and labor to
	Install	S.F.	1.51	.79	.01	2.31	install 4″ concrete slab-on-grade
	Demolish and Install	S.F.	1.51	2.49	.27	4.27	poured by chute, including forms,
	Clean	S.F.	.04	.28		.32	vapor barrier, 3000 PSI concrete, float
	Paint	S.F.	.11	.66		.77	finish, and curing.
	Minimum Charge	Job		160		160	
6″							
	Demolish	S.F.		2.40	.37	2.77	Cost includes material and labor to
	Install	S.F.	2.21	.81	.01	3.03	install 6″ concrete slab-on-grade
	Demolish and Install	S.F.	2.21	3.21	.38	5.80	poured by chute, including forms,
	Clean	S.F.	.04	.28		.32	vapor barrier, 3000 PSI concrete, float
	Paint	S.F.	.11	.66		.77	finish, and curing.
	Minimum Charge	Job		160		160	
8″							
	Demolish	S.F.		3.95	.60	4.55	Cost includes material and labor to
	Install	S.F.	3.03	.85	.01	3.89	install 8″ concrete slab-on-grade
	Demolish and Install	S.F.	3.03	4.80	.61	8.44	poured by chute, including forms,
	Clean	S.F.	.04	.28		.32	vapor barrier, 3000 PSI concrete, float
	Paint	S.F.	.11	.66		.77	finish, and curing.
	Minimum Charge	Job		160		160	
Lightweight							
	Demolish	S.F.		.75	.18	.93	Includes material and labor to install
	Install	S.F.	1.89	2.39		4.28	lightweight concrete, 3″ - 4″ thick.
	Demolish and Install	S.F.	1.89	3.14	.18	5.21	
	Clean	S.F.	.04	.28		.32	
	Paint	S.F.	.11	.66		.77	
	Minimum Charge	Job		160		160	

Foundation

Concrete Slab

	Unit	Material	Labor	Equip.	Total	Specification
Gunite						Includes material and labor to install
Demolish	S.F.		.94	.30	1.24	gunite on flat plane per inch of
Install	S.F.	.32	1.02	.39	1.73	thickness using 1" as base price. No
Demolish and Install	S.F.	.32	1.96	.69	2.97	forms or reinforcing are included.
Clean	S.F.	.04	.28		.32	
Paint	S.F.	.11	.66		.77	
Minimum Charge	Job		895		895	
Epoxy Inject						
Install	L.F.	5.80	3.20		9	Includes labor and material to inject
Minimum Charge	Job		160		160	epoxy into cracks.
Pressure Grout						
Install	S.F.	2.29	2.37		4.66	Includes labor, material and
Minimum Charge	Job		160		160	equipment to install pressure grout
						under an existing building.
Scrape and Paint						
Paint	S.F.	.20	.76		.96	Includes labor and materials to scrape
Minimum Charge	Job		153		153	and paint a concrete floor.
Vapor Barrier						
Install	S.F.	.03	.09		.12	Cost includes material and labor to
Minimum Charge	Job		174		174	install 4 mil polyethylene vapor barrier
						10' wide sheets with 6" overlaps.

Rough Framing

Beam / Girder		Unit	Material	Labor	Equip.	Total	Specification
Solid Wood Beam							
2″ x 8″							
	Demolish	L.F.		.68		.68	Cost includes material and labor to
	Install	L.F.	.92	1.07		1.99	install 2″ x 8″ beam.
	Demolish and Install	L.F.	.92	1.75		2.67	
	Clean	L.F.	.10	.23		.33	
	Paint	L.F.	.12	1.65		1.77	
	Minimum Charge	Job		174		174	
2″ x 10″							
	Demolish	L.F.		.85		.85	Cost includes material and labor to
	Install	L.F.	1.31	1.16		2.47	install 2″ x 10″ beam.
	Demolish and Install	L.F.	1.31	2.01		3.32	
	Clean	L.F.	.12	.27		.39	
	Paint	L.F.	.19	1.02		1.21	
	Minimum Charge	Job		174		174	
2″ x 12″							
	Demolish	L.F.		1.03		1.03	Cost includes material and labor to
	Install	L.F.	1.57	1.27		2.84	install 2″ x 12″ beam.
	Demolish and Install	L.F.	1.57	2.30		3.87	
	Clean	L.F.	.14	.32		.46	
	Paint	L.F.	.18	2.44		2.62	
	Minimum Charge	Job		174		174	
4″ x 8″							
	Demolish	L.F.		2.12		2.12	Cost includes material and labor to
	Install	L.F.	4.11	1.57	.80	6.48	install 4″ x 8″ beam.
	Demolish and Install	L.F.	4.11	3.69	.80	8.60	
	Clean	L.F.	.12	.27		.39	
	Paint	L.F.	.19	1.02		1.21	
	Minimum Charge	Job		174		174	
4″ x 10″							
	Demolish	L.F.		2.66		2.66	Cost includes material and labor to
	Install	L.F.	6.05	1.65	.84	8.54	install 4″ x 10″ beam.
	Demolish and Install	L.F.	6.05	4.31	.84	11.20	
	Clean	L.F.	.14	.32		.46	
	Paint	L.F.	.21	1.19		1.40	
	Minimum Charge	Job		174		174	
4″ x 12″							
	Demolish	L.F.		3.19		3.19	Cost includes material and labor to
	Install	L.F.	5.60	1.74	.88	8.22	install 4″ x 12″ beam.
	Demolish and Install	L.F.	5.60	4.93	.88	11.41	
	Clean	L.F.	.17	.36		.53	
	Paint	L.F.	.18	2.44		2.62	
	Minimum Charge	Job		174		174	

Rough Frame / Structure

Beam / Girder		Unit	Material	Labor	Equip.	Total	Specification
6" x 8"							
	Demolish	L.F.		3.19		3.19	Cost includes material and labor to
	Install	L.F.	8.35	2.99	1.52	12.86	install 6" x 8" beam.
	Demolish and Install	L.F.	8.35	6.18	1.52	16.05	
	Clean	L.F.	.14	.32		.46	
	Paint	L.F.	.21	1.19		1.40	
	Minimum Charge	Job		174		174	
6" x 10"							
	Demolish	L.F.		3.99		3.99	Cost includes material and labor to
	Install	L.F.	3.93	1.40		5.33	install 6" x 10" triple beam.
	Demolish and Install	L.F.	3.93	5.39		9.32	
	Clean	L.F.	.17	.36		.53	
	Paint	L.F.	.24	1.36		1.60	
	Minimum Charge	Job		174		174	
6" x 12"							
	Demolish	L.F.		4.74		4.74	Cost includes material and labor to
	Install	L.F.	4.72	1.47		6.19	install 6" x 12" triple beam.
	Demolish and Install	L.F.	4.72	6.21		10.93	
	Clean	L.F.	.19	.41		.60	
	Paint	L.F.	.26	1.53		1.79	
	Minimum Charge	Job		174		174	
Steel I-type							
W8 x 31							
	Demolish	L.F.		7.35	3.62	10.97	Includes material and labor to install
	Install	L.F.	38.50	2.37	1.58	42.45	W shaped steel beam / girder.
	Demolish and Install	L.F.	38.50	9.72	5.20	53.42	
	Minimum Charge	Job		430		430	
W8 x 48							
	Demolish	L.F.		7.35	3.62	10.97	Includes material and labor to install
	Install	L.F.	59.50	2.48	1.65	63.63	W shaped steel beam / girder.
	Demolish and Install	L.F.	59.50	9.83	5.27	74.60	
	Minimum Charge	Job		430		430	
W8 x 67							
	Demolish	L.F.		7.35	3.62	10.97	Includes material and labor to install
	Install	L.F.	83	2.60	1.73	87.33	W shaped steel beam / girder.
	Demolish and Install	L.F.	83	9.95	5.35	98.30	
	Minimum Charge	Job		430		430	
W10 x 45							
	Demolish	L.F.		7.35	3.62	10.97	Includes material and labor to install
	Install	L.F.	55.50	2.48	1.65	59.63	W shaped steel beam / girder.
	Demolish and Install	L.F.	55.50	9.83	5.27	70.60	
	Minimum Charge	Job		430		430	
W10 x 68							
	Demolish	L.F.		7.35	3.62	10.97	Includes material and labor to install
	Install	L.F.	84	2.60	1.73	88.33	W shaped steel beam / girder.
	Demolish and Install	L.F.	84	9.95	5.35	99.30	
	Minimum Charge	Job		430		430	
W12 x 50							
	Demolish	L.F.		7.35	3.62	10.97	Includes material and labor to install
	Install	L.F.	62	2.48	1.65	66.13	W shaped steel beam / girder.
	Demolish and Install	L.F.	62	9.83	5.27	77.10	
	Minimum Charge	Job		430		430	
W12 x 87							
	Demolish	L.F.		10.85	5.30	16.15	Includes material and labor to install
	Install	L.F.	108	2.60	1.73	112.33	W shaped steel beam / girder.
	Demolish and Install	L.F.	108	13.45	7.03	128.48	
	Minimum Charge	Job		430		430	

Rough Frame / Structure

Beam / Girder

	Unit	Material	Labor	Equip.	Total	Specification
W12 x 120						
Demolish	L.F.		10.85	5.30	16.15	Includes material and labor to install
Install	L.F.	149	2.66	1.77	153.43	W shaped steel beam / girder.
Demolish and Install	L.F.	149	13.51	7.07	169.58	
Minimum Charge	Job		430		430	
W14 x 74						
Demolish	L.F.		10.85	5.30	16.15	Includes material and labor to install
Install	L.F.	92	2.60	1.73	96.33	W shaped steel beam / girder.
Demolish and Install	L.F.	92	13.45	7.03	112.48	
Minimum Charge	Job		430		430	
W14 x 120						
Demolish	L.F.		10.85	5.30	16.15	Includes material and labor to install
Install	L.F.	149	2.66	1.77	153.43	W shaped steel beam / girder.
Demolish and Install	L.F.	149	13.51	7.07	169.58	
Minimum Charge	Job		430		430	

Column

	Unit	Material	Labor	Equip.	Total	Specification
Concrete						
Small Diameter						
Demolish	L.F.		12	1.83	13.83	Includes material and labor to install
Install	L.F.	67.50	10.95	2.99	81.44	small diameter concrete column.
Demolish and Install	L.F.	67.50	22.95	4.82	95.27	
Minimum Charge	Job		550	149	699	
Large Diameter						
Demolish	L.F.		36	5.50	41.50	Includes material and labor to install
Install	L.F.	88	12.90	3.52	104.42	large diameter concrete column.
Demolish and Install	L.F.	88	48.90	9.02	145.92	
Minimum Charge	Job		550	149	699	

Floor Framing System

	Unit	Material	Labor	Equip.	Total	Specification
12" O.C.						
2" x 6" Joists						
Demolish	S.F.		.58		.58	Cost includes material and labor to
Install	S.F.	.74	.28		1.02	install 2" x 6" joists, 12" O.C.
Demolish and Install	S.F.	.74	.86		1.60	including box or band joist. Blocking
Reinstall	S.F.		.22		.22	or bridging not included.
Minimum Charge	Job		174		174	
2" x 8" Joists						
Demolish	S.F.		.60		.60	Cost includes material and labor to
Install	S.F.	1.08	.32		1.40	install 2" x 8" joists, 12" O.C.
Demolish and Install	S.F.	1.08	.92		2	including box or band joist. Blocking
Reinstall	S.F.		.25		.25	or bridging not included.
Minimum Charge	Job		174		174	
2" x 10" Joists						
Demolish	S.F.		.62		.62	Cost includes material and labor to
Install	S.F.	1.53	.39		1.92	install 2" x 10" joists, 12" O.C.
Demolish and Install	S.F.	1.53	1.01		2.54	including box or band joist. Blocking
Reinstall	S.F.		.31		.31	or bridging not included.
Minimum Charge	Job		174		174	

Rough Frame / Structure

Floor Framing System		Unit	Material	Labor	Equip.	Total	Specification
2″ x 12″ Joists							
	Demolish	S.F.		.64		.64	Cost includes material and labor to
	Install	S.F.	1.84	.40		2.24	install 2″ x 12″ joists, 12″ O.C.
	Demolish and Install	S.F.	1.84	1.04		2.88	including box or band joist. Blocking
	Reinstall	S.F.		.32		.32	or bridging not included.
	Minimum Charge	Job		174		174	
Block / Bridge							
	Demolish	Ea.		.88		.88	Includes material and labor to install
	Install	Ea.	1.75	1.96		3.71	set of cross bridging or per block of
	Demolish and Install	Ea.	1.75	2.84		4.59	solid bridging for 2″ x 10″ joists cut to
	Minimum Charge	Job		174		174	size on site.

16″ O.C.

		Unit	Material	Labor	Equip.	Total	Specification
2″ x 6″ Joists							
	Demolish	S.F.		.44		.44	Cost includes material and labor to
	Install	S.F.	.57	.42		.99	install 2″ x 6″ joists including box or
	Demolish and Install	S.F.	.57	.86		1.43	band joist installed 16″ O.C. Does not
	Reinstall	S.F.		.33		.33	include beams, blocking or bridging.
	Minimum Charge	Job		174		174	
2″ x 8″ Joists							
	Demolish	S.F.		.45		.45	Cost includes material and labor to
	Install	S.F.	.84	.48		1.32	install 2″ x 8″ joists including box or
	Demolish and Install	S.F.	.84	.93		1.77	band joist installed 16″ O.C. Does not
	Reinstall	S.F.		.38		.38	include beams, blocking or bridging.
	Minimum Charge	Job		174		174	
2″ x 10″ Joists							
	Demolish	S.F.		.46		.46	Cost includes material and labor to
	Install	S.F.	1.18	.58		1.76	install 2″ x 10″ joists including box or
	Demolish and Install	S.F.	1.18	1.04		2.22	band joist 16″ O.C. Does not include
	Reinstall	S.F.		.47		.47	beams, blocking or bridging.
	Minimum Charge	Job		174		174	
2″ x 12″ Joists							
	Demolish	S.F.		.48		.48	Cost includes material and labor to
	Install	S.F.	1.42	.60		2.02	install 2″ x 12″ joists including box or
	Demolish and Install	S.F.	1.42	1.08		2.50	band joist installed 16″ O.C. Does not
	Reinstall	S.F.		.48		.48	include beams, blocking or bridging.
	Minimum Charge	Job		174		174	
Block / Bridge							
	Demolish	Ea.		.88		.88	Includes material and labor to install
	Install	Ea.	1.75	1.96		3.71	set of cross bridging or per block of
	Demolish and Install	Ea.	1.75	2.84		4.59	solid bridging for 2″ x 10″ joists cut to
	Minimum Charge	Job		174		174	size on site.

24″ O.C.

		Unit	Material	Labor	Equip.	Total	Specification
2″ x 6″ Joists							
	Demolish	S.F.		.29		.29	Cost includes material and labor to
	Install	S.F.	.41	.14		.55	install 2″ x 6″ joists including box or
	Demolish and Install	S.F.	.41	.43		.84	band joist installed 24″ O.C. Does not
	Reinstall	S.F.		.11		.11	include beams, blocking or bridging.
	Minimum Charge	Job		174		174	
2″ x 8″ Joists							
	Demolish	S.F.		.30		.30	Cost includes material and labor to
	Install	S.F.	.59	.16		.75	install 2″ x 8″ joists including box or
	Demolish and Install	S.F.	.59	.46		1.05	band joist installed 24″ O.C. Does not
	Reinstall	S.F.		.13		.13	include beams, blocking or bridging.
	Minimum Charge	Job		174		174	

Rough Frame / Structure

Floor Framing System		Unit	Material	Labor	Equip.	Total	Specification
2″ x 10″ Joists							Cost includes material and labor to
	Demolish	S.F.		.31		.31	install 2″ x 10″ joists including box or
	Install	S.F.	.84	.19		1.03	band joist installed 24″ O.C. Does not
	Demolish and Install	S.F.	.84	.50		1.34	include beams, blocking or bridging.
	Reinstall	S.F.		.16		.16	
	Minimum Charge	Job		174		174	
2″ x 12″ Joists							Cost includes material and labor to
	Demolish	S.F.		.32		.32	install 2″ x 12″ joists including box or
	Install	S.F.	1.01	.20		1.21	band joist installed 24″ O.C. Does not
	Demolish and Install	S.F.	1.01	.52		1.53	include beams, blocking or bridging.
	Reinstall	S.F.		.16		.16	
	Minimum Charge	Job		174		174	
Block / Bridge							Includes material and labor to install
	Demolish	Ea.		.88		.88	set of cross bridging or per block of
	Install	Ea.	1.75	1.96		3.71	solid bridging for 2″ x 10″ joists cut to
	Demolish and Install	Ea.	1.75	2.84		4.59	size on site.
	Minimum Charge	Job		174		174	
Engineered Lumber, Joist							
9-1/2″							Includes material and labor to install
	Demolish	S.F.		.46		.46	engineered lumber truss / joists per
	Install	S.F.	1.84	.35		2.19	S.F. of floor area based on joists 16″
	Demolish and Install	S.F.	1.84	.81		2.65	O.C. Beams, supports and bridging
	Reinstall	S.F.		.28		.28	are not included.
	Minimum Charge	Job		174		174	
11-7/8″							Includes material and labor to install
	Demolish	S.F.		.48		.48	engineered lumber truss / joists per
	Install	S.F.	1.96	.36		2.32	S.F. of floor area based on joists 16″
	Demolish and Install	S.F.	1.96	.84		2.80	O.C. Beams, supports and bridging
	Reinstall	S.F.		.29		.29	are not included.
	Minimum Charge	Job		174		174	
14″							Includes material and labor to install
	Demolish	S.F.		.50		.50	engineered lumber truss / joists per
	Install	S.F.	2.30	.38		2.68	S.F. of floor area based on joists 16″
	Demolish and Install	S.F.	2.30	.88		3.18	O.C. Beams, supports and bridging
	Reinstall	S.F.		.31		.31	are not included.
	Minimum Charge	Job		174		174	
16″							Includes material and labor to install
	Demolish	S.F.		.52		.52	engineered lumber truss / joists per
	Install	S.F.	2.57	.40		2.97	S.F. of floor area based on joists 16″
	Demolish and Install	S.F.	2.57	.92		3.49	O.C. Beams, supports and bridging
	Reinstall	S.F.		.32		.32	are not included.
	Minimum Charge	Job		174		174	
Block / Bridge							Includes material and labor to install
	Demolish	Pr.		.88		.88	set of steel, one-nail type cross
	Install	Pr.	1.41	2.58		3.99	bridging for trusses placed 16″ O.C.
	Demolish and Install	Pr.	1.41	3.46		4.87	
	Minimum Charge	Job		174		174	
Block / Bridge							
2″ x 6″							Includes material and labor to install
	Demolish	Ea.		.88		.88	set of cross bridging or per block of
	Install	Ea.	.85	1.57		2.42	solid bridging for 2″ x 6″ joists cut to
	Demolish and Install	Ea.	.85	2.45		3.30	size on site.
	Minimum Charge	Job		174		174	

Rough Frame / Structure

Floor Framing System		Unit	Material	Labor	Equip.	Total	Specification
2" x 8"							
	Demolish	Ea.		.88		.88	Includes material and labor to install
	Install	Ea.	1.24	1.74		2.98	set of cross bridging or per block of
	Demolish and Install	Ea.	1.24	2.62		3.86	solid bridging for 2" x 8" joists cut to
	Minimum Charge	Job		174		174	size on site.
2" x 10"							
	Demolish	Ea.		.88		.88	Includes material and labor to install
	Install	Ea.	1.75	1.96		3.71	set of cross bridging or per block of
	Demolish and Install	Ea.	1.75	2.84		4.59	solid bridging for 2" x 10" joists cut to
	Minimum Charge	Job		174		174	size on site.
2" x 12"							
	Demolish	Ea.		.88		.88	Includes material and labor to install
	Install	Ea.	2.11	2.31		4.42	set of cross bridging or per block of
	Demolish and Install	Ea.	2.11	3.19		5.30	solid bridging for 2" x 12" joists cut to
	Minimum Charge	Job		174		174	size on site.
Ledger Strips							
1" x 2"							
	Demolish	L.F.		.24		.24	Cost includes material and labor to
	Install	L.F.	.13	1.01		1.14	install 1" x 2" ledger strip nailed to the
	Demolish and Install	L.F.	.13	1.25		1.38	face of studs, beams or joist.
	Minimum Charge	Job		174		174	
1" x 3"							
	Demolish	L.F.		.24		.24	Includes material and labor to install
	Install	L.F.	.51	1.16		1.67	up to 1" x 4" ledger strip nailed to the
	Demolish and Install	L.F.	.51	1.40		1.91	face of studs, beams or joist.
	Minimum Charge	Job		174		174	
1" x 4"							
	Demolish	L.F.		.24		.24	Includes material and labor to install
	Install	L.F.	.51	1.16		1.67	up to 1" x 4" ledger strip nailed to the
	Demolish and Install	L.F.	.51	1.40		1.91	face of studs, beams or joist.
	Minimum Charge	Job		174		174	
2" x 2"							
	Demolish	L.F.		.26		.26	Cost includes material and labor to
	Install	L.F.	.33	1.06		1.39	install 2" x 2" ledger strip nailed to the
	Demolish and Install	L.F.	.33	1.32		1.65	face of studs, beams or joist.
	Minimum Charge	Job		174		174	
2" x 4"							
	Demolish	L.F.		.28		.28	Cost includes material and labor to
	Install	L.F.	.41	1.40		1.81	install 2" x 4" ledger strip nailed to the
	Demolish and Install	L.F.	.41	1.68		2.09	face of studs, beams or joist.
	Reinstall	L.F.		1.12		1.12	
	Minimum Charge	Job		174		174	
Ledger Boards							
2" x 6"							
	Demolish	L.F.		.28		.28	Cost includes material and labor to
	Install	L.F.	1.89	2.18		4.07	install 2" x 6" ledger board fastened to
	Demolish and Install	L.F.	1.89	2.46		4.35	a wall, joists or studs.
	Reinstall	L.F.		1.74		1.74	
	Minimum Charge	Job		174		174	
4" x 6"							
	Demolish	L.F.		.47		.47	Cost includes material and labor to
	Install	L.F.	4.62	2.18		6.80	install 4" x 6" ledger board fastened to
	Demolish and Install	L.F.	4.62	2.65		7.27	a wall, joists or studs.
	Reinstall	L.F.		1.74		1.74	
	Minimum Charge	Job		174		174	

Rough Frame / Structure

Floor Framing System

	Unit	Material	Labor	Equip.	Total	Specification
4" x 8"						
Demolish	L.F.		.63		.63	Cost includes material and labor to
Install	L.F.	5.65	2.18		7.83	install 4" x 8" ledger board fastened to
Demolish and Install	L.F.	5.65	2.81		8.46	a wall, joists or studs.
Reinstall	L.F.		1.74		1.74	
Minimum Charge	Job		174		174	

Sill Plate Per L.F.

	Unit	Material	Labor	Equip.	Total	Specification
2" x 4"						
Demolish	L.F.		.24		.24	Cost includes material and labor to
Install	L.F.	.57	1.27		1.84	install 2" x 4" pressure treated lumber,
Demolish and Install	L.F.	.57	1.51		2.08	drilled and installed on foundation
Reinstall	L.F.		1.01		1.01	bolts 48" O.C. Bolts, nuts and washers
Minimum Charge	Job		174		174	are not included.
2" x 6"						
Demolish	L.F.		.36		.36	Cost includes material and labor to
Install	L.F.	.91	1.40		2.31	install 2" x 6" pressure treated lumber,
Demolish and Install	L.F.	.91	1.76		2.67	drilled and installed on foundation
Reinstall	L.F.		1.12		1.12	bolts 48" O.C. Bolts, nuts and washers
Minimum Charge	Job		174		174	not included.
2" x 8"						
Demolish	L.F.		.48		.48	Cost includes material and labor to
Install	L.F.	1.27	1.55		2.82	install 2" x 8" pressure treated lumber,
Demolish and Install	L.F.	1.27	2.03		3.30	drilled and installed on foundation
Reinstall	L.F.		1.24		1.24	bolts 48" O.C. Bolts, nuts and washers
Minimum Charge	Job		174		174	not included.

Earthquake Strapping

	Unit	Material	Labor	Equip.	Total	Specification
Install	Ea.	2.19	2.18		4.37	Includes labor and material to replace
Minimum Charge	Job		174		174	an earthquake strap.

Subflooring

	Unit	Material	Labor	Equip.	Total	Specification
Plywood						
1/2"						
Demolish	S.F.		.74		.74	Cost includes material and labor to
Install	Ea.	.39	.37		.76	install 1/2" plywood subfloor, CD
Demolish and Install	Ea.	.39	1.11		1.50	standard interior grade, plugged and
Minimum Charge	Job		174		174	touch sanded.
5/8"						
Demolish	S.F.		.74		.74	Cost includes material and labor to
Install	SF Flr.	.49	.42		.91	install 5/8" plywood subfloor, CD
Demolish and Install	SF Flr.	.49	1.16		1.65	standard interior grade, plugged and
Reinstall	SF Flr.		.33		.33	touch sanded.
Minimum Charge	Job		174		174	
3/4"						
Demolish	S.F.		.75		.75	Cost includes material and labor to
Install	SF Flr.	.59	.45		1.04	install 3/4" plywood subfloor, CD
Demolish and Install	SF Flr.	.59	1.20		1.79	standard interior grade, plugged and
Reinstall	SF Flr.		.36		.36	touch sanded.
Minimum Charge	Job		174		174	

Subflooring

Subflooring		Unit	Material	Labor	Equip.	Total	Specification
Particle Board							
1/2"							
	Demolish	S.F.		.37		.37	Cost includes material and labor to
	Install	SF Flr.	.41	.47		.88	install 1/2" particle board subfloor.
	Demolish and Install	SF Flr.	.41	.84		1.25	
	Minimum Charge	Job		174		174	
5/8"							
	Demolish	S.F.		.37		.37	Cost includes material and labor to
	Install	S.F.	.45	.51		.96	install 5/8" particle board subfloor.
	Demolish and Install	S.F.	.45	.88		1.33	
	Minimum Charge	Job		174		174	
3/4"							
	Demolish	S.F.		.38		.38	Cost includes material and labor to
	Install	SF Flr.	.70	.56		1.26	install 3/4" particle board subfloor.
	Demolish and Install	SF Flr.	.70	.94		1.64	
	Minimum Charge	Job		174		174	
Plank Board							
	Demolish	S.F.		1.19		1.19	Cost includes material and labor to
	Install	S.F.	1.56	.39		1.95	install 1" x 6" standard grade plank
	Demolish and Install	S.F.	1.56	1.58		3.14	flooring.
	Reinstall	S.F.		.39		.39	
	Minimum Charge	Job		174		174	
Felt Underlay							
	Demolish	S.F.		.07		.07	Includes material and labor to install
	Install	S.F.	.04	.09		.13	#15 felt building paper.
	Demolish and Install	S.F.	.04	.16		.20	
	Minimum Charge	Job		174		174	
Prep (for flooring)							
	Install	S.F.		.30		.30	Preparation of subflooring for
	Minimum Charge	Job		160		160	installation of new finished flooring. Cost reflects average time to prep area.

Wall Framing System

Wall Framing System		Unit	Material	Labor	Equip.	Total	Specification
2" x 4"							
8' High							
	Demolish	L.F.		3.76		3.76	Cost includes material and labor to
	Install	L.F.	4.28	4.10		8.38	install 2" x 4" x 8' high wall system,
	Demolish and Install	L.F.	4.28	7.86		12.14	including studs, treated bottom plate,
	Reinstall	L.F.		3.28		3.28	double top plate and one row of fire
	Minimum Charge	Job		174		174	blocking.
9' High							
	Demolish	L.F.		4.21		4.21	Cost includes material and labor to
	Install	L.F.	4.59	4.10		8.69	install 2" x 4" x 9' wall system
	Demolish and Install	L.F.	4.59	8.31		12.90	including studs, treated bottom plate,
	Reinstall	L.F.		3.28		3.28	double top plate and one row of fire
	Minimum Charge	Job		174		174	blocking.
10' High							
	Demolish	L.F.		4.70		4.70	Cost includes material and labor to
	Install	L.F.	4.90	4.10		9	install 2" x 4" x 10' wall system
	Demolish and Install	L.F.	4.90	8.80		13.70	including studs, treated bottom plate,
	Reinstall	L.F.		3.28		3.28	double top plate and one row of fire
	Minimum Charge	Job		174		174	blocking.

Rough Frame / Structure

Wall Framing System		Unit	Material	Labor	Equip.	Total	Specification
12' High							
	Demolish	L.F.		5.65		5.65	Cost includes material and labor to
	Install	L.F.	5.50	4.98		10.48	install 2" x 4" x 12' wall system
	Demolish and Install	L.F.	5.50	10.63		16.13	including studs, treated bottom plate,
	Reinstall	L.F.		3.99		3.99	double top plate and one row of fire
	Minimum Charge	Job		174		174	blocking.
2" x 6"							
8' High							
	Demolish	L.F.		3.53		3.53	Cost includes material and labor to
	Install	L.F.	6.60	7.55		14.15	install 2" x 6" x 8' wall system
	Demolish and Install	L.F.	6.60	11.08		17.68	including studs, treated bottom plate,
	Reinstall	L.F.		6.03		6.03	double top plate and one row of fire
	Minimum Charge	Job		174		174	blocking.
9' High							
	Demolish	L.F.		3.98		3.98	Cost includes material and labor to
	Install	L.F.	7.10	7.55		14.65	install 2" x 6" x 9' wall system
	Demolish and Install	L.F.	7.10	11.53		18.63	including studs, treated bottom plate,
	Reinstall	L.F.		6.03		6.03	double top plate and one row of fire
	Minimum Charge	Job		174		174	blocking.
10' High							
	Demolish	L.F.		4.34		4.34	Cost includes material and labor to
	Install	L.F.	7.55	7.55		15.10	install 2" x 6" x 10' wall system
	Demolish and Install	L.F.	7.55	11.89		19.44	including studs, treated bottom plate,
	Reinstall	L.F.		6.03		6.03	double top plate and one row of fire
	Minimum Charge	Job		174		174	blocking.
12' High							
	Demolish	L.F.		5.15		5.15	Cost includes material and labor to
	Install	L.F.	8.55	9.30		17.85	install 2" x 6" x 12' wall system
	Demolish and Install	L.F.	8.55	14.45		23	including studs, treated bottom plate,
	Reinstall	L.F.		7.42		7.42	double top plate and one row of fire
	Minimum Charge	Job		174		174	blocking.
Fireblock							
2" x 4", 16" O.C. System							
	Demolish	L.F.		.47		.47	Cost includes material and labor to
	Install	L.F.	.41	1.16		1.57	install 2" x 4" fireblocks in wood frame
	Demolish and Install	L.F.	.41	1.63		2.04	walls per L.F. of wall to be blocked.
	Reinstall	L.F.		.93		.93	
	Minimum Charge	Job		174		174	
2" x 4", 24" O.C. System							
	Demolish	L.F.		.47		.47	Cost includes material and labor to
	Install	L.F.	.41	1.16		1.57	install 2" x 4" fireblocks in wood frame
	Demolish and Install	L.F.	.41	1.63		2.04	walls per L.F. of wall to be blocked.
	Reinstall	L.F.		.93		.93	
	Minimum Charge	Job		174		174	
2" x 6", 16" O.C. System							
	Demolish	L.F.		.71		.71	Cost includes material and labor to
	Install	L.F.	.64	1.16		1.80	install 2" x 6" fireblocks in wood frame
	Demolish and Install	L.F.	.64	1.87		2.51	walls per L.F. of wall to be blocked.
	Reinstall	L.F.		.93		.93	
	Minimum Charge	Job		174		174	
2" x 6", 24" O.C. System							
	Demolish	L.F.		.71		.71	Cost includes material and labor to
	Install	L.F.	.64	1.16		1.80	install 2" x 6" fireblocks in wood frame
	Demolish and Install	L.F.	.64	1.87		2.51	walls per L.F. of wall to be blocked.
	Reinstall	L.F.		.93		.93	
	Minimum Charge	Job		174		174	

Rough Frame / Structure

Wall Framing System		Unit	Material	Labor	Equip.	Total	Specification
Bracing							
1″ x 3″							
	Demolish	L.F.		.27		.27	Cost includes material and labor to
	Install	L.F.	.24	2.18		2.42	install 1″ x 3″ let-in bracing.
	Demolish and Install	L.F.	.24	2.45		2.69	
	Minimum Charge	Job		174		174	
1″ x 4″							
	Demolish	L.F.		.27		.27	Cost includes material and labor to
	Install	L.F.	.43	2.18		2.61	install 1″ x 4″ let-in bracing.
	Demolish and Install	L.F.	.43	2.45		2.88	
	Minimum Charge	Job		174		174	
1″ x 6″							
	Demolish	L.F.		.27		.27	Cost includes material and labor to
	Install	L.F.	.72	2.33		3.05	install 1″ x 6″ let-in bracing.
	Demolish and Install	L.F.	.72	2.60		3.32	
	Minimum Charge	Job		174		174	
2″ x 3″							
	Demolish	L.F.		.35		.35	Cost includes material and labor to
	Install	L.F.	.33	2.33		2.66	install 2″ x 3″ let-in bracing.
	Demolish and Install	L.F.	.33	2.68		3.01	
	Reinstall	L.F.		1.86		1.86	
	Minimum Charge	Job		174		174	
2″ x 4″							
	Demolish	L.F.		.35		.35	Cost includes material and labor to
	Install	L.F.	.41	2.33		2.74	install 2″ x 4″ let-in bracing.
	Demolish and Install	L.F.	.41	2.68		3.09	
	Reinstall	L.F.		1.86		1.86	
	Minimum Charge	Job		174		174	
2″ x 6″							
	Demolish	L.F.		.35		.35	Cost includes material and labor to
	Install	L.F.	.64	2.49		3.13	install 2″ x 6″ let-in bracing.
	Demolish and Install	L.F.	.64	2.84		3.48	
	Reinstall	L.F.		1.99		1.99	
	Minimum Charge	Job		174		174	
2″ x 8″							
	Demolish	L.F.		.35		.35	Cost includes material and labor to
	Install	L.F.	.92	2.49		3.41	install 2″ x 8″ let-in bracing.
	Demolish and Install	L.F.	.92	2.84		3.76	
	Reinstall	L.F.		1.99		1.99	
	Minimum Charge	Job		174		174	
Earthquake Strapping							
	Install	Ea.	2.19	2.18		4.37	Includes labor and material to replace
	Minimum Charge	Job		174		174	an earthquake strap.
Hurricane Clips							
	Install	Ea.	1.93	2.41		4.34	Includes labor and material to install a
	Minimum Charge	Job		174		174	hurricane clip.
Stud							
2″ x 4″							
	Demolish	L.F.		.28		.28	Cost includes material and labor to
	Install	L.F.	.41	.63		1.04	install 2″ x 4″ wall stud per L.F. of stud.
	Demolish and Install	L.F.	.41	.91		1.32	
	Reinstall	L.F.		.51		.51	
	Minimum Charge	Job		174		174	

Rough Frame / Structure

Wall Framing System

	Unit	Material	Labor	Equip.	Total	Specification
2" x 6"						
Demolish	L.F.		.35		.35	Cost includes material and labor to
Install	L.F.	.64	.70		1.34	install 2" x 6" stud per L.F. of stud.
Demolish and Install	L.F.	.64	1.05		1.69	
Reinstall	L.F.		.56		.56	
Minimum Charge	Job		174		174	
Plates						
2" x 4"						
Demolish	L.F.		.26		.26	Cost includes material and labor to
Install	L.F.	.41	.87		1.28	install 2" x 4" plate per L.F. of plate.
Demolish and Install	L.F.	.41	1.13		1.54	
Reinstall	L.F.		.70		.70	
Minimum Charge	Job		174		174	
2" x 6"						
Demolish	L.F.		.27		.27	Cost includes material and labor to
Install	L.F.	.64	.93		1.57	install 2" x 6" plate per L.F. of plate.
Demolish and Install	L.F.	.64	1.20		1.84	
Reinstall	L.F.		.74		.74	
Minimum Charge	Job		174		174	
Headers						
Demolish	L.F.		.25		.25	Includes material and labor to install
Install	L.F.	.64	1.94		2.58	header over wall openings and
Demolish and Install	L.F.	.64	2.19		2.83	around floor, ceiling and roof
Clean	L.F.	.01	.32		.33	openings or flush beam.
Paint	L.F.	.09	1.22		1.31	
Minimum Charge	Job		174		174	
2" x 8"						
Demolish	L.F.		.27		.27	Includes material and labor to install
Install	L.F.	.92	2.05		2.97	header over wall openings and
Demolish and Install	L.F.	.92	2.32		3.24	around floor, ceiling and roof
Clean	L.F.	.02	.32		.34	openings or flush beam.
Paint	L.F.	.12	1.65		1.77	
Minimum Charge	Job		174		174	
2" x 10"						
Demolish	L.F.		.27		.27	Includes material and labor to install
Install	L.F.	1.31	2.18		3.49	header over wall openings and
Demolish and Install	L.F.	1.31	2.45		3.76	around floor, ceiling and roof
Clean	L.F.	.02	.32		.34	openings or flush beam.
Paint	L.F.	.14	2.03		2.17	
Minimum Charge	Job		174		174	
2" x 12"						
Demolish	L.F.		.28		.28	Includes material and labor to install
Install	L.F.	1.57	2.33		3.90	header over wall openings and
Demolish and Install	L.F.	1.57	2.61		4.18	around floor, ceiling and roof
Clean	L.F.	.03	.33		.36	openings or flush beam.
Paint	L.F.	.18	2.44		2.62	
Minimum Charge	Job		174		174	
4" x 8"						
Demolish	L.F.		.50		.50	Includes material and labor to install
Install	L.F.	4.11	2.68		6.79	header over wall openings and
Demolish and Install	L.F.	4.11	3.18		7.29	around floor, ceiling and roof
Clean	L.F.	.02	.32		.34	openings or flush beam.
Paint	L.F.	.12	1.65		1.77	
Minimum Charge	Job		174		174	

Wall Framing System

Wall Framing System	Unit	Material	Labor	Equip.	Total	Specification
4" x 10"						
Demolish	L.F.		.54		.54	Includes material and labor to install
Install	L.F.	6.05	2.91		8.96	header over wall openings and
Demolish and Install	L.F.	6.05	3.45		9.50	around floor, ceiling and roof
Clean	L.F.	.02	.32		.34	openings or flush beam.
Paint	L.F.	.14	2.03		2.17	
Minimum Charge	Job		174		174	
4" x 12"						
Demolish	L.F.		.56		.56	Includes material and labor to install
Install	L.F.	5.60	3.67		9.27	header over wall openings and
Demolish and Install	L.F.	5.60	4.23		9.83	around floor, ceiling and roof
Clean	L.F.	.03	.33		.36	openings or flush beam.
Paint	L.F.	.18	2.44		2.62	
Minimum Charge	Job		174		174	
6" x 8"						
Demolish	L.F.		.50		.50	Includes material and labor to install
Install	L.F.	8.35	1.94		10.29	header over wall openings and
Demolish and Install	L.F.	8.35	2.44		10.79	around floor, ceiling and roof
Clean	L.F.	.02	.32		.34	openings or flush beam.
Paint	L.F.	.12	1.65		1.77	
Minimum Charge	Job		174		174	
6" x 10"						
Demolish	L.F.		.54		.54	Includes material and labor to install
Install	L.F.	12	4.23		16.23	header over wall openings and
Demolish and Install	L.F.	12	4.77		16.77	around floor, ceiling and roof
Clean	L.F.	.02	.32		.34	openings or flush beam.
Paint	L.F.	.14	2.03		2.17	
Minimum Charge	Job		174		174	
6" x 12"						
Demolish	L.F.		.56		.56	Includes material and labor to install
Install	L.F.	10.05	4.98		15.03	header over wall openings and
Demolish and Install	L.F.	10.05	5.54		15.59	around floor, ceiling and roof
Clean	L.F.	.03	.33		.36	openings or flush beam.
Paint	L.F.	.18	.85		1.03	
Minimum Charge	Job		174		174	

Rough-in Opening

Rough-in Opening	Unit	Material	Labor	Equip.	Total	Specification
Door w / 2" x 4" Lumber						
3' Wide						
Demolish	Ea.		8.80		8.80	Includes material and labor to install
Install	Ea.	17.25	10.90		28.15	header, double studs each side,
Demolish and Install	Ea.	17.25	19.70		36.95	cripples, blocking and nails, up to 3'
Reinstall	Ea.		8.72		8.72	opening in 2" x 4" stud wall 8' high.
Minimum Charge	Job		174		174	
4' Wide						
Demolish	Ea.		8.80		8.80	Includes material and labor to install
Install	Ea.	18.55	10.90		29.45	header, double studs each side,
Demolish and Install	Ea.	18.55	19.70		38.25	cripples, blocking and nails, up to 4'
Reinstall	Ea.		8.72		8.72	opening in 2" x 4" stud wall 8' high.
Minimum Charge	Job		174		174	
5' Wide						
Demolish	Ea.		8.80		8.80	Includes material and labor to install
Install	Ea.	23	10.90		33.90	header, double studs each side,
Demolish and Install	Ea.	23	19.70		42.70	cripples, blocking and nails, up to 5'
Reinstall	Ea.		8.72		8.72	opening in 2" x 4" stud wall 8' high.
Minimum Charge	Job		174		174	

Rough Frame / Structure

Rough-in Opening		Unit	Material	Labor	Equip.	Total	Specification
6' Wide							
	Demolish	Ea.		8.80		8.80	Includes material and labor to install
	Install	Ea.	25	10.90		35.90	header, double studs each side,
	Demolish and Install	Ea.	25	19.70		44.70	cripples, blocking and nails, up to 6'
	Reinstall	Ea.		8.72		8.72	opening in 2" x 4" stud wall 8' high.
	Minimum Charge	Job		174		174	
8' Wide							
	Demolish	Ea.		9.40		9.40	Includes material and labor to install
	Install	Ea.	34.50	11.65		46.15	header, double studs each side,
	Demolish and Install	Ea.	34.50	21.05		55.55	cripples, blocking and nails, up to 8'
	Reinstall	Ea.		9.30		9.30	opening in 2" x 4" stud wall 8' high.
	Minimum Charge	Job		174		174	
10' Wide							
	Demolish	Ea.		9.40		9.40	Includes material and labor to install
	Install	Ea.	45.50	11.65		57.15	header, double studs each side,
	Demolish and Install	Ea.	45.50	21.05		66.55	cripples, blocking and nails, up to 10'
	Reinstall	Ea.		9.30		9.30	opening in 2" x 4" stud wall 8' high.
	Minimum Charge	Job		174		174	
12' Wide							
	Demolish	Ea.		9.40		9.40	Includes material and labor to install
	Install	Ea.	72	11.65		83.65	header, double studs each side,
	Demolish and Install	Ea.	72	21.05		93.05	cripples, blocking and nails, up to 12'
	Reinstall	Ea.		9.30		9.30	opening in 2" x 4" stud wall 8' high.
	Minimum Charge	Job		174		174	
Door w / 2" x 6" Lumber							
3' Wide							
	Demolish	Ea.		8.80		8.80	Includes material and labor to install
	Install	Ea.	24	10.90		34.90	header, double studs each side,
	Demolish and Install	Ea.	24	19.70		43.70	cripples, blocking and nails, up to 3'
	Reinstall	Ea.		8.72		8.72	opening in 2" x 6" stud wall 8' high.
	Minimum Charge	Job		174		174	
4' Wide							
	Demolish	Ea.		8.80		8.80	Includes material and labor to install
	Install	Ea.	26	10.90		36.90	header, double studs each side,
	Demolish and Install	Ea.	26	19.70		45.70	cripples, blocking and nails, up to 4'
	Reinstall	Ea.		8.72		8.72	opening in 2" x 6" stud wall 8' high.
	Minimum Charge	Job		174		174	
5' Wide							
	Demolish	Ea.		8.80		8.80	Includes material and labor to install
	Install	Ea.	29.50	10.90		40.40	header, double studs each side,
	Demolish and Install	Ea.	29.50	19.70		49.20	cripples, blocking and nails up to 5'
	Reinstall	Ea.		8.72		8.72	opening in 2" x 6" stud wall 8' high.
	Minimum Charge	Job		174		174	
6' Wide							
	Demolish	Ea.		8.80		8.80	Includes material and labor to install
	Install	Ea.	32	10.90		42.90	header, double studs each side,
	Demolish and Install	Ea.	32	19.70		51.70	cripples, blocking and nails up to 6'
	Reinstall	Ea.		8.72		8.72	opening in 2" x 6" stud wall 8' high.
	Minimum Charge	Job		174		174	
8' Wide							
	Demolish	Ea.		9.40		9.40	Includes material and labor to install
	Install	Ea.	42	11.65		53.65	header, double studs each side,
	Demolish and Install	Ea.	42	21.05		63.05	cripples, blocking and nails up to 8'
	Reinstall	Ea.		9.30		9.30	opening in 2" x 6" stud wall 8' high.
	Minimum Charge	Job		174		174	
10' Wide							
	Demolish	Ea.		9.40		9.40	Includes material and labor to install
	Install	Ea.	52.50	11.65		64.15	header, double studs each side,
	Demolish and Install	Ea.	52.50	21.05		73.55	cripples, blocking and nails up to 10'
	Reinstall	Ea.		9.30		9.30	opening in 2" x 6" stud wall 8' high.
	Minimum Charge	Job		174		174	

Rough Frame / Structure

Rough-in Opening

	Unit	Material	Labor	Equip.	Total	Specification
12' Wide						
Demolish	Ea.		9.40		9.40	Includes material and labor to install
Install	Ea.	79	11.65		90.65	header, double studs each side,
Demolish and Install	Ea.	79	21.05		100.05	cripples, blocking and nails, up to 12'
Reinstall	Ea.		9.30		9.30	opening in 2" x 6" stud wall 8' high.
Minimum Charge	Job		174		174	

Rough-in Opening

	Unit	Material	Labor	Equip.	Total	Specification
Window w / 2" x 4" Lumber						
2' Wide						
Demolish	Ea.		11.75		11.75	Includes material and labor to install
Install	Ea.	18.30	14.55		32.85	header, double studs each side of
Demolish and Install	Ea.	18.30	26.30		44.60	opening, cripples, blocking, nails and
Reinstall	Ea.		11.63		11.63	sub-sills for opening up to 2' in a 2" x
Minimum Charge	Job		174		174	4" stud wall 8' high.
3' Wide						
Demolish	Ea.		11.75		11.75	Includes material and labor to install
Install	Ea.	21.50	14.55		36.05	header, double studs each side of
Demolish and Install	Ea.	21.50	26.30		47.80	opening, cripples, blocking, nails and
Reinstall	Ea.		11.63		11.63	sub-sills for opening up to 3' in a 2" x
Minimum Charge	Job		174		174	4" stud wall 8' high.
4' Wide						
Demolish	Ea.		11.75		11.75	Includes material and labor to install
Install	Ea.	23.50	14.55		38.05	header, double studs each side of
Demolish and Install	Ea.	23.50	26.30		49.80	opening, cripples, blocking, nails and
Reinstall	Ea.		11.63		11.63	sub-sills for opening up to 4' in a 2" x
Minimum Charge	Job		174		174	4" stud wall 8' high.
5' Wide						
Demolish	Ea.		11.75		11.75	Includes material and labor to install
Install	Ea.	28	14.55		42.55	header, double studs each side of
Demolish and Install	Ea.	28	26.30		54.30	opening, cripples, blocking, nails and
Reinstall	Ea.		11.63		11.63	sub-sills for opening up to 5' in a 2" x
Minimum Charge	Job		174		174	4" stud wall 8' high.
6' Wide						
Demolish	Ea.		11.75		11.75	Includes material and labor to install
Install	Ea.	31	14.55		45.55	header, double studs each side of
Demolish and Install	Ea.	31	26.30		57.30	opening, cripples, blocking, nails and
Reinstall	Ea.		11.63		11.63	sub-sills for opening up to 6' in a 2" x
Minimum Charge	Job		174		174	4" stud wall 8' high.
7' Wide						
Demolish	Ea.		11.75		11.75	Includes material and labor to install
Install	Ea.	39.50	14.55		54.05	header, double studs each side of
Demolish and Install	Ea.	39.50	26.30		65.80	opening, cripples, blocking, nails and
Reinstall	Ea.		11.63		11.63	sub-sills for opening up to 7' in a 2" x
Minimum Charge	Job		174		174	4" stud wall 8' high.
8' Wide						
Demolish	Ea.		12.85		12.85	Includes material and labor to install
Install	Ea.	43.50	15.85		59.35	header, double studs each side of
Demolish and Install	Ea.	43.50	28.70		72.20	opening, cripples, blocking, nails and
Reinstall	Ea.		12.68		12.68	sub-sills for opening up to 8' in a 2" x
Minimum Charge	Job		174		174	4" stud wall 8' high.

Rough Frame / Structure

Rough-in Opening

	Unit	Material	Labor	Equip.	Total	Specification
10′ Wide						
Demolish	Ea.		12.85		12.85	Includes material and labor to install
Install	Ea.	55.50	15.85		71.35	header, double studs each side of
Demolish and Install	Ea.	55.50	28.70		84.20	opening, cripples, blocking, nails and
Reinstall	Ea.		12.68		12.68	sub-sills for opening up to 10′ in a 2″
Minimum Charge	Job		174		174	x 4″ stud wall 8′ high.
12′ Wide						
Demolish	Ea.		12.85		12.85	Includes material and labor to install
Install	Ea.	84	15.85		99.85	header, double studs each side of
Demolish and Install	Ea.	84	28.70		112.70	opening, cripples, blocking, nails and
Reinstall	Ea.		12.68		12.68	sub-sills for opening up to 12′ in a 2″
Minimum Charge	Job		174		174	x 4″ stud wall 8′ high.

Window w / 2″ x 6″ Lumber

	Unit	Material	Labor	Equip.	Total	Specification
2′ Wide						
Demolish	Ea.		11.75		11.75	Includes material and labor to install
Install	Ea.	27	14.55		41.55	header, double studs each side of
Demolish and Install	Ea.	27	26.30		53.30	opening, cripples, blocking, nails and
Reinstall	Ea.		11.63		11.63	sub-sills for opening up to 2′ in a 2″ x
Minimum Charge	Job		174		174	6″ stud wall 8′ high.
3′ Wide						
Demolish	Ea.		11.75		11.75	Includes material and labor to install
Install	Ea.	31	14.55		45.55	header, double studs each side of
Demolish and Install	Ea.	31	26.30		57.30	opening, cripples, blocking, nails and
Reinstall	Ea.		11.63		11.63	sub-sills for opening up to 3′ in a 2″ x
Minimum Charge	Job		174		174	6″ stud wall 8′ high.
4′ Wide						
Demolish	Ea.		11.75		11.75	Includes material and labor to install
Install	Ea.	33	14.55		47.55	header, double studs each side of
Demolish and Install	Ea.	33	26.30		59.30	opening, cripples, blocking, nails and
Reinstall	Ea.		11.63		11.63	sub-sills for opening up to 4′ in a 2″ x
Minimum Charge	Job		174		174	6″ stud wall 8′ high.
5′ Wide						
Demolish	Ea.		11.75		11.75	Includes material and labor to install
Install	Ea.	37.50	14.55		52.05	header, double studs each side of
Demolish and Install	Ea.	37.50	26.30		63.80	opening, cripples, blocking, nails and
Reinstall	Ea.		11.63		11.63	sub-sills for opening up to 5′ in a 2″ x
Minimum Charge	Job		174		174	6″ stud wall 8′ high.
6′ Wide						
Demolish	Ea.		11.75		11.75	Includes material and labor to install
Install	Ea.	41.50	14.55		56.05	header, double studs each side of
Demolish and Install	Ea.	41.50	26.30		67.80	opening, cripples, blocking, nails and
Reinstall	Ea.		11.63		11.63	sub-sills for opening up to 6′ in a 2″ x
Minimum Charge	Job		174		174	6″ stud wall 8′ high.
7′ Wide						
Demolish	Ea.		11.75		11.75	Includes material and labor to install
Install	Ea.	50.50	14.55		65.05	header, double studs each side of
Demolish and Install	Ea.	50.50	26.30		76.80	opening, cripples, blocking, nails and
Reinstall	Ea.		11.63		11.63	sub-sills for opening up to 7′ in a 2″ x
Minimum Charge	Job		174		174	6″ stud wall 8′ high.

Rough Frame / Structure

Rough-in Opening

	Unit	Material	Labor	Equip.	Total	Specification
8' Wide						
Demolish	Ea.		12.85		12.85	Includes material and labor to install
Install	Ea.	55	15.85		70.85	header, double studs each side of
Demolish and Install	Ea.	55	28.70		83.70	opening, cripples, blocking, nails and
Reinstall	Ea.		12.68		12.68	sub-sills for opening up to 8' in a 2" x
Minimum Charge	Job		174		174	6" stud wall 8' high.
10' Wide						
Demolish	Ea.		12.85		12.85	Includes material and labor to install
Install	Ea.	67.50	15.85		83.35	header, double studs each side of
Demolish and Install	Ea.	67.50	28.70		96.20	opening, cripples, blocking, nails and
Reinstall	Ea.		12.68		12.68	sub-sills for opening up to 10' in a 2"
Minimum Charge	Job		174		174	x 6" stud wall 8' high.
12' Wide						
Demolish	Ea.		12.85		12.85	Includes material and labor to install
Install	Ea.	98	15.85		113.85	header, double studs each side of
Demolish and Install	Ea.	98	28.70		126.70	opening, cripples, blocking, nails and
Reinstall	Ea.		12.68		12.68	sub-sills for opening up to 12' in a 2"
Minimum Charge	Job		174		174	x 6" stud wall 8' high.

Glue-Laminated Beams

	Unit	Material	Labor	Equip.	Total	Specification
3-1/2" x 6"						
Demolish	L.F.		2.12		2.12	Includes material and labor to install
Install	L.F.	5.40	2.61	1.33	9.34	glue-laminated wood beam.
Demolish and Install	L.F.	5.40	4.73	1.33	11.46	
Reinstall	L.F.		2.09	1.06	3.15	
Clean	L.F.	.10	.32		.42	
Paint	L.F.	.32	.90		1.22	
Minimum Charge	Job		174		174	
3-1/2" x 9"						
Demolish	L.F.		2.12		2.12	Includes material and labor to install
Install	L.F.	7.20	2.61	1.33	11.14	glue-laminated wood beam.
Demolish and Install	L.F.	7.20	4.73	1.33	13.26	
Reinstall	L.F.		2.09	1.06	3.15	
Clean	L.F.	.13	.33		.46	
Paint	L.F.	.41	1.16		1.57	
Minimum Charge	Job		174		174	
3-1/2" x 12"						
Demolish	L.F.		3.19		3.19	Includes material and labor to install
Install	L.F.	9.65	2.61	1.33	13.59	glue-laminated wood beam.
Demolish and Install	L.F.	9.65	5.80	1.33	16.78	
Reinstall	L.F.		2.09	1.06	3.15	
Clean	L.F.	.17	.34		.51	
Paint	L.F.	.51	1.45		1.96	
Minimum Charge	Job		174		174	
3-1/2" x 15"						
Demolish	L.F.		3.19		3.19	Includes material and labor to install
Install	L.F.	12	2.70	1.37	16.07	glue-laminated wood beam.
Demolish and Install	L.F.	12	5.89	1.37	19.26	
Reinstall	L.F.		2.16	1.10	3.26	
Clean	L.F.	.20	.35		.55	
Paint	L.F.	.59	1.70		2.29	
Minimum Charge	Job		174		174	

Rough Frame / Structure

Glue-Laminated Beams	Unit	Material	Labor	Equip.	Total	Specification
3-1/2″ x 18″						
Demolish	L.F.		3.99		3.99	Includes material and labor to install
Install	L.F.	16.10	2.70	1.37	20.17	glue-laminated wood beam.
Demolish and Install	L.F.	16.10	6.69	1.37	24.16	
Reinstall	L.F.		2.16	1.10	3.26	
Clean	L.F.	.23	.36		.59	
Paint	L.F.	.70	1.97		2.67	
Minimum Charge	Job		174		174	
5-1/8″ x 6″						
Demolish	L.F.		3.19		3.19	Includes material and labor to install
Install	L.F.	7.85	2.61	1.33	11.79	glue-laminated wood beam.
Demolish and Install	L.F.	7.85	5.80	1.33	14.98	
Reinstall	L.F.		2.09	1.06	3.15	
Clean	L.F.	.11	.32		.43	
Paint	L.F.	.34	.98		1.32	
Minimum Charge	Job		174		174	
5-1/8″ x 9″						
Demolish	L.F.		3.99		3.99	Includes material and labor to install
Install	L.F.	11.85	2.61	1.33	15.79	glue-laminated wood beam.
Demolish and Install	L.F.	11.85	6.60	1.33	19.78	
Reinstall	L.F.		2.09	1.06	3.15	
Clean	L.F.	.14	.33		.47	
Paint	L.F.	.44	1.26		1.70	
Minimum Charge	Job		174		174	
5-1/8″ x 12″						
Demolish	L.F.		4.74		4.74	Includes material and labor to install
Install	L.F.	15.75	2.61	1.33	19.69	glue-laminated wood beam.
Demolish and Install	L.F.	15.75	7.35	1.33	24.43	
Reinstall	L.F.		2.09	1.06	3.15	
Clean	L.F.	.18	.34		.52	
Paint	L.F.	.54	1.53		2.07	
Minimum Charge	Job		174		174	
5-1/8″ x 18″						
Demolish	L.F.		7.95		7.95	Includes material and labor to install
Install	L.F.	23.50	2.70	1.37	27.57	glue-laminated wood beam.
Demolish and Install	L.F.	23.50	10.65	1.37	35.52	
Reinstall	L.F.		2.16	1.10	3.26	
Clean	L.F.	.24	.36		.60	
Paint	L.F.	.63	2.06		2.69	
Minimum Charge	Job		174		174	
6-3/4″ x 12″						
Demolish	L.F.		4.74		4.74	Includes material and labor to install
Install	L.F.	20.50	2.70	1.37	24.57	glue-laminated wood beam.
Demolish and Install	L.F.	20.50	7.44	1.37	29.31	
Reinstall	L.F.		2.16	1.10	3.26	
Clean	L.F.	.20	.35		.55	
Paint	L.F.	.59	1.70		2.29	
Minimum Charge	Job		174		174	
6-3/4″ x 15″						
Demolish	L.F.		7.95		7.95	Includes material and labor to install
Install	L.F.	26	2.70	1.37	30.07	glue-laminated wood beam.
Demolish and Install	L.F.	26	10.65	1.37	38.02	
Reinstall	L.F.		2.16	1.10	3.26	
Clean	L.F.	.23	.36		.59	
Paint	L.F.	.70	1.97		2.67	
Minimum Charge	Job		174		174	

Rough Frame / Structure

Glue-Laminated Beams

Glue-Laminated Beams		Unit	Material	Labor	Equip.	Total	Specification
6-3/4" x 18"							
	Demolish	L.F.		7.95		7.95	Includes material and labor to install
	Install	L.F.	31.50	2.80	1.42	35.72	glue-laminated wood beam.
	Demolish and Install	L.F.	31.50	10.75	1.42	43.67	
	Reinstall	L.F.		2.24	1.14	3.38	
	Clean	L.F.	.26	.38		.64	
	Paint	L.F.	.79	2.24		3.03	
	Minimum Charge	Job		174		174	
Hardware							
5-1/4" Glue-Lam Seat							
	Install	Ea.	63.50	1.94		65.44	Includes labor and material to install
	Minimum Charge	Job		174		174	beam hangers.
6-3/4" Glue-Lam Seat							
	Install	Ea.	65.50	1.94		67.44	Includes labor and material to install
	Minimum Charge	Job		174		174	beam hangers.
8-3/4" Glue-Lam Seat							
	Install	Ea.	75	1.94		76.94	Includes labor and material to install
	Minimum Charge	Job		174		174	beam hangers.
Earthquake Strapping							
	Install	Ea.	2.19	2.18		4.37	Includes labor and material to replace
	Minimum Charge	Job		174		174	an earthquake strap.
Hurricane Clips							
	Install	Ea.	1.93	2.41		4.34	Includes labor and material to install a
	Minimum Charge	Job		174		174	hurricane clip.

Metal Stud Framing

Metal Stud Framing		Unit	Material	Labor	Equip.	Total	Specification
16" O.C. System							
4", 16 Ga.							
	Demolish	S.F.		1.04		1.04	Includes material and labor to install
	Install	S.F.	1.40	1.07		2.47	load bearing cold rolled metal stud
	Demolish and Install	S.F.	1.40	2.11		3.51	walls, to 10' high, including studs, top
	Reinstall	S.F.		.86		.86	and bottom track and screws.
	Minimum Charge	Job		174		174	
6", 16 Ga.							
	Demolish	S.F.		1.04		1.04	Includes material and labor to install
	Install	S.F.	1.76	1.09		2.85	load bearing cold rolled metal stud
	Demolish and Install	S.F.	1.76	2.13		3.89	walls, to 10' high, including studs, top
	Reinstall	S.F.		.87		.87	and bottom track and screws.
	Minimum Charge	Job		174		174	
4", 25 Ga.							
	Demolish	S.F.		.72		.72	Includes material and labor to install
	Install	S.F.	.52	.73		1.25	cold rolled metal stud walls, to 10'
	Demolish and Install	S.F.	.52	1.45		1.97	high, including studs, top and bottom
	Reinstall	S.F.		.59		.59	track and screws.
	Minimum Charge	Job		174		174	
6", 25 Ga.							
	Demolish	S.F.		.72		.72	Includes material and labor to install
	Install	S.F.	.64	.74		1.38	cold rolled metal stud walls, to 10'
	Demolish and Install	S.F.	.64	1.46		2.10	high, including studs, top and bottom
	Reinstall	S.F.		.59		.59	track and screws.
	Minimum Charge	Job		174		174	

Rough Frame / Structure

Metal Stud Framing

Metal Stud Framing		Unit	Material	Labor	Equip.	Total	Specification
24″ O.C. System							
4″, 25 Ga.							
	Demolish	S.F.		.62		.62	Includes material and labor to install
	Install	S.F.	.39	.47		.86	cold rolled metal stud walls, to 10′
	Demolish and Install	S.F.	.39	1.09		1.48	high, including studs, top and bottom
	Reinstall	S.F.		.38		.38	track and screws.
	Minimum Charge	Job		174		174	
6″, 25 Ga.							
	Demolish	S.F.		.62		.62	Includes material and labor to install
	Install	S.F.	.47	.48		.95	cold rolled metal stud walls, to 10′
	Demolish and Install	S.F.	.47	1.10		1.57	high, including studs, top and bottom
	Reinstall	S.F.		.38		.38	track and screws.
	Minimum Charge	Job		174		174	
4″, 16 Ga.							
	Demolish	L.F.		7		7	Includes material and labor to install
	Install	S.F.	1.01	.78		1.79	load bearing cold rolled metal stud
	Demolish and Install	S.F.	1.01	7.78		8.79	walls, to 10′ high, including studs, top
	Reinstall	S.F.		.62		.62	and bottom track and screws.
	Minimum Charge	Job		174		174	
6″, 16 Ga.							
	Demolish	L.F.		7		7	Includes material and labor to install
	Install	S.F.	1.28	.79		2.07	load bearing cold rolled metal stud
	Demolish and Install	S.F.	1.28	7.79		9.07	walls, to 10′ high, including studs, top
	Reinstall	S.F.		.63		.63	and bottom track and screws.
	Minimum Charge	Job		174		174	

Metal Joist

Metal Joist		Unit	Material	Labor	Equip.	Total	Specification
Bar Joist							
18K9							
	Demolish	L.F.		.84	.41	1.25	Includes material, labor and
	Install	L.F.	8	1.93	.98	10.91	equipment to install open web joist.
	Demolish and Install	L.F.	8	2.77	1.39	12.16	
	Reinstall	L.F.		1.54	.78	2.32	
	Minimum Charge	Job		1575	795	2370	
16K6							
	Demolish	L.F.		.84	.41	1.25	Includes material, labor and
	Install	L.F.	5.40	1.30	.09	6.79	equipment to install open web joist.
	Demolish and Install	L.F.	5.40	2.14	.50	8.04	
	Reinstall	L.F.		1.04	.08	1.12	
	Minimum Charge	Job		216		216	

Exterior Sheathing

Exterior Sheathing		Unit	Material	Labor	Equip.	Total	Specification
CDX Plywood							
5/16″							
	Demolish	S.F.		.48		.48	Cost includes material and labor to
	Install	S.F.	.68	.44		1.12	install 5/16″ CDX plywood sheathing.
	Demolish and Install	S.F.	.68	.92		1.60	
	Minimum Charge	Job		174		174	
3/8″							
	Demolish	S.F.		.49		.49	Cost includes material and labor to
	Install	S.F.	.48	.58		1.06	install 3/8″ CDX plywood sheathing.
	Demolish and Install	S.F.	.48	1.07		1.55	
	Minimum Charge	Job		174		174	

Rough Frame / Structure

Exterior Sheathing

	Unit	Material	Labor	Equip.	Total	Specification
1/2″						
Demolish	S.F.		.50		.50	Cost includes material and labor to
Install	S.F.	.52	.62		1.14	install 1/2″ CDX plywood sheathing.
Demolish and Install	S.F.	.52	1.12		1.64	
Minimum Charge	Job		174		174	
5/8″						
Demolish	S.F.		.51		.51	Cost includes material and labor to
Install	S.F.	.65	.66		1.31	install 5/8″ CDX plywood sheathing.
Demolish and Install	S.F.	.65	1.17		1.82	
Minimum Charge	Job		174		174	
3/4″						
Demolish	S.F.		.53		.53	Cost includes material and labor to
Install	S.F.	.78	.72		1.50	install 3/4″ CDX plywood sheathing.
Demolish and Install	S.F.	.78	1.25		2.03	
Minimum Charge	Job		174		174	
OSB						
1/2″						
Demolish	S.F.		.50		.50	Cost includes material and labor to
Install	S.F.	.77	.26		1.03	install 4′ x 8′ x 1/2″ OSB sheathing.
Demolish and Install	S.F.	.77	.76		1.53	
Minimum Charge	Job		174		174	
5/8″						
Demolish	S.F.		.51		.51	Cost includes material and labor to
Install	S.F.	.88	.28		1.16	install 4′ x 8′ x 5/8″ OSB sheathing.
Demolish and Install	S.F.	.88	.79		1.67	
Minimum Charge	Job		174		174	
Vapor Barrier						
Black Paper						
Install	S.F.	.04	.09		.13	Includes material and labor to install
Minimum Charge	Job		174		174	#15 felt building paper.

Plywood Sheathing

	Unit	Material	Labor	Equip.	Total	Specification
Finish Plywood						
5/16″						
Demolish	S.F.		.48		.48	Includes material and labor to install
Install	S.F.	.86	.22		1.08	exterior 5/16″ AC plywood on walls.
Demolish and Install	S.F.	.86	.70		1.56	
Minimum Charge	Job		174		174	
3/8″						
Demolish	S.F.		.49		.49	Includes material and labor to install
Install	S.F.	.86	.29		1.15	exterior 3/8″ AC plywood on walls.
Demolish and Install	S.F.	.86	.78		1.64	
Minimum Charge	Job		174		174	
1/2″						
Demolish	S.F.		.50		.50	Includes material and labor to install
Install	S.F.	.99	.31		1.30	exterior 1/2″ AC plywood on walls.
Demolish and Install	S.F.	.99	.81		1.80	
Minimum Charge	Job		174		174	
5/8″						
Demolish	S.F.		.51		.51	Includes material and labor to install
Install	S.F.	1.20	.31		1.51	exterior 5/8″ AC plywood on walls.
Demolish and Install	S.F.	1.20	.82		2.02	
Reinstall	S.F.		.31		.31	
Minimum Charge	Job		174		174	

Rough Frame / Structure

Plywood Sheathing

	Unit	Material	Labor	Equip.	Total	Specification
3/4″						Includes material and labor to install
Demolish	S.F.		.53		.53	exterior 3/4″ AC plywood on walls.
Install	S.F.	1.22	.31		1.53	
Demolish and Install	S.F.	1.22	.84		2.06	
Reinstall	S.F.		.31		.31	
Minimum Charge	Job		174		174	

Stairs

	Unit	Material	Labor	Equip.	Total	Specification
Job-built						
Treads and Risers						Includes material and labor to install
Demolish	Ea.		35.50		35.50	three 2″ x 12″ stringers, treads and
Install	Ea.	500	43.50		543.50	risers of 3/4″ CDX plywood, installed
Demolish and Install	Ea.	500	79		579	in a straight or ″L″ shaped run
Clean	Flight	3.11	22		25.11	including carpet.
Paint	Ea.	.10	61		61.10	
Minimum Charge	Job		174		174	
Landing						Includes material and labor to install
Demolish	S.F.		.28		.28	landing framing, 3/4″ CDX plywood
Install	S.F.	59	2.42		61.42	surface and carpet.
Demolish and Install	S.F.	59	2.70		61.70	
Clean	S.F.		.32		.32	
Paint	S.F.	.29	.75		1.04	
Minimum Charge	Job		174		174	

Ceiling Framing

	Unit	Material	Labor	Equip.	Total	Specification
16″ O.C. System						
2″ x 6″ Joists						Cost includes material and labor to
Demolish	S.F.		.44		.44	install 2″ x 6″ joists including end and
Install	S.F.	.48	.57		1.05	header joist installed 16″ O.C. Does
Demolish and Install	S.F.	.48	1.01		1.49	not include beams, ledger strips,
Reinstall	S.F.		.46		.46	blocking or bridging.
Minimum Charge	Job		174		174	
2″ x 8″ Joists						Cost includes material and labor to
Demolish	S.F.		.45		.45	install 2″ x 8″ joists including end and
Install	S.F.	.70	.71		1.41	header joist installed 16″ O.C. Does
Demolish and Install	S.F.	.70	1.16		1.86	not include beams, ledger strips,
Reinstall	S.F.		.56		.56	blocking or bridging.
Minimum Charge	Job		174		174	
2″ x 10″ Joists						Cost includes material and labor to
Demolish	S.F.		.46		.46	install 2″ x 10″ joists including end
Install	S.F.	.99	.83		1.82	and header joist installed 16″ O.C.
Demolish and Install	S.F.	.99	1.29		2.28	Does not include beams, ledger strips,
Reinstall	S.F.		.67		.67	blocking or bridging.
Minimum Charge	Job		174		174	
2″ x 12″ Joists						Cost includes material and labor to
Demolish	S.F.		.48		.48	install 2″ x 12″ joists including end
Install	S.F.	1.20	1.02		2.22	and header joist installed 16″ O.C.
Demolish and Install	S.F.	1.20	1.50		2.70	Does not include beams, ledger strips,
Reinstall	S.F.		.82		.82	blocking or bridging.
Minimum Charge	Job		174		174	

Rough Frame / Structure

Ceiling Framing		Unit	Material	Labor	Equip.	Total	Specification
Block / Bridge							
	Demolish	Ea.		.88		.88	Includes material and labor to install
	Install	Ea.	1.75	1.96		3.71	set of cross bridging or per block of
	Demolish and Install	Ea.	1.75	2.84		4.59	solid bridging for 2" x 10" joists cut to
	Minimum Charge	Job		174		174	size on site.
24" O.C. System							
2" x 6" Joists							
	Demolish	S.F.		.29		.29	Cost includes material and labor to
	Install	S.F.	.32	.38		.70	install 2" x 6" joists including end and
	Demolish and Install	S.F.	.32	.67		.99	header joist installed 24" O.C. Does
	Reinstall	S.F.		.31		.31	not include beams, ledger strips,
	Minimum Charge	Job		174		174	blocking or bridging.
2" x 8" Joists							
	Demolish	S.F.		.30		.30	Cost includes material and labor to
	Install	S.F.	.47	.47		.94	install 2" x 8" joists including end and
	Demolish and Install	S.F.	.47	.77		1.24	header joist installed 24" O.C. Does
	Reinstall	S.F.		.38		.38	not include beams, ledger strips,
	Minimum Charge	Job		174		174	blocking or bridging.
2" x 10" Joists							
	Demolish	S.F.		.31		.31	Cost includes material and labor to
	Install	S.F.	.67	.56		1.23	install 2" x 10" joists including end
	Demolish and Install	S.F.	.67	.87		1.54	and header joist installed 24" O.C.
	Reinstall	S.F.		.45		.45	Does not include beams, ledger strips,
	Minimum Charge	Job		174		174	blocking or bridging.
2" x 12" Joists							
	Demolish	S.F.		.32		.32	Cost includes material and labor to
	Install	S.F.	.80	.68		1.48	install 2" x 12" joists including end
	Demolish and Install	S.F.	.80	1		1.80	and header joist installed 24" O.C.
	Reinstall	S.F.		.54		.54	Does not include beams, ledger strips,
	Minimum Charge	Job		174		174	blocking or bridging.
Block / Bridge							
	Demolish	Ea.		.88		.88	Includes material and labor to install
	Install	Ea.	1.75	1.96		3.71	set of cross bridging or per block of
	Demolish and Install	Ea.	1.75	2.84		4.59	solid bridging for 2" x 10" joists cut to
	Minimum Charge	Job		174		174	size on site.
Ledger Strips							
1" x 2"							
	Demolish	L.F.		.24		.24	Cost includes material and labor to
	Install	L.F.	.13	1.01		1.14	install 1" x 2" ledger strip nailed to the
	Demolish and Install	L.F.	.13	1.25		1.38	face of studs, beams or joist.
	Minimum Charge	Job		174		174	
1" x 3"							
	Demolish	L.F.		.24		.24	Includes material and labor to install
	Install	L.F.	.51	1.16		1.67	up to 1" x 4" ledger strip nailed to the
	Demolish and Install	L.F.	.51	1.40		1.91	face of studs, beams or joist.
	Minimum Charge	Job		174		174	
1" x 4"							
	Demolish	L.F.		.24		.24	Includes material and labor to install
	Install	L.F.	.51	1.16		1.67	up to 1" x 4" ledger strip nailed to the
	Demolish and Install	L.F.	.51	1.40		1.91	face of studs, beams or joist.
	Minimum Charge	Job		174		174	
2" x 2"							
	Demolish	L.F.		.26		.26	Cost includes material and labor to
	Install	L.F.	.33	1.06		1.39	install 2" x 2" ledger strip nailed to the
	Demolish and Install	L.F.	.33	1.32		1.65	face of studs, beams or joist.
	Minimum Charge	Job		174		174	

Rough Frame / Structure

Ceiling Framing

	Unit	Material	Labor	Equip.	Total	Specification
2″ x 4″						
Demolish	L.F.		.28		.28	Cost includes material and labor to
Install	L.F.	.41	1.40		1.81	install 2″ x 4″ ledger strip nailed to the
Demolish and Install	L.F.	.41	1.68		2.09	face of studs, beams or joist.
Reinstall	L.F.		1.12		1.12	
Minimum Charge	Job		174		174	
Ledger Boards						
Demolish	L.F.		.28		.28	Cost includes material and labor to
Install	L.F.	1.87	1.94		3.81	install 2″ x 4″ ledger board fastened to
Demolish and Install	L.F.	1.87	2.22		4.09	a wall, joists or studs.
Reinstall	L.F.		1.55		1.55	
Minimum Charge	Job		174		174	
4″ x 6″						
Demolish	L.F.		.47		.47	Cost includes material and labor to
Install	L.F.	4.62	2.18		6.80	install 4″ x 6″ ledger board fastened to
Demolish and Install	L.F.	4.62	2.65		7.27	a wall, joists or studs.
Reinstall	L.F.		1.74		1.74	
Minimum Charge	Job		174		174	
4″ x 8″						
Demolish	L.F.		.63		.63	Cost includes material and labor to
Install	L.F.	5.65	2.18		7.83	install 4″ x 8″ ledger board fastened to
Demolish and Install	L.F.	5.65	2.81		8.46	a wall, joists or studs.
Reinstall	L.F.		1.74		1.74	
Minimum Charge	Job		174		174	
Earthquake Strapping						
Install	Ea.	2.19	2.18		4.37	Includes labor and material to replace
Minimum Charge	Job		174		174	an earthquake strap.
Hurricane Clips						
Install	Ea.	1.93	2.41		4.34	Includes labor and material to install a
Minimum Charge	Job		174		174	hurricane clip.

Roof Framing

	Unit	Material	Labor	Equip.	Total	Specification
16″ O.C. System						
2″ x 4″ Rafters						
Demolish	S.F.		.64		.64	Cost includes material and labor to
Install	S.F.	.31	.31		.62	install 2″ x 4″, 16″ O.C. rafter framing
Demolish and Install	S.F.	.31	.95		1.26	for flat, shed or gable roofs, 25' span,
Reinstall	S.F.		.25		.25	up to 5/12 slope, per S.F. of roof.
Minimum Charge	Job		174		174	
2″ x 6″ Rafters						
Demolish	S.F.		.67		.67	Cost includes material and labor to
Install	S.F.	.47	.33		.80	install 2″ x 6″, 16″ O.C. rafter framing
Demolish and Install	S.F.	.47	1		1.47	for flat, shed or gable roofs, 25' span,
Reinstall	S.F.		.26		.26	up to 5/12 slope, per S.F. of roof.
Minimum Charge	Job		174		174	
2″ x 8″ Rafters						
Demolish	S.F.		.69		.69	Cost includes material and labor to
Install	S.F.	.69	.35		1.04	install 2″ x 8″, 16″ O.C. rafter framing
Demolish and Install	S.F.	.69	1.04		1.73	for flat, shed or gable roofs, 25' span,
Reinstall	S.F.		.28		.28	up to 5/12 slope, per S.F. of roof.
Minimum Charge	Job		174		174	

Rough Frame / Structure

Roof Framing		Unit	Material	Labor	Equip.	Total	Specification
2" x 10" Rafters							
	Demolish	S.F.		.69		.69	Cost includes material and labor to
	Install	S.F.	.98	.53		1.51	install 2" x 10", 16" O.C. rafter
	Demolish and Install	S.F.	.98	1.22		2.20	framing for flat, shed or gable roofs,
	Reinstall	S.F.		.42		.42	25' span, up to 5/12 slope, per S.F.
	Minimum Charge	Job		174		174	of roof.
2" x 12" Rafters							
	Demolish	S.F.		.70		.70	Cost includes material and labor to
	Install	S.F.	1.18	.57		1.75	install 2" x 12", 16" O.C. rafter
	Demolish and Install	S.F.	1.18	1.27		2.45	framing for flat, shed or gable roofs,
	Reinstall	S.F.		.46		.46	25' span, up to 5/12 slope, per S.F.
	Minimum Charge	Job		174		174	of roof.
24" O.C. System							
2" x 4" Rafters							
	Demolish	S.F.		.48		.48	Cost includes material and labor to
	Install	S.F.	.21	.23		.44	install 2" x 4", 24" O.C. rafter framing
	Demolish and Install	S.F.	.21	.71		.92	for flat, shed or gable roofs, 25' span,
	Reinstall	S.F.		.19		.19	up to 5/12 slope, per S.F. of roof.
	Minimum Charge	Job		174		174	
2" x 6" Rafters							
	Demolish	S.F.		.51		.51	Cost includes material and labor to
	Install	S.F.	.32	.25		.57	install 2" x 6", 24" O.C. rafter framing
	Demolish and Install	S.F.	.32	.76		1.08	for flat, shed or gable roofs, 25' span,
	Reinstall	S.F.		.20		.20	up to 5/12 slope, per S.F. of roof.
	Minimum Charge	Job		174		174	
2" x 8" Rafters							
	Demolish	S.F.		.52		.52	Cost includes material and labor to
	Install	S.F.	.46	.26		.72	install 2" x 8", 24" O.C. rafter framing
	Demolish and Install	S.F.	.46	.78		1.24	for flat, shed or gable roofs, 25' span,
	Reinstall	S.F.		.21		.21	up to 5/12 slope, per S.F. of roof.
	Minimum Charge	Job		174		174	
2" x 10" Rafters							
	Demolish	S.F.		.52		.52	Cost includes material and labor to
	Install	S.F.	.66	.40		1.06	install 2" x 10", 24" O.C. rafter
	Demolish and Install	S.F.	.66	.92		1.58	framing for flat, shed or gable roofs,
	Reinstall	S.F.		.32		.32	25' span, up to 5/12 slope, per S.F.
	Minimum Charge	Job		174		174	of roof.
2" x 12" Rafters							
	Demolish	S.F.		.52		.52	Cost includes material and labor to
	Install	S.F.	.79	.43		1.22	install 2" x 12", 24" O.C. rafter
	Demolish and Install	S.F.	.79	.95		1.74	framing for flat, shed or gable roofs,
	Reinstall	S.F.		.35		.35	25' span, up to 5/12 slope, per S.F.
	Minimum Charge	Job		174		174	of roof.
Rafter							
2" x 4"							
	Demolish	L.F.		.65		.65	Cost includes material and labor to
	Install	L.F.	.41	1.01		1.42	install 2" x 4" rafters for flat, shed or
	Demolish and Install	L.F.	.41	1.66		2.07	gable roofs, up to 5/12 slope, 25'
	Reinstall	L.F.		.81		.81	span, per L.F.
	Minimum Charge	Job		174		174	
2" x 6"							
	Demolish	L.F.		.66		.66	Cost includes material and labor to
	Install	L.F.	.64	.87		1.51	install 2" x 6" rafters for flat, shed or
	Demolish and Install	L.F.	.64	1.53		2.17	gable roofs, up to 5/12 slope, 25'
	Reinstall	L.F.		.70		.70	span, per L.F.
	Minimum Charge	Job		174		174	

50

Rough Frame / Structure

Roof Framing

Roof Framing		Unit	Material	Labor	Equip.	Total	Specification
2″ x 8″							
	Demolish	L.F.		.67		.67	Cost includes material and labor to
	Install	L.F.	.92	.93		1.85	install 2″ x 8″ rafters for flat, shed or
	Demolish and Install	L.F.	.92	1.60		2.52	gable roofs, up to 5/12 slope, 25′
	Reinstall	L.F.		.74		.74	span, per L.F.
	Minimum Charge	Job		174		174	
2″ x 10″							
	Demolish	L.F.		.68		.68	Cost includes material and labor to
	Install	L.F.	1.31	1.41		2.72	install 2″ x 10″ rafters for flat, shed or
	Demolish and Install	L.F.	1.31	2.09		3.40	gable roofs, up to 5/12 slope, 25′
	Reinstall	L.F.		1.13		1.13	span, per L.F.
	Minimum Charge	Job		174		174	
2″ x 12″							
	Demolish	L.F.		.70		.70	Cost includes material and labor to
	Install	L.F.	1.57	1.53		3.10	install 2″ x 12″ rafters for flat, shed or
	Demolish and Install	L.F.	1.57	2.23		3.80	gable roofs, up to 5/12 slope, 25′
	Reinstall	L.F.		1.23		1.23	span, per L.F.
	Minimum Charge	Job		174		174	
2″x 4″ Valley / Jack							
	Demolish	L.F.		.65		.65	Includes material and labor to install
	Install	L.F.	.64	1.47		2.11	up to 2″ x 6″ valley/jack rafters for
	Demolish and Install	L.F.	.64	2.12		2.76	flat, shed or gable roofs, up to 5/12
	Reinstall	L.F.		1.17		1.17	slope, 25′ span, per L.F.
	Minimum Charge	Job		174		174	
2″x 6″ Valley / Jack							
	Demolish	L.F.		.66		.66	Includes material and labor to install
	Install	L.F.	.64	1.47		2.11	up to 2″ x 6″ valley/jack rafters for
	Demolish and Install	L.F.	.64	2.13		2.77	flat, shed or gable roofs, up to 5/12
	Reinstall	L.F.		1.17		1.17	slope, 25′ span, per L.F.
	Minimum Charge	Job		174		174	

Ridgeboard

Ridgeboard		Unit	Material	Labor	Equip.	Total	Specification
2″ x 4″							
	Demolish	L.F.		.63		.63	Cost includes material and labor to
	Install	L.F.	.41	.63		1.04	install 2″ x 4″ ridgeboard for flat, shed
	Demolish and Install	L.F.	.41	1.26		1.67	or gable roofs, up to 5/12 slope, 25′
	Reinstall	L.F.		.51		.51	span, per L.F.
	Minimum Charge	Job		174		174	
2″ x 6″							
	Demolish	L.F.		.65		.65	Cost includes material and labor to
	Install	L.F.	.64	1.40		2.04	install 2″ x 6″ ridgeboard for flat, shed
	Demolish and Install	L.F.	.64	2.05		2.69	or gable roofs, up to 5/12 slope, 25′
	Reinstall	L.F.		1.12		1.12	span, per L.F.
	Minimum Charge	Job		174		174	
2″ x 8″							
	Demolish	L.F.		.66		.66	Cost includes material and labor to
	Install	L.F.	.92	1.55		2.47	install 2″ x 8″ ridgeboard for flat, shed
	Demolish and Install	L.F.	.92	2.21		3.13	or gable roofs, up to 5/12 slope, 25′
	Reinstall	L.F.		1.24		1.24	span, per L.F.
	Minimum Charge	Job		174		174	
2″ x 10″							
	Demolish	L.F.		.68		.68	Cost includes material and labor to
	Install	L.F.	1.31	1.74		3.05	install 2″ x 10″ ridgeboard for flat,
	Demolish and Install	L.F.	1.31	2.42		3.73	shed or gable roofs, up to 5/12
	Reinstall	L.F.		1.40		1.40	slope, 25′ span, per L.F.
	Minimum Charge	Job		174		174	

Rough Frame / Structure

Roof Framing		Unit	Material	Labor	Equip.	Total	Specification
2" x 12"							
	Demolish	L.F.		.71		.71	Cost includes material and labor to
	Install	L.F.	1.57	1		2.57	install 2" x 12" ridgeboard for flat,
	Demolish and Install	L.F.	1.57	1.71		3.28	shed or gable roofs, up to 5/12
	Reinstall	L.F.		.80		.80	slope, 25' span, per L.F.
	Minimum Charge	Job		174		174	
Collar Beam							
1" x 6"							
	Demolish	L.F.		.50		.50	Cost includes material and labor to
	Install	L.F.	.72	.44		1.16	install 1" x 6" collar beam or tie for
	Demolish and Install	L.F.	.72	.94		1.66	roof framing.
	Reinstall	L.F.		.35		.35	
	Minimum Charge	Job		174		174	
2" x 6"							
	Demolish	L.F.		.58		.58	Cost includes material and labor to
	Install	L.F.	.64	.44		1.08	install 2" x 6" collar beam or tie for
	Demolish and Install	L.F.	.64	1.02		1.66	roof framing.
	Reinstall	L.F.		.35		.35	
	Minimum Charge	Job		174		174	
Purlins							
	Demolish	L.F.		.56		.56	Cost includes material and labor to
	Install	L.F.	.64	.39		1.03	install 2" x 6" purlins below roof
	Demolish and Install	L.F.	.64	.95		1.59	rafters.
	Reinstall	L.F.		.31		.31	
	Minimum Charge	Job		174		174	
2" x 8"							
	Demolish	L.F.		.57		.57	Cost includes material and labor to
	Install	L.F.	.92	.39		1.31	install 2" x 8" purlins below roof
	Demolish and Install	L.F.	.92	.96		1.88	rafters.
	Reinstall	L.F.		.31		.31	
	Minimum Charge	Job		174		174	
2" x 10"							
	Demolish	L.F.		.58		.58	Cost includes material and labor to
	Install	L.F.	1.31	.40		1.71	install 2" x 10" purlins below roof
	Demolish and Install	L.F.	1.31	.98		2.29	rafters.
	Reinstall	L.F.		.32		.32	
	Minimum Charge	Job		174		174	
2" x 12"							
	Demolish	L.F.		.58		.58	Cost includes material and labor to
	Install	L.F.	1.57	.40		1.97	install 2" x 12" purlins below roof
	Demolish and Install	L.F.	1.57	.98		2.55	rafters.
	Reinstall	L.F.		.32		.32	
	Minimum Charge	Job		174		174	
4" x 6"							
	Demolish	L.F.		.58		.58	Cost includes material and labor to
	Install	L.F.	3.06	.40		3.46	install 4" x 6" purlins below roof
	Demolish and Install	L.F.	3.06	.98		4.04	rafters.
	Reinstall	L.F.		.32		.32	
	Minimum Charge	Job		174		174	
4" x 8"							
	Demolish	L.F.		.59		.59	Cost includes material and labor to
	Install	L.F.	4.11	.41		4.52	install 4" x 8" purlins below roof
	Demolish and Install	L.F.	4.11	1		5.11	rafters.
	Reinstall	L.F.		.33		.33	
	Minimum Charge	Job		174		174	

Rough Frame / Structure

Roof Framing		Unit	Material	Labor	Equip.	Total	Specification
Ledger Board							
2″ x 6″							
	Demolish	L.F.		.28		.28	Cost includes material and labor to
	Install	L.F.	.64	1.57		2.21	install 2″ x 6″ ledger board nailed to
	Demolish and Install	L.F.	.64	1.85		2.49	the face of a rafter.
	Reinstall	L.F.		1.26		1.26	
	Minimum Charge	Job		174		174	
4″ x 6″							
	Demolish	L.F.		.47		.47	Cost includes material and labor to
	Install	L.F.	3.06	2.31		5.37	install 4″ x 6″ ledger board nailed to
	Demolish and Install	L.F.	3.06	2.78		5.84	the face of a rafter.
	Reinstall	L.F.		1.85		1.85	
	Minimum Charge	Job		174		174	
4″ x 8″							
	Demolish	L.F.		.63		.63	Cost includes material and labor to
	Install	L.F.	4.11	2.66		6.77	install 4″ x 8″ ledger board nailed to
	Demolish and Install	L.F.	4.11	3.29		7.40	the face of a rafter.
	Reinstall	L.F.		2.13		2.13	
	Minimum Charge	Job		174		174	
Outriggers							
2″ x 4″							
	Demolish	L.F.		.19		.19	Cost includes material and labor to
	Install	L.F.	.41	1.40		1.81	install 2″ x 4″ outrigger rafters for flat,
	Demolish and Install	L.F.	.41	1.59		2	shed or gabled roofs, up to 5/12
	Reinstall	L.F.		1.12		1.12	slope, per L.F.
	Minimum Charge	Job		174		174	
2″ x 6″							
	Demolish	L.F.		.29		.29	Cost includes material and labor to
	Install	L.F.	.64	1.40		2.04	install 2″ x 6″ outrigger rafters for flat,
	Demolish and Install	L.F.	.64	1.69		2.33	shed or gabled roofs, up to 5/12
	Reinstall	L.F.		1.12		1.12	slope, per L.F.
	Minimum Charge	Job		174		174	
Lookout Rafter							
2″ x 4″							
	Demolish	L.F.		.65		.65	Cost includes material and labor to
	Install	L.F.	.41	.58		.99	install 2″ x 4″ lookout rafters for flat,
	Demolish and Install	L.F.	.41	1.23		1.64	shed or gabled roofs, up to 5/12
	Reinstall	L.F.		.47		.47	slope, per L.F.
	Minimum Charge	Job		174		174	
2″ x 6″							
	Demolish	L.F.		.66		.66	Cost includes material and labor to
	Install	L.F.	.64	.59		1.23	install 2″ x 6″ lookout rafters for flat,
	Demolish and Install	L.F.	.64	1.25		1.89	shed or gabled roofs, up to 5/12
	Reinstall	L.F.		.47		.47	slope, per L.F.
	Minimum Charge	Job		174		174	
Fly Rafter							
2″ x 4″							
	Demolish	L.F.		.65		.65	Cost includes material and labor to
	Install	L.F.	.41	.58		.99	install 2″ x 4″ fly rafters for flat, shed
	Demolish and Install	L.F.	.41	1.23		1.64	or gabled roofs, up to 5/12 slope,
	Reinstall	L.F.		.47		.47	per L.F.
	Minimum Charge	Job		174		174	

Rough Frame / Structure

Roof Framing

	Unit	Material	Labor	Equip.	Total	Specification
2" x 6"						
Demolish	L.F.		.66		.66	Cost includes material and labor to
Install	L.F.	.64	.59		1.23	install 2" x 6" fly rafters for flat, shed
Demolish and Install	L.F.	.64	1.25		1.89	or gabled roofs, up to 5/12 slope,
Reinstall	L.F.		.47		.47	per L.F.
Minimum Charge	Job		174		174	

Residential Trusses

	Unit	Material	Labor	Equip.	Total	Specification
W or Fink Truss						
24' Span						
Demolish	Ea.		26.50	13.40	39.90	Includes material, labor and
Install	Ea.	82.50	26	13.25	121.75	equipment to install wood gang-nailed
Demolish and Install	Ea.	82.50	52.50	26.65	161.65	residential truss, up to 24' span.
Reinstall	Ea.		26.12	13.27	39.39	
Minimum Charge	Job		174		174	
28' Span						
Demolish	Ea.		28	14.25	42.25	Includes material, labor and
Install	Ea.	76	29.50	15.05	120.55	equipment to install wood gang-nailed
Demolish and Install	Ea.	76	57.50	29.30	162.80	residential truss, up to 28' span.
Reinstall	Ea.		29.57	15.03	44.60	
Minimum Charge	Job		174		174	
32' Span						
Demolish	Ea.		31	15.80	46.80	Includes material, labor and
Install	Ea.	110	31.50	15.95	157.45	equipment to install wood gang-nailed
Demolish and Install	Ea.	110	62.50	31.75	204.25	residential truss, up to 32' span.
Reinstall	Ea.		31.34	15.93	47.27	
Minimum Charge	Job		174		174	
36' Span						
Demolish	Ea.		33.50	17	50.50	Includes material, labor and
Install	Ea.	146	34	17.30	197.30	equipment to install wood gang-nailed
Demolish and Install	Ea.	146	67.50	34.30	247.80	residential truss, up to 36' span.
Reinstall	Ea.		34.07	17.31	51.38	
Minimum Charge	Job		174		174	
40' Span						
Demolish	Ea.		36.50	18.45	54.95	Includes material, labor and
Install	Ea.	173	36.50	18.50	228	equipment to install wood gang-nailed
Demolish and Install	Ea.	173	73	36.95	282.95	residential truss, up to 40' span.
Reinstall	Ea.		36.45	18.52	54.97	
Minimum Charge	Job		174		174	
Gable End						
24' Span						
Demolish	Ea.		26.50	13.40	39.90	Includes material, labor and
Install	Ea.	116	28	14.20	158.20	equipment to install wood gang-nailed
Demolish and Install	Ea.	116	54.50	27.60	198.10	residential gable end truss, up to 24'
Reinstall	Ea.		27.99	14.22	42.21	span.
Clean	Ea.	1.30	34		35.30	
Paint	Ea.	3.96	40.50		44.46	
Minimum Charge	Job		174		174	
28' Span						
Demolish	Ea.		28	14.25	42.25	Includes material, labor and
Install	Ea.	110	33.50	16.95	160.45	equipment to install wood gang-nailed
Demolish and Install	Ea.	110	61.50	31.20	202.70	residential gable end truss, up to 30'
Reinstall	Ea.		33.34	16.94	50.28	span.
Clean	Ea.	1.35	46		47.35	
Paint	Ea.	4.13	55.50		59.63	
Minimum Charge	Job		174		174	

Rough Frame / Structure

Residential Trusses

	Unit	Material	Labor	Equip.	Total	Specification
32' Span						
Demolish	Ea.		31	15.80	46.80	Includes material, labor and
Install	Ea.	123	34	17.30	174.30	equipment to install wood gang-nailed
Demolish and Install	Ea.	123	65	33.10	221.10	residential gable end truss, up to 32'
Reinstall	Ea.		34.07	17.31	51.38	span.
Clean	Ea.	1.42	60.50		61.92	
Paint	Ea.	4.32	72.50		76.82	
Minimum Charge	Job		174		174	
36' Span						
Demolish	Ea.		33.50	17	50.50	Includes material, labor and
Install	Ea.	146	39	19.90	204.90	equipment to install wood gang-nailed
Demolish and Install	Ea.	146	72.50	36.90	255.40	residential gable end truss, up to 36'
Reinstall	Ea.		39.18	19.91	59.09	span.
Clean	Ea.	1.49	76.50		77.99	
Paint	Ea.	4.52	91.50		96.02	
Minimum Charge	Job		174		174	
40' Span						
Demolish	Ea.		36.50	18.45	54.95	Includes material, labor and
Install	Ea.	164	41	21	226	equipment to install wood gang-nailed
Demolish and Install	Ea.	164	77.50	39.45	280.95	residential gable end truss, up to 40'
Reinstall	Ea.		41.24	20.96	62.20	span.
Clean	Ea.	4.75	94		98.75	
Paint	Ea.	4.75	113		117.75	
Minimum Charge	Job		174		174	
Truss Hardware						
5-1/4" Glue-Lam Seat						
Install	Ea.	63.50	1.94		65.44	Includes labor and material to install
Minimum Charge	Job		174		174	beam hangers.
6-3/4" Glue-Lam Seat						
Install	Ea.	65.50	1.94		67.44	Includes labor and material to install
Minimum Charge	Job		174		174	beam hangers.
8-3/4" Glue-Lam Seat						
Install	Ea.	75	1.94		76.94	Includes labor and material to install
Minimum Charge	Job		174		174	beam hangers.
2" x 4" Joist Hanger						
Install	Ea.	.78	1.99		2.77	Includes labor and material to install
Minimum Charge	Job		174		174	joist and beam hangers.
2" x 12" Joist Hanger						
Install	Ea.	.89	2.11		3	Cost includes labor and material to
Minimum Charge	Job		174		174	install 2" x 12" joist hanger.

Commercial Trusses

	Unit	Material	Labor	Equip.	Total	Specification
Engineered Lumber, Truss / Joist						
9-1/2"						
Demolish	S.F.		.46		.46	Includes material and labor to install
Install	S.F.	1.84	.35		2.19	engineered lumber truss / joists per
Demolish and Install	S.F.	1.84	.81		2.65	S.F. of floor area based on joists 16"
Reinstall	S.F.		.28		.28	O.C. Beams, supports and bridging
Minimum Charge	Job		174		174	are not included.

Rough Frame / Structure

Commercial Trusses

		Unit	Material	Labor	Equip.	Total	Specification
11-7/8"							
	Demolish	S.F.		.48		.48	Includes material and labor to install
	Install	S.F.	1.96	.36		2.32	engineered lumber truss / joists per
	Demolish and Install	S.F.	1.96	.84		2.80	S.F. of floor area based on joists 16"
	Reinstall	S.F.		.29		.29	O.C. Beams, supports and bridging
	Minimum Charge	Job		174		174	are not included.
14"							
	Demolish	S.F.		.50		.50	Includes material and labor to install
	Install	S.F.	2.30	.38		2.68	engineered lumber truss / joists per
	Demolish and Install	S.F.	2.30	.88		3.18	S.F. of floor area based on joists 16"
	Reinstall	S.F.		.31		.31	O.C. Beams, supports and bridging
	Minimum Charge	Job		174		174	are not included.
16"							
	Demolish	S.F.		.52		.52	Includes material and labor to install
	Install	S.F.	2.57	.40		2.97	engineered lumber truss / joists per
	Demolish and Install	S.F.	2.57	.92		3.49	S.F. of floor area based on joists 16"
	Reinstall	S.F.		.32		.32	O.C. Beams, supports and bridging
	Minimum Charge	Job		174		174	are not included.

Bowstring Truss

		Unit	Material	Labor	Equip.	Total	Specification
100' Clear Span							
	Demolish	SF Flr.		.48	.37	.85	Includes material, labor and
	Install	SF Flr.	7.05	.39	.29	7.73	equipment to install bow string truss
	Demolish and Install	SF Flr.	7.05	.87	.66	8.58	system up to 100' clear span per S.F.
	Minimum Charge	Job		174		174	of floor area.
120' Clear Span							
	Demolish	SF Flr.		.53	.41	.94	Includes material, labor and
	Install	SF Flr.	7.60	.44	.32	8.36	equipment to install bow string truss
	Demolish and Install	SF Flr.	7.60	.97	.73	9.30	system up to 120' clear span per S.F.
	Minimum Charge	Job		174		174	of floor area.

Roof Sheathing

		Unit	Material	Labor	Equip.	Total	Specification
Plywood							
3/8"							
	Demolish	S.F.		.37		.37	Cost includes material and labor to
	Install	S.F.	.48	.46		.94	install 3/8" CDX plywood roof
	Demolish and Install	S.F.	.48	.83		1.31	sheathing.
	Reinstall	S.F.		.46		.46	
	Minimum Charge	Job		174		174	
1/2"							
	Demolish	S.F.		.37		.37	Cost includes material and labor to
	Install	S.F.	.52	.50		1.02	install 1/2" CDX plywood roof
	Demolish and Install	S.F.	.52	.87		1.39	sheathing.
	Minimum Charge	Job		174		174	
5/8"							
	Demolish	S.F.		.37		.37	Cost includes material and labor to
	Install	S.F.	.65	.54		1.19	install 5/8" CDX plywood roof
	Demolish and Install	S.F.	.65	.91		1.56	sheathing.
	Minimum Charge	Job		174		174	
3/4"							
	Demolish	S.F.		.37		.37	Cost includes material and labor to
	Install	S.F.	.78	.58		1.36	install 3/4" CDX plywood roof
	Demolish and Install	S.F.	.78	.95		1.73	sheathing.
	Minimum Charge	Job		174		174	

Rough Frame / Structure

Roof Sheathing

Roof Sheathing	Unit	Material	Labor	Equip.	Total	Specification
OSB						
1/2″						
Demolish	S.F.		.37		.37	Cost includes material and labor to
Install	S.F.	.77	.26		1.03	install 4′ x 8′ x 1/2″ OSB sheathing.
Demolish and Install	S.F.	.77	.63		1.40	
Minimum Charge	Job		174		174	
5/8″						
Demolish	S.F.		.37		.37	Cost includes material and labor to
Install	S.F.	.88	.28		1.16	install 4′ x 8′ x 5/8″ OSB sheathing.
Demolish and Install	S.F.	.88	.65		1.53	
Minimum Charge	Job		174		174	
Plank						
Demolish	S.F.		.37		.37	Cost includes material and labor to
Install	S.F.	1.41	.96		2.37	install 1″ x 6″ or 1″ x 8″ utility T&G
Demolish and Install	S.F.	1.41	1.33		2.74	board sheathing laid diagonal.
Reinstall	S.F.		.96		.96	
Minimum Charge	Job		174		174	
Skip-type						
1″ x 4″						
Demolish	S.F.		.56		.56	Includes material and labor to install
Install	S.F.	.54	.29		.83	sheathing board material 1″ x 4″, 7″
Demolish and Install	S.F.	.54	.85		1.39	O.C.
Minimum Charge	Job		174		174	
1″ x 6″						
Demolish	S.F.		.56		.56	Includes material and labor to install
Install	S.F.	.72	.24		.96	sheathing board material 1″ x 6″, 7″
Demolish and Install	S.F.	.72	.80		1.52	or 9″ O.C.
Minimum Charge	Job		174		174	

Wood Deck

Wood Deck	Unit	Material	Labor	Equip.	Total	Specification
Treated						
Demolish	S.F.		1.18		1.18	Includes material and labor to install
Install	S.F.	4.91	8.70		13.61	unfinished deck using 2″ x 4″ treated
Demolish and Install	S.F.	4.91	9.88		14.79	decking, with 2″ x 6″ double beams
Reinstall	S.F.		6.98		6.98	24″ O.C., 4″ x 4″ posts 5′ O.C. set in
Clean	S.F.	.07	.26		.33	concrete.
Paint	S.F.	.06	.23		.29	
Minimum Charge	Job		174		174	
Redwood						
Demolish	S.F.		1.18		1.18	Includes material and labor to install
Install	S.F.	16.85	8.70		25.55	unfinished deck using 2″ x 4″ select
Demolish and Install	S.F.	16.85	9.88		26.73	redwood decking, with 2″ x 6″ double
Reinstall	S.F.		6.98		6.98	beams 24″ O.C., 4″ x 4″ posts 5′
Clean	S.F.	.07	.26		.33	O.C. set in concrete.
Paint	S.F.	.06	.23		.29	
Minimum Charge	Job		174		174	
Cedar						
Demolish	S.F.		1.18		1.18	Includes material and labor to install
Install	S.F.	7.80	8.70		16.50	unfinished deck using 2″ x 4″ select
Demolish and Install	S.F.	7.80	9.88		17.68	cedar decking, with 2″ x 6″ double
Reinstall	S.F.		6.98		6.98	beams 24″ O.C., 4″ x 4″ posts 5′
Clean	S.F.	.07	.26		.33	O.C. set in concrete.
Paint	S.F.	.06	.23		.29	
Minimum Charge	Job		174		174	

Rough Frame / Structure

Wood Deck

Wood Deck		Unit	Material	Labor	Equip.	Total	Specification
Stairs							
	Demolish	Ea.		335		335	Includes material and labor to install
	Install	Ea.	122	605		727	wood steps. 5 - 7 risers included.
	Demolish and Install	Ea.	122	940		1062	
	Reinstall	Ea.		482.43		482.43	
	Clean	Flight	3.11	22		25.11	
	Paint	Flight	12.05	40.50		52.55	
	Minimum Charge	Job		174		174	
Railing							
	Demolish	L.F.		4.34		4.34	Includes material and labor to install
	Install	L.F.	17.95	15.85		33.80	porch rail with balusters.
	Demolish and Install	L.F.	17.95	20.19		38.14	
	Clean	L.F.	.63	1.59		2.22	
	Paint	L.F.	.55	2.54		3.09	
	Minimum Charge	Job		174		174	
Bench Seating							
	Demolish	L.F.		4.70		4.70	Includes material and labor to install
	Install	L.F.	2.41	17.45		19.86	pressure treated bench seating.
	Demolish and Install	L.F.	2.41	22.15		24.56	
	Reinstall	L.F.		13.95		13.95	
	Clean	L.F.	.12	1.27		1.39	
	Paint	L.F.	.56	2.44		3	
	Minimum Charge	Job		174		174	
Privacy Wall							
	Demolish	SF Flr.		1.13		1.13	Includes material, labor and
	Install	SF Flr.	.50	.59	.12	1.21	equipment to install privacy wall.
	Demolish and Install	SF Flr.	.50	1.72	.12	2.34	
	Reinstall	SF Flr.		.47	.10	.57	
	Clean	SF Flr.	.07	.26		.33	
	Paint	S.F.	.06	.23		.29	
	Minimum Charge	Job		174		174	

Porch

Porch		Unit	Material	Labor	Equip.	Total	Specification
Treated							
	Demolish	S.F.		2.35		2.35	Includes material and labor to install
	Install	S.F.	5.80	17.45		23.25	floor and roof frame and decking
	Demolish and Install	S.F.	5.80	19.80		25.60	using 2" x 4" treated decking, with 2"
	Reinstall	S.F.		13.95		13.95	x 6" double beams 24" O.C., 4" x 4"
	Clean	S.F.	.12	.52		.64	posts 5' O.C. set in concrete.
	Paint	S.F.	.19	.61		.80	
	Minimum Charge	Job		174		174	
Redwood							
	Demolish	S.F.		2.35		2.35	Includes material and labor to install
	Install	S.F.	18	17.45		35.45	floor and roof frame and decking
	Demolish and Install	S.F.	18	19.80		37.80	using 2" x 4" redwood decking, with
	Reinstall	S.F.		13.95		13.95	2" x 6" double beams 24" O.C., 4" x
	Clean	S.F.	.12	.52		.64	4" posts 5' O.C. set in concrete.
	Paint	S.F.	.19	.61		.80	
	Minimum Charge	Job		174		174	
Cedar							
	Demolish	S.F.		2.35		2.35	Includes material and labor to install
	Install	S.F.	8.95	17.45		26.40	floor and roof frame and cedar
	Demolish and Install	S.F.	8.95	19.80		28.75	decking using 2" x 4" select cedar,
	Reinstall	S.F.		13.95		13.95	with 2" x 6" double beams 24" O.C.,
	Clean	S.F.	.12	.52		.64	4" x 4" posts 5' O.C. set in concrete.
	Paint	S.F.	.19	.61		.80	
	Minimum Charge	Job		174		174	

Rough Frame / Structure

Porch	Unit	Material	Labor	Equip.	Total	Specification
Stairs						
Demolish	Ea.		335		335	Includes material and labor to install
Install	Ea.	122	605		727	wood steps. 5 - 7 risers included.
Demolish and Install	Ea.	122	940		1062	
Reinstall	Ea.		482.43		482.43	
Clean	Flight	3.11	22		25.11	
Paint	Flight	12.05	40.50		52.55	
Minimum Charge	Job		174		174	
Railing						
Demolish	L.F.		4.34		4.34	Includes material and labor to install
Install	L.F.	17.95	15.85		33.80	porch rail with balusters.
Demolish and Install	L.F.	17.95	20.19		38.14	
Clean	L.F.	.63	1.98		2.61	
Paint	L.F.	.55	2.54		3.09	
Minimum Charge	Job		174		174	
Bench Seating						
Demolish	L.F.		4.70		4.70	Includes material and labor to install
Install	L.F.	2.41	17.45		19.86	pressure treated bench seating.
Demolish and Install	L.F.	2.41	22.15		24.56	
Reinstall	L.F.		13.95		13.95	
Clean	L.F.	.12	1.27		1.39	
Paint	L.F.	.56	2.44		3	
Minimum Charge	Job		174		174	
Privacy Wall						
Demolish	SF Flr.		1.13		1.13	Includes material, labor and
Install	SF Flr.	.50	.59	.12	1.21	equipment to install privacy wall.
Demolish and Install	SF Flr.	.50	1.72	.12	2.34	
Reinstall	SF Flr.		.47	.10	.57	
Clean	SF Flr.	.07	.26		.33	
Paint	S.F.	.06	.23		.29	
Minimum Charge	Job		174		174	
Wood Board Ceiling						
Demolish	S.F.		.47		.47	Includes material and labor to install
Install	S.F.	1.99	1.55		3.54	wood board on ceiling.
Demolish and Install	S.F.	1.99	2.02		4.01	
Clean	S.F.	.02	.15		.17	
Paint	S.F.	.17	.38		.55	
Minimum Charge	Job		174		174	
Re-screen						
Install	S.F.	1.16	1.40		2.56	Includes labor and material to
Minimum Charge	Job		174		174	re-screen wood frame.

Exterior Trim

Exterior Trim		Unit	Material	Labor	Equip.	Total	Specification
Stock Lumber							
1″ x 2″							
	Demolish	L.F.		.24		.24	Cost includes material and labor to
	Install	L.F.	.35	1.06		1.41	install 1″ x 2″ pine trim.
	Demolish and Install	L.F.	.35	1.30		1.65	
	Clean	L.F.	.01	.17		.18	
	Paint	L.F.	.02	.48		.50	
	Minimum Charge	Job		174		174	
1″ x 3″							
	Demolish	L.F.		.24		.24	Cost includes material and labor to
	Install	L.F.	.24	1.16		1.40	install 1″ x 3″ pine trim.
	Demolish and Install	L.F.	.24	1.40		1.64	
	Clean	L.F.	.01	.17		.18	
	Paint	L.F.	.02	.48		.50	
	Minimum Charge	Job		174		174	
1″ x 4″							
	Demolish	L.F.		.24		.24	Cost includes material and labor to
	Install	L.F.	.68	1.40		2.08	install 1″ x 4″ pine trim.
	Demolish and Install	L.F.	.68	1.64		2.32	
	Clean	L.F.	.01	.17		.18	
	Paint	L.F.	.02	.48		.50	
	Minimum Charge	Job		174		174	
1″ x 6″							
	Demolish	L.F.		.24		.24	Cost includes material and labor to
	Install	L.F.	1.18	1.40		2.58	install 1″ x 6″ pine trim.
	Demolish and Install	L.F.	1.18	1.64		2.82	
	Clean	L.F.	.01	.17		.18	
	Paint	L.F.	.07	.48		.55	
	Minimum Charge	Job		174		174	
1″ x 8″							
	Demolish	L.F.		.26		.26	Cost includes material and labor to
	Install	L.F.	1.53	1.74		3.27	install 1″ x 8″ pine trim.
	Demolish and Install	L.F.	1.53	2		3.53	
	Reinstall	L.F.		1.40		1.40	
	Clean	L.F.	.02	.18		.20	
	Paint	L.F.	.07	.48		.55	
	Minimum Charge	Job		174		174	

Exterior Trim

Exterior Trim	Unit	Material	Labor	Equip.	Total	Specification
1″ x 10″						
Demolish	L.F.		.39		.39	Cost includes material and labor to
Install	L.F.	1.89	1.94		3.83	install 1″ x 10″ pine trim.
Demolish and Install	L.F.	1.89	2.33		4.22	
Minimum Charge	Job		174		174	
1″ x 12″						
Demolish	L.F.		.39		.39	Cost includes material and labor to
Install	L.F.	2.08	1.94		4.02	install 1″ x 12″ pine trim.
Demolish and Install	L.F.	2.08	2.33		4.41	
Minimum Charge	Job		174		174	

Fascia	Unit	Material	Labor	Equip.	Total	Specification
Aluminum						
Demolish	L.F.		.49		.49	Includes material and labor to install
Install	S.F.	1.54	1.66		3.20	aluminum fascia.
Demolish and Install	S.F.	1.54	2.15		3.69	
Reinstall	S.F.		1.33		1.33	
Clean	L.F.		.32		.32	
Minimum Charge	Job		174		174	
Vinyl						
Demolish	L.F.		.49		.49	Cost includes material and labor to
Install	L.F.	3.06	1.99		5.05	install 6″ vinyl fascia with 12″ vinyl
Demolish and Install	L.F.	3.06	2.48		5.54	soffit and J-channel.
Reinstall	L.F.		1.59		1.59	
Clean	L.F.		.32		.32	
Minimum Charge	Job		174		174	
Plywood						
Demolish	L.F.		.49		.49	Cost includes material and labor to
Install	L.F.	1.31	.78		2.09	install 12″ wide plywood fascia.
Demolish and Install	L.F.	1.31	1.27		2.58	
Reinstall	L.F.		.62		.62	
Clean	L.F.	.03	.21		.24	
Paint	L.F.	.14	.76		.90	
Minimum Charge	Job		174		174	
Cedar						
Demolish	L.F.		.49		.49	Cost includes material and labor to
Install	L.F.	2.18	.78		2.96	install 12″ wide cedar fascia.
Demolish and Install	L.F.	2.18	1.27		3.45	
Reinstall	L.F.		.62		.62	
Clean	L.F.	.03	.21		.24	
Paint	L.F.	.14	.76		.90	
Minimum Charge	Job		174		174	
Redwood						
1″ x 6″						
Demolish	L.F.		.38		.38	Cost includes material and labor to
Install	L.F.	1.75	1.40		3.15	install 1″ x 6″ redwood fascia board.
Demolish and Install	L.F.	1.75	1.78		3.53	
Reinstall	L.F.		1.12		1.12	
Minimum Charge	Job		174		174	
1″ x 8″						
Demolish	L.F.		.43		.43	Cost includes material and labor to
Install	L.F.	3.82	1.55		5.37	install 1″ x 8″ redwood fascia board.
Demolish and Install	L.F.	3.82	1.98		5.80	
Reinstall	L.F.		1.24		1.24	
Minimum Charge	Job		174		174	

Exterior Trim

Fascia

	Unit	Material	Labor	Equip.	Total	Specification
2″ x 6″						
Demolish	L.F.		.38		.38	Cost includes material and labor to
Install	L.F.	2.70	1.55		4.25	install 2″ x 6″ redwood fascia board.
Demolish and Install	L.F.	2.70	1.93		4.63	
Reinstall	L.F.		1.24		1.24	
Minimum Charge	Job		174		174	
2″ x 8″						
Demolish	L.F.		.43		.43	Cost includes material and labor to
Install	L.F.	3.65	1.74		5.39	install 2″ x 8″ redwood fascia board.
Demolish and Install	L.F.	3.65	2.17		5.82	
Reinstall	L.F.		1.40		1.40	
Minimum Charge	Job		174		174	

Hem-fir Std & Better

	Unit	Material	Labor	Equip.	Total	Specification
2″ x 6″						
Demolish	L.F.		.38		.38	Cost includes material and labor to
Install	L.F.	.64	2.79		3.43	install 2″ x 6″ hem-fir exterior trim
Demolish and Install	L.F.	.64	3.17		3.81	board.
Reinstall	L.F.		2.23		2.23	
Clean	L.F.	.01	.21		.22	
Paint	L.F.	.14	.76		.90	
Minimum Charge	Job		174		174	
2″ x 8″						
Demolish	L.F.		.43		.43	Cost includes material and labor to
Install	L.F.	.92	3.10		4.02	install 2″ x 8″ hem-fir exterior trim
Demolish and Install	L.F.	.92	3.53		4.45	board.
Reinstall	L.F.		2.48		2.48	
Clean	L.F.	.02	.21		.23	
Paint	L.F.	.14	.76		.90	
Minimum Charge	Job		174		174	
2″ x 10″						
Demolish	L.F.		.49		.49	Cost includes material and labor to
Install	L.F.	1.31	3.88		5.19	install 2″ x 10″ hem-fir exterior trim
Demolish and Install	L.F.	1.31	4.37		5.68	board.
Reinstall	L.F.		3.10		3.10	
Clean	L.F.	.03	.23		.26	
Paint	L.F.	.14	.76		.90	
Minimum Charge	Job		174		174	

Barge Rafter

	Unit	Material	Labor	Equip.	Total	Specification
2″ x 6″						
Demolish	L.F.		.75		.75	Cost includes material and labor to
Install	L.F.	.64	.39		1.03	install 2″ x 6″ barge rafters.
Demolish and Install	L.F.	.64	1.14		1.78	
Reinstall	L.F.		.31		.31	
Clean	L.F.	.01	.21		.22	
Paint	L.F.	.14	.76		.90	
Minimum Charge	Job		174		174	
2″ x 8″						
Demolish	L.F.		.77		.77	Cost includes material and labor to
Install	L.F.	.92	.41		1.33	install 2″ x 8″ barge rafters.
Demolish and Install	L.F.	.92	1.18		2.10	
Reinstall	L.F.		.33		.33	
Clean	L.F.	.02	.21		.23	
Paint	L.F.	.14	.76		.90	
Minimum Charge	Job		174		174	

Exterior Trim

Fascia

	Unit	Material	Labor	Equip.	Total	Specification
2" x 10"						
Demolish	L.F.		.78		.78	Cost includes material and labor to
Install	L.F.	1.31	.42		1.73	install 2" x 10" barge rafters.
Demolish and Install	L.F.	1.31	1.20		2.51	
Reinstall	L.F.		.34		.34	
Clean	L.F.	.03	.23		.26	
Paint	L.F.	.14	.76		.90	
Minimum Charge	Job		174		174	

Soffit

	Unit	Material	Labor	Equip.	Total	Specification
Aluminum						
12" w / Fascia						
Demolish	L.F.		.49		.49	Cost includes material and labor to
Install	L.F.	3.75	1.99		5.74	install 12" aluminum soffit with 6"
Demolish and Install	L.F.	3.75	2.48		6.23	fascia and J-channel.
Reinstall	L.F.		1.59		1.59	
Clean	L.F.		.32		.32	
Paint	L.F.	.19	.76		.95	
Minimum Charge	Job		174		174	
18" w / Fascia						
Demolish	L.F.		.49		.49	Cost includes material and labor to
Install	L.F.	4.48	2.33		6.81	install 18" aluminum soffit with 6"
Demolish and Install	L.F.	4.48	2.82		7.30	fascia and J-channel.
Reinstall	L.F.		1.86		1.86	
Clean	L.F.		.32		.32	
Paint	L.F.	.19	.76		.95	
Minimum Charge	Job		174		174	
24" w / Fascia						
Demolish	L.F.		.49		.49	Cost includes material and labor to
Install	L.F.	5.25	2.58		7.83	install 24" aluminum soffit with 6"
Demolish and Install	L.F.	5.25	3.07		8.32	fascia and J-channel.
Reinstall	L.F.		2.07		2.07	
Clean	L.F.		.46		.46	
Paint	L.F.	.19	.76		.95	
Minimum Charge	Job		174		174	
Vinyl						
Demolish	L.F.		.49		.49	Cost includes material and labor to
Install	L.F.	3.06	1.99		5.05	install 6" vinyl fascia with 12" vinyl
Demolish and Install	L.F.	3.06	2.48		5.54	soffit and J-channel.
Reinstall	L.F.		1.59		1.59	
Clean	L.F.		.32		.32	
Paint	L.F.	.19	.76		.95	
Minimum Charge	Job		174		174	
Plywood						
Demolish	L.F.		.54		.54	Cost includes material and labor to
Install	L.F.	1.31	1		2.31	install 12" wide plywood soffit.
Demolish and Install	L.F.	1.31	1.54		2.85	
Reinstall	L.F.		.80		.80	
Clean	L.F.	.03	.21		.24	
Paint	L.F.	.14	.76		.90	
Minimum Charge	Job		174		174	

Exterior Trim

Soffit	Unit	Material	Labor	Equip.	Total	Specification
Redwood						
Demolish	L.F.		.54		.54	Cost includes material and labor to
Install	L.F.	3.50	1		4.50	install 12″ wide redwood soffit.
Demolish and Install	L.F.	3.50	1.54		5.04	
Reinstall	L.F.		.80		.80	
Clean	L.F.	.03	.21		.24	
Paint	L.F.	.14	.76		.90	
Minimum Charge	Job		174		174	
Cedar						
Demolish	L.F.		.54		.54	Cost includes material and labor to
Install	L.F.	2.18	1		3.18	install 12″ wide cedar soffit.
Demolish and Install	L.F.	2.18	1.54		3.72	
Reinstall	L.F.		.80		.80	
Clean	L.F.	.03	.21		.24	
Paint	L.F.	.14	.76		.90	
Minimum Charge	Job		174		174	
Douglas Fir						
Demolish	L.F.		.54		.54	Cost includes material and labor to
Install	L.F.	2.08	1		3.08	install 12″ wide Douglas fir soffit.
Demolish and Install	L.F.	2.08	1.54		3.62	
Reinstall	L.F.		.80		.80	
Clean	L.F.	.03	.21		.24	
Paint	L.F.	.14	.76		.90	
Minimum Charge	Job		174		174	

Roofing and Flashing   Tile Roofing

Metal Flashing		Unit	Material	Labor	Equip.	Total	Specification
Galvanized							
6"							
	Demolish	L.F.		.19		.19	Includes material and labor to install
	Install	L.F.	.55	.81		1.36	flat galvanized steel flashing, 6" wide.
	Demolish and Install	L.F.	.55	1		1.55	
	Paint	L.F.	.10	.38		.48	
	Minimum Charge	Job		161		161	
14"							
	Demolish	L.F.		.19		.19	Includes material and labor to install
	Install	L.F.	1.29	.81		2.10	flat galvanized steel flashing, 14"
	Demolish and Install	L.F.	1.29	1		2.29	wide.
	Paint	L.F.	.10	.38		.48	
	Minimum Charge	Job		161		161	
20"							
	Demolish	L.F.		.19		.19	Includes material and labor to install
	Install	L.F.	1.82	.81		2.63	flat galvanized steel flashing, 20"
	Demolish and Install	L.F.	1.82	1		2.82	wide.
	Paint	L.F.	.10	.38		.48	
	Minimum Charge	Job		161		161	
Aluminum							
6"							
	Demolish	L.F.		.19		.19	Includes material and labor to install
	Install	L.F.	.31	.81		1.12	flat aluminum flashing, 6" wide.
	Demolish and Install	L.F.	.31	1		1.31	
	Paint	L.F.	.10	.38		.48	
	Minimum Charge	Job		161		161	
14"							
	Demolish	L.F.		.19		.19	Includes material and labor to install
	Install	L.F.	.54	.81		1.35	flat aluminum flashing, 14" wide.
	Demolish and Install	L.F.	.54	1		1.54	
	Paint	L.F.	.10	.38		.48	
	Minimum Charge	Job		161		161	
20"							
	Demolish	L.F.		.19		.19	Includes material and labor to install
	Install	L.F.	.78	.81		1.59	flat aluminum flashing, 20" wide.
	Demolish and Install	L.F.	.78	1		1.78	
	Paint	L.F.	.10	.38		.48	
	Minimum Charge	Job		161		161	

Roofing

Metal Flashing		Unit	Material	Labor	Equip.	Total	Specification
Copper							
6"							
	Demolish	L.F.		.19		.19	Includes material and labor to install
	Install	L.F.	2.77	1.07		3.84	flat copper flashing, 6" wide.
	Demolish and Install	L.F.	2.77	1.26		4.03	
	Minimum Charge	Job		193		193	
14"							
	Demolish	L.F.		.19		.19	Includes material and labor to install
	Install	L.F.	6.50	1.07		7.57	flat copper flashing, 14" wide.
	Demolish and Install	L.F.	6.50	1.26		7.76	
	Minimum Charge	Job		193		193	
18"							
	Demolish	L.F.		.19		.19	Includes material and labor to install
	Install	L.F.	9.25	1.07		10.32	flat copper flashing, 18" wide.
	Demolish and Install	L.F.	9.25	1.26		10.51	
	Minimum Charge	Job		193		193	
Aluminum Valley							
	Demolish	L.F.		.19		.19	Includes material and labor to install
	Install	L.F.	2.92	2.05		4.97	aluminum valley flashing, .024" thick.
	Demolish and Install	L.F.	2.92	2.24		5.16	
	Paint	L.F.	.10	.38		.48	
	Minimum Charge	Job		174		174	
Gravel Stop							
	Demolish	L.F.		.17		.17	Includes material and labor to install
	Install	L.F.	5.15	2.67		7.82	aluminum gravel stop, .050" thick, 4"
	Demolish and Install	L.F.	5.15	2.84		7.99	high, mill finish.
	Reinstall	L.F.		2.13		2.13	
	Paint	L.F.	.17	.38		.55	
	Minimum Charge	Job		174		174	
Drip Edge							
	Demolish	L.F.		.17		.17	Includes material and labor to install
	Install	L.F.	.46	.87		1.33	aluminum drip edge, .016" thick, 5"
	Demolish and Install	L.F.	.46	1.04		1.50	wide, mill finish.
	Reinstall	L.F.		.70		.70	
	Paint	L.F.	.17	.38		.55	
	Minimum Charge	Job		174		174	
End Wall							
	Demolish	L.F.		.87		.87	Includes material and labor to install
	Install	L.F.	1.88	2.05		3.93	aluminum end wall flashing, .024"
	Demolish and Install	L.F.	1.88	2.92		4.80	thick.
	Minimum Charge	Job		174		174	
Side Wall							
	Demolish	L.F.		.17		.17	Includes material and labor to install
	Install	L.F.	1.88	2.05		3.93	aluminum side wall flashing, .024"
	Demolish and Install	L.F.	1.88	2.22		4.10	thick.
	Minimum Charge	Job		174		174	
Coping and Wall Cap							
	Demolish	L.F.		1.18		1.18	Includes material and labor to install
	Install	L.F.	1.95	2.22		4.17	aluminum flashing, mill finish, 0.40
	Demolish and Install	L.F.	1.95	3.40		5.35	thick, 12" wide.
	Clean	L.F.	.08	.79		.87	
	Minimum Charge	Job		161		161	

Roofing

Roof Shingles

	Unit	Material	Labor	Equip.	Total	Specification
Underlayment						
15#						
Demolish	Sq.		9.40		9.40	Includes material and labor to install
Install	Sq.	4.65	5.05		9.70	#15 felt underlayment.
Demolish and Install	Sq.	4.65	14.45		19.10	
Minimum Charge	Job		161		161	
30#						
Demolish	Sq.		9.40		9.40	Includes material and labor to install
Install	Sq.	7.75	5.55		13.30	#30 felt underlayment.
Demolish and Install	Sq.	7.75	14.95		22.70	
Minimum Charge	Job		161		161	
Self Adhering						
Demolish	Sq.		23.50		23.50	Includes material and labor to install
Install	Sq.	54.50	14.65		69.15	self adhering ice barrier roofing
Demolish and Install	Sq.	54.50	38.15		92.65	underlayment.
Minimum Charge	Job		161		161	

Rolled Roofing

	Unit	Material	Labor	Equip.	Total	Specification
90 lb						
Demolish	Sq.		23.50		23.50	Cost includes material and labor to
Install	Sq.	19.85	21.50		41.35	install 90 lb mineral surface rolled
Demolish and Install	Sq.	19.85	45		64.85	roofing.
Minimum Charge	Job		161		161	
110 lb						
Demolish	Sq.		23.50		23.50	Cost includes material and labor to
Install	Sq.	53	21.50		74.50	install 110 lb mineral surface rolled
Demolish and Install	Sq.	53	45		98	roofing.
Minimum Charge	Job		161		161	
140 lb						
Demolish	Sq.		23.50		23.50	Cost includes material and labor to
Install	Sq.	53	32		85	install 140 lb mineral surface rolled
Demolish and Install	Sq.	53	55.50		108.50	roofing.
Minimum Charge	Job		161		161	

Composition Shingles

	Unit	Material	Labor	Equip.	Total	Specification
Fiberglass (3-tab)						
225 lb, 20 Year						
Demolish	Sq.		25.50		25.50	Cost includes material and labor to
Install	Sq.	38.50	63		101.50	install 225 lb, 20 year fiberglass
Demolish and Install	Sq.	38.50	88.50		127	shingles.
Minimum Charge	Job		161		161	
300 lb, 25 Year						
Demolish	Sq.		25.50		25.50	Cost includes material and labor to
Install	Sq.	38.50	63		101.50	install 300 lb, 25 year fiberglass
Demolish and Install	Sq.	38.50	88.50		127	shingles.
Minimum Charge	Job		161		161	
Emergency Repairs						
Install	Ea.	79	161		240	Includes labor and material for
Minimum Charge	Job		161		161	emergency repairs.

Roofing

Composition Shingles

	Unit	Material	Labor	Equip.	Total	Specification
Architectural						
Laminated, 25 Year						
Demolish	Sq.		25.50		25.50	Cost includes material and labor to
Install	Sq.	52.50	63		115.50	install 25 year fiberglass roof shingles.
Demolish and Install	Sq.	52.50	88.50		141	
Minimum Charge	Job		161		161	
Laminated, 30 Year						
Demolish	Sq.		25.50		25.50	Cost includes material and labor to
Install	Sq.	53.50	63		116.50	install 30 year fiberglass roof shingles.
Demolish and Install	Sq.	53.50	88.50		142	
Minimum Charge	Job		161		161	
Laminated, 40 Year						
Demolish	Sq.		25.50		25.50	Cost includes material and labor to
Install	Sq.	71.50	63		134.50	install 40 year fiberglass roof shingles.
Demolish and Install	Sq.	71.50	88.50		160	
Minimum Charge	Job		161		161	
Laminated Shake, 40 Year						
Demolish	Sq.		25.50		25.50	Cost includes material and labor to
Install	Sq.	88	84		172	install 40 year fiberglass roof shakes.
Demolish and Install	Sq.	88	109.50		197.50	
Minimum Charge	Job		161		161	
Emergency Repairs						
Install	Ea.	79	161		240	Includes labor and material for
Minimum Charge	Job		161		161	emergency repairs.
Roof Jack						
Demolish	Ea.		5.15		5.15	Includes material and labor to install
Install	Ea.	14.65	35		49.65	residential roof jack, w/bird screen,
Demolish and Install	Ea.	14.65	40.15		54.80	backdraft damper, 3″ & 4″ dia. round
Paint	Ea.	.06	3.82		3.88	duct.
Minimum Charge	Job		192		192	
Ridge Vent						
Demolish	L.F.		.91		.91	Includes material and labor to install a
Install	L.F.	2.95	2.50		5.45	mill finish aluminum ridge vent strip.
Demolish and Install	L.F.	2.95	3.41		6.36	
Minimum Charge	Job		174		174	
Shingle Molding						
1″ x 2″						
Demolish	L.F.		.18		.18	Cost includes material and labor to
Install	L.F.	.35	1.06		1.41	install 1″ x 2″ pine trim.
Demolish and Install	L.F.	.35	1.24		1.59	
Paint	L.F.	.02	.48		.50	
Minimum Charge	Job		174		174	
1″ x 3″						
Demolish	L.F.		.18		.18	Cost includes material and labor to
Install	L.F.	.40	1.20		1.60	install 1″ x 3″ pine trim.
Demolish and Install	L.F.	.40	1.38		1.78	
Paint	L.F.	.02	.48		.50	
Minimum Charge	Job		174		174	
1″ x 4″						
Demolish	L.F.		.18		.18	Cost includes material and labor to
Install	L.F.	.68	1.40		2.08	install 1″ x 4″ pine trim.
Demolish and Install	L.F.	.68	1.58		2.26	
Paint	L.F.	.02	.48		.50	
Minimum Charge	Job		174		174	

Roofing

Composition Shingles		Unit	Material	Labor	Equip.	Total	Specification
Add for Steep Pitch							
	Install	Sq.		20		20	Includes labor and material added for 2nd story or steep roofs.
Add for Additional Story							
	Install	Sq.		20		20	Includes labor and material added for 2nd story or steep roofs.

Wood Shingles		Unit	Material	Labor	Equip.	Total	Specification
Cedar Wood Shingles							
16" Red Label (#2)							
	Demolish	Sq.		62.50		62.50	Includes material and labor to install
	Install	Sq.	135	140		275	wood shingles/shakes.
	Demolish and Install	Sq.	135	202.50		337.50	
	Minimum Charge	Job		174		174	
16" Blue Label (#1)							
	Demolish	Sq.		62.50		62.50	Includes material and labor to install
	Install	Sq.	208	107		315	wood shingles/shakes.
	Demolish and Install	Sq.	208	169.50		377.50	
	Minimum Charge	Job		174		174	
18" Red Label (#2)							
	Demolish	Sq.		62.50		62.50	Includes material and labor to install
	Install	Sq.	264	89.50		353.50	wood shingles/shakes.
	Demolish and Install	Sq.	264	152		416	
	Minimum Charge	Job		174		174	
18" Blue Label (#1)							
	Demolish	Sq.		62.50		62.50	Includes material and labor to install
	Install	Sq.	184	97.50		281.50	wood shingles/shakes.
	Demolish and Install	Sq.	184	160		344	
	Minimum Charge	Job		174		174	
Replace Shingle							
	Install	Ea.	13.20	8.70		21.90	Includes labor and material to replace
	Minimum Charge	Job		174		174	individual wood shingle/shake. Cost is per shingle.
Cedar Shake							
18" Medium Handsplit							
	Demolish	Sq.		62.50		62.50	Includes material and labor to install
	Install	Sq.	210	134		344	wood shingles/shakes.
	Demolish and Install	Sq.	210	196.50		406.50	
	Minimum Charge	Job		174		174	
18" Heavy Handsplit							
	Demolish	Sq.		62.50		62.50	Includes material and labor to install
	Install	Sq.	210	149		359	wood shingles/shakes.
	Demolish and Install	Sq.	210	211.50		421.50	
	Minimum Charge	Job		174		174	
24" Heavy Handsplit							
	Demolish	Sq.		62.50		62.50	Includes material and labor to install
	Install	Sq.	182	119		301	wood shingles/shakes.
	Demolish and Install	Sq.	182	181.50		363.50	
	Minimum Charge	Job		174		174	
24" Medium Handsplit							
	Demolish	Sq.		62.50		62.50	Includes material and labor to install
	Install	Sq.	182	107		289	wood shingles/shakes.
	Demolish and Install	Sq.	182	169.50		351.50	
	Minimum Charge	Job		174		174	

Roofing

Wood Shingles

	Unit	Material	Labor	Equip.	Total	Specification
24″ Straight Split						
Demolish	Sq.		62.50		62.50	Includes material and labor to install
Install	Sq.	182	159		341	wood shingles/shakes.
Demolish and Install	Sq.	182	221.50		403.50	
Minimum Charge	Job		174		174	
24″ Tapersplit						
Demolish	Sq.		62.50		62.50	Includes material and labor to install
Install	Sq.	182	159		341	wood shingles/shakes.
Demolish and Install	Sq.	182	221.50		403.50	
Minimum Charge	Job		174		174	
18″ Straight Split						
Demolish	Sq.		62.50		62.50	Includes material and labor to install
Install	Sq.	238	159		397	wood shingles/shakes.
Demolish and Install	Sq.	238	221.50		459.50	
Minimum Charge	Job		174		174	
Replace Shingle						
Install	Ea.	13.20	8.70		21.90	Includes labor and material to replace
Minimum Charge	Job		174		174	individual wood shingle/shake. Cost is per shingle.

Slate Tile Roofing

	Unit	Material	Labor	Equip.	Total	Specification
Unfading Green						
Demolish	Sq.		56.50		56.50	Includes material and labor to install
Install	Sq.	480	185		665	clear Vermont slate tile.
Demolish and Install	Sq.	480	241.50		721.50	
Reinstall	Sq.		148.11		148.11	
Minimum Charge	Job		162		162	
Unfading Purple						
Demolish	Sq.		56.50		56.50	Includes material and labor to install
Install	Sq.	430	185		615	clear Vermont slate tile.
Demolish and Install	Sq.	430	241.50		671.50	
Reinstall	Sq.		148.11		148.11	
Minimum Charge	Job		162		162	
Replace Slate Tiles						
Install	Ea.	6.25	17.05		23.30	Includes labor and material to replace
Minimum Charge	Job		162		162	individual slate tiles. Cost is per tile.
Variegated Purple						
Demolish	Sq.		56.50		56.50	Includes material and labor to install
Install	Sq.	450	185		635	clear Vermont slate tile.
Demolish and Install	Sq.	450	241.50		691.50	
Reinstall	Sq.		148.11		148.11	
Minimum Charge	Job		162		162	
Unfading Grey / Black						
Demolish	Sq.		56.50		56.50	Includes material and labor to install
Install	Sq.	490	185		675	clear Vermont slate tile.
Demolish and Install	Sq.	490	241.50		731.50	
Reinstall	Sq.		148.11		148.11	
Minimum Charge	Job		162		162	
Unfading Red						
Demolish	Sq.		56.50		56.50	Includes material and labor to install
Install	Sq.	1250	185		1435	clear Vermont slate tile.
Demolish and Install	Sq.	1250	241.50		1491.50	
Reinstall	Sq.		148.11		148.11	
Minimum Charge	Job		162		162	

Roofing

Slate Tile Roofing

	Unit	Material	Labor	Equip.	Total	Specification
Weathering Black						
Demolish	Sq.		56.50		56.50	Includes material and labor to install
Install	Sq.	525	185		710	clear Pennsylvania slate tile.
Demolish and Install	Sq.	525	241.50		766.50	
Reinstall	Sq.		148.11		148.11	
Minimum Charge	Job		162		162	
Weathering Green						
Demolish	Sq.		56.50		56.50	Includes material and labor to install
Install	Sq.	380	185		565	clear Vermont slate tile.
Demolish and Install	Sq.	380	241.50		621.50	
Reinstall	Sq.		148.11		148.11	
Minimum Charge	Job		162		162	

Flat Clay Tile Roofing

	Unit	Material	Labor	Equip.	Total	Specification
Glazed						
Demolish	Sq.		62.50		62.50	Includes material and labor to install
Install	Sq.	595	130		725	tile roofing. Cost based on terra cotta
Demolish and Install	Sq.	595	192.50		787.50	red flat clay tile and includes ridge
Reinstall	Sq.		103.68		103.68	and rake tiles.
Minimum Charge	Job		162		162	
Terra Cotta Red						
Demolish	Sq.		62.50		62.50	Includes material and labor to install
Install	Sq.	595	130		725	tile roofing. Cost based on terra cotta
Demolish and Install	Sq.	595	192.50		787.50	red flat clay tile and includes ridge
Reinstall	Sq.		103.68		103.68	and rake tiles.
Minimum Charge	Job		162		162	

Mission Tile Roofing

	Unit	Material	Labor	Equip.	Total	Specification
Glazed Red						
Demolish	Sq.		62.50		62.50	Includes material and labor to install
Install	Sq.	785	216		1001	tile roofing. Cost based on glazed red
Demolish and Install	Sq.	785	278.50		1063.50	tile and includes birdstop, ridge and
Reinstall	Sq.		172.80		172.80	rake tiles.
Minimum Charge	Job		162		162	
Unglazed Red						
Demolish	Sq.		62.50		62.50	Includes material and labor to install
Install	Sq.	785	216		1001	tile roofing. Cost based on unglazed
Demolish and Install	Sq.	785	278.50		1063.50	red tile and includes birdstop, ridge
Reinstall	Sq.		172.80		172.80	and rake tiles.
Minimum Charge	Job		162		162	
Glazed Blue						
Demolish	Sq.		62.50		62.50	Includes material and labor to install
Install	Sq.	785	216		1001	tile roofing. Cost based on glazed
Demolish and Install	Sq.	785	278.50		1063.50	blue tile and includes birdstop, ridge
Reinstall	Sq.		172.80		172.80	and rake tiles.
Minimum Charge	Job		162		162	
Glazed White						
Demolish	Sq.		62.50		62.50	Includes material and labor to install
Install	Sq.	785	216		1001	tile roofing. Cost based on glazed
Demolish and Install	Sq.	785	278.50		1063.50	white tile and includes birdstop, ridge
Reinstall	Sq.		172.80		172.80	and rake tiles.
Minimum Charge	Job		162		162	

Mission Tile Roofing	Unit	Material	Labor	Equip.	Total	Specification
Unglazed White						
Demolish	Sq.		62.50		62.50	Includes material and labor to install
Install	Sq.	785	216		1001	tile roofing. Cost based on unglazed
Demolish and Install	Sq.	785	278.50		1063.50	white tile and includes birdstop, ridge
Reinstall	Sq.		172.80		172.80	and rake tiles.
Minimum Charge	Job		162		162	
Color Blend						
Demolish	Sq.		62.50		62.50	Includes material and labor to install
Install	Sq.	785	216		1001	tile roofing. Cost based on color blend
Demolish and Install	Sq.	785	278.50		1063.50	tile and includes birdstop, ridge and
Reinstall	Sq.		172.80		172.80	rake tiles.
Minimum Charge	Job		162		162	
Accessories / Extras						
Add for Cap Furring Strips						
Install	Sq.	92.50	203		295.50	Includes labor and material to install vertically placed wood furring strips under cap tiles.
Add for Tile Adhesive						
Install	Sq.	15.85	11.80		27.65	Includes material and labor for adhesive material placed between tiles.
Add for Wire Attachment						
Install	Sq.	15.60	141		156.60	Includes labor and material to install single wire anchors for attaching tile.
Add for Braided Runners						
Install	Sq.	28	135		163	Includes labor and material to install braided wire runners for secure attachment in high wind areas.
Add for Stainless Steel Nails						
Install	Sq.	21.50			21.50	Includes material only for stainless steel nails.
Add for Brass Nails						
Install	Sq.	28.50			28.50	Includes material only for brass nails.
Add for Hurricane Clips						
Install	Sq.	18.90	191		209.90	Includes material and labor for Hurricane or wind clips for high wind areas.
Add for Copper Nails						
Install	Sq.	28.50			28.50	Includes material only for copper nails.
Add for Vertical Furring Strips						
Install	Sq.	18.25	37.50		55.75	Includes labor and material to install vertically placed wood furring strips under tiles.
Add for Horizontal Furring Strips						
Install	Sq.	11.70	16.90		28.60	Includes labor and material to install horizontally placed wood furring strips under tiles.

Roofing

Spanish Tile Roofing		Unit	Material	Labor	Equip.	Total	Specification
Glazed Red							
	Demolish	Sq.		62.50		62.50	Includes material and labor to install
	Install	Sq.	615	135		750	tile roofing. Cost based on glazed red
	Demolish and Install	Sq.	615	197.50		812.50	tile and includes birdstop, ridge and
	Reinstall	Sq.		108		108	rake tiles.
	Minimum Charge	Job		162		162	
Glazed White							
	Demolish	Sq.		62.50		62.50	Includes material and labor to install
	Install	Sq.	625	135		760	tile roofing. Cost based on glazed
	Demolish and Install	Sq.	625	197.50		822.50	white tile and includes birdstop, ridge
	Reinstall	Sq.		108		108	and rake tiles.
	Minimum Charge	Job		162		162	
Glazed Blue							
	Demolish	Sq.		62.50		62.50	Includes material and labor to install
	Install	Sq.	625	135		760	tile roofing. Cost based on glazed
	Demolish and Install	Sq.	625	197.50		822.50	blue tile and includes birdstop, ridge
	Reinstall	Sq.		108		108	and rake tiles.
	Minimum Charge	Job		162		162	
Unglazed Red							
	Demolish	Sq.		62.50		62.50	Includes material and labor to install
	Install	Sq.	625	135		760	tile roofing. Cost based on unglazed
	Demolish and Install	Sq.	625	197.50		822.50	red tile and includes birdstop, ridge
	Reinstall	Sq.		108		108	and rake tiles.
	Minimum Charge	Job		162		162	
Unglazed White							
	Demolish	Sq.		62.50		62.50	Includes material and labor to install
	Install	Sq.	625	135		760	tile roofing. Cost based on unglazed
	Demolish and Install	Sq.	625	197.50		822.50	white tile and includes birdstop, ridge
	Reinstall	Sq.		108		108	and rake tiles.
	Minimum Charge	Job		162		162	
Color Blend							
	Demolish	Sq.		62.50		62.50	Includes material and labor to install
	Install	Sq.	625	135		760	tile roofing. Cost based on color blend
	Demolish and Install	Sq.	625	197.50		822.50	tile and includes birdstop, ridge and
	Reinstall	Sq.		108		108	rake tiles.
	Minimum Charge	Job		162		162	
Accessories / Extras							
Add for Vertical Furring Strips							
	Install	Sq.	18.25	37.50		55.75	Includes labor and material to install vertically placed wood furring strips under tiles.
Add for Wire Attachment							
	Install	Sq.	13.05	108		121.05	Includes labor and material to install single wire ties.
Add for Braided Runners							
	Install	Sq.	23	92.50		115.50	Includes labor and material to install braided wire runners for secure attachment in high wind areas.
Add for Stainless Steel Nails							
	Install	Sq.	18.25			18.25	Includes material only for stainless steel nails.

Roofing

Spanish Tile Roofing		Unit	Material	Labor	Equip.	Total	Specification
Add for Copper Nails							
	Install	Sq.	24			24	Includes material only for copper nails.
Add for Brass Nails							
	Install	Sq.	24			24	Includes material only for brass nails.
Add for Hurricane Clips							
	Install	Sq.	14.45	112		126.45	Includes material and labor to install hurricane or wind clips for high wind areas.
Add for Mortar Set Tiles							
	Install	Sq.	45	162		207	Includes material and labor to install setting tiles by using mortar material placed between tiles.
Add for Tile Adhesive							
	Install	Sq.	14.70	10.50		25.20	Includes material and labor for adhesive placed between tiles.
Add for Cap Furring Strips							
	Install	Sq.	92.50	203		295.50	Includes labor and material to install vertically placed wood furring strips under cap tiles.
Add for Horizontal Furring Strips							
	Install	Sq.	11.70	16.90		28.60	Includes labor and material to install horizontally placed wood furring strips under tiles.

Concrete Tile Roofing		Unit	Material	Labor	Equip.	Total	Specification
Corrugated Red / Brown							
	Demolish	Sq.		62.50		62.50	Includes material and labor to install tile roofing. Cost includes birdstop, booster, ridge and rake tiles.
	Install	Sq.	229	240		469	
	Demolish and Install	Sq.	229	302.50		531.50	
	Reinstall	Sq.		192		192	
	Minimum Charge	Job		162		162	
Corrugated Gray							
	Demolish	Sq.		62.50		62.50	Includes material and labor to install tile roofing. Cost includes birdstop, booster, ridge and rake tiles.
	Install	Sq.	229	240		469	
	Demolish and Install	Sq.	229	302.50		531.50	
	Reinstall	Sq.		192		192	
	Minimum Charge	Job		162		162	
Corrugated Bright Red							
	Demolish	Sq.		62.50		62.50	Includes material and labor to install tile roofing. Cost includes birdstop, booster, ridge and rake tiles.
	Install	Sq.	229	240		469	
	Demolish and Install	Sq.	229	302.50		531.50	
	Reinstall	Sq.		192		192	
	Minimum Charge	Job		162		162	
Corrugated Black							
	Demolish	Sq.		62.50		62.50	Includes material and labor to install tile roofing. Cost includes birdstop, booster, ridge and rake tiles.
	Install	Sq.	229	240		469	
	Demolish and Install	Sq.	229	302.50		531.50	
	Reinstall	Sq.		192		192	
	Minimum Charge	Job		162		162	
Corrugated Green							
	Demolish	Sq.		62.50		62.50	Includes material and labor to install tile roofing. Cost includes birdstop, booster, ridge and rake tiles.
	Install	Sq.	229	240		469	
	Demolish and Install	Sq.	229	302.50		531.50	
	Reinstall	Sq.		192		192	
	Minimum Charge	Job		162		162	

Roofing

Concrete Tile Roofing

Concrete Tile Roofing	Unit	Material	Labor	Equip.	Total	Specification
Corrugated Blue						
Demolish	Sq.		62.50		62.50	Includes material and labor to install
Install	Sq.	229	240		469	tile roofing. Cost includes birdstop,
Demolish and Install	Sq.	229	302.50		531.50	booster, ridge and rake tiles.
Reinstall	Sq.		192		192	
Minimum Charge	Job		162		162	
Flat Natural Gray						
Demolish	Sq.		62.50		62.50	Includes material and labor to install
Install	Sq.	221	240		461	tile roofing. Cost includes birdstop,
Demolish and Install	Sq.	221	302.50		523.50	booster, ridge and rake tiles.
Reinstall	Sq.		192		192	
Minimum Charge	Job		162		162	
Flat Red / Brown						
Demolish	Sq.		62.50		62.50	Includes material and labor to install
Install	Sq.	221	240		461	tile roofing. Cost includes birdstop,
Demolish and Install	Sq.	221	302.50		523.50	booster, ridge and rake tiles.
Reinstall	Sq.		192		192	
Minimum Charge	Job		162		162	
Flat Bright Red						
Demolish	Sq.		62.50		62.50	Includes material and labor to install
Install	Sq.	221	240		461	tile roofing. Cost includes birdstop,
Demolish and Install	Sq.	221	302.50		523.50	booster, ridge and rake tiles.
Reinstall	Sq.		192		192	
Minimum Charge	Job		162		162	
Flat Green						
Demolish	Sq.		62.50		62.50	Includes material and labor to install
Install	Sq.	221	240		461	tile roofing. Cost includes birdstop,
Demolish and Install	Sq.	221	302.50		523.50	booster, ridge and rake tiles.
Minimum Charge	Job		162		162	
Flat Blue						
Demolish	Sq.		62.50		62.50	Includes material and labor to install
Install	Sq.	221	240		461	tile roofing. Cost includes birdstop,
Demolish and Install	Sq.	221	302.50		523.50	booster, ridge and rake tiles.
Reinstall	Sq.		192		192	
Minimum Charge	Job		162		162	
Flat Black						
Demolish	Sq.		62.50		62.50	Includes material and labor to install
Install	Sq.	221	240		461	tile roofing. Cost includes birdstop,
Demolish and Install	Sq.	221	302.50		523.50	booster, ridge and rake tiles.
Reinstall	Sq.		192		192	
Minimum Charge	Job		162		162	

Aluminum Sheet

Aluminum Sheet	Unit	Material	Labor	Equip.	Total	Specification
Corrugated						
.019″ Thick Natural						
Demolish	S.F.		.67		.67	Includes material and labor to install
Install	S.F.	1.05	1.61		2.66	aluminum sheet metal roofing.
Demolish and Install	S.F.	1.05	2.28		3.33	
Reinstall	S.F.		1.29		1.29	
Minimum Charge	Job		385		385	
.016″ Thick Natural						
Demolish	S.F.		.67		.67	Includes material and labor to install
Install	S.F.	.79	1.61		2.40	aluminum sheet metal roofing.
Demolish and Install	S.F.	.79	2.28		3.07	
Reinstall	S.F.		1.29		1.29	
Minimum Charge	Job		385		385	

Roofing

Aluminum Sheet

	Unit	Material	Labor	Equip.	Total	Specification
.016" Colored Finish						
Demolish	S.F.		.67		.67	Includes material and labor to install
Install	S.F.	1.16	1.61		2.77	aluminum sheet metal roofing.
Demolish and Install	S.F.	1.16	2.28		3.44	
Reinstall	S.F.		1.29		1.29	
Minimum Charge	Job		385		385	
.019" Colored Finish						
Demolish	S.F.		.67		.67	Includes material and labor to install
Install	S.F.	1.29	1.61		2.90	aluminum sheet metal roofing.
Demolish and Install	S.F.	1.29	2.28		3.57	
Reinstall	S.F.		1.29		1.29	
Minimum Charge	Job		385		385	

Ribbed

	Unit	Material	Labor	Equip.	Total	Specification
.019" Thick Natural						
Demolish	S.F.		.67		.67	Includes material and labor to install
Install	S.F.	1.05	1.61		2.66	aluminum sheet metal roofing.
Demolish and Install	S.F.	1.05	2.28		3.33	
Reinstall	S.F.		1.29		1.29	
Minimum Charge	Job		385		385	
.016" Thick Natural						
Demolish	S.F.		.67		.67	Includes material and labor to install
Install	S.F.	.79	1.61		2.40	aluminum sheet metal roofing.
Demolish and Install	S.F.	.79	2.28		3.07	
Reinstall	S.F.		1.29		1.29	
Minimum Charge	Job		385		385	
.050" Colored Finish						
Demolish	S.F.		.67		.67	Includes material and labor to install
Install	S.F.	3.59	1.61		5.20	aluminum sheet metal roofing.
Demolish and Install	S.F.	3.59	2.28		5.87	
Reinstall	S.F.		1.29		1.29	
Minimum Charge	Job		385		385	
.032" Thick Natural						
Demolish	S.F.		.67		.67	Includes material and labor to install
Install	S.F.	2.04	1.61		3.65	aluminum sheet metal roofing.
Demolish and Install	S.F.	2.04	2.28		4.32	
Reinstall	S.F.		1.29		1.29	
Minimum Charge	Job		385		385	
.040" Thick Natural						
Demolish	S.F.		.67		.67	Includes material and labor to install
Install	S.F.	2.48	1.61		4.09	aluminum sheet metal roofing.
Demolish and Install	S.F.	2.48	2.28		4.76	
Reinstall	S.F.		1.29		1.29	
Minimum Charge	Job		385		385	
.050" Thick Natural						
Demolish	S.F.		.67		.67	Includes material and labor to install
Install	S.F.	2.97	1.61		4.58	aluminum sheet metal roofing.
Demolish and Install	S.F.	2.97	2.28		5.25	
Reinstall	S.F.		1.29		1.29	
Minimum Charge	Job		385		385	
.016" Colored Finish						
Demolish	S.F.		.67		.67	Includes material and labor to install
Install	S.F.	1.16	1.61		2.77	aluminum sheet metal roofing.
Demolish and Install	S.F.	1.16	2.28		3.44	
Reinstall	S.F.		1.29		1.29	
Minimum Charge	Job		385		385	

Roofing

Aluminum Sheet

Aluminum Sheet	Unit	Material	Labor	Equip.	Total	Specification
.019″ Colored Finish						
Demolish	S.F.		.67		.67	Includes material and labor to install
Install	S.F.	1.29	1.61		2.90	aluminum sheet metal roofing.
Demolish and Install	S.F.	1.29	2.28		3.57	
Reinstall	S.F.		1.29		1.29	
Minimum Charge	Job		385		385	
.032″ Colored Finish						
Demolish	S.F.		.67		.67	Includes material and labor to install
Install	S.F.	2.50	1.61		4.11	aluminum sheet metal roofing.
Demolish and Install	S.F.	2.50	2.28		4.78	
Reinstall	S.F.		1.29		1.29	
Minimum Charge	Job		385		385	
.040″ Colored Finish						
Demolish	S.F.		.67		.67	Includes material and labor to install
Install	S.F.	2.98	1.61		4.59	aluminum sheet metal roofing.
Demolish and Install	S.F.	2.98	2.28		5.26	
Reinstall	S.F.		1.29		1.29	
Minimum Charge	Job		385		385	

Fiberglass Sheet

Fiberglass Sheet	Unit	Material	Labor	Equip.	Total	Specification
12 Ounce Corrugated						
Demolish	S.F.		1.20		1.20	Cost includes material and labor to
Install	S.F.	3.66	1.28		4.94	install 12 ounce corrugated fiberglass
Demolish and Install	S.F.	3.66	2.48		6.14	sheet roofing.
Reinstall	S.F.		1.03		1.03	
Minimum Charge	Job		161		161	
8 Ounce Corrugated						
Demolish	S.F.		1.20		1.20	Cost includes material and labor to
Install	S.F.	1.68	1.28		2.96	install 8 ounce corrugated fiberglass
Demolish and Install	S.F.	1.68	2.48		4.16	roofing.
Reinstall	S.F.		1.03		1.03	
Minimum Charge	Job		161		161	

Galvanized Steel

Galvanized Steel	Unit	Material	Labor	Equip.	Total	Specification
Corrugated						
24 Gauge						
Demolish	S.F.		.67		.67	Includes material and labor to install
Install	S.F.	2.86	1.35		4.21	galvanized steel sheet metal roofing.
Demolish and Install	S.F.	2.86	2.02		4.88	Cost based on 24 gauge corrugated
Reinstall	S.F.		1.08		1.08	or ribbed steel roofing.
Minimum Charge	Job		161		161	
26 Gauge						
Demolish	S.F.		.67		.67	Includes material and labor to install
Install	S.F.	1.23	1.28		2.51	galvanized steel sheet metal roofing.
Demolish and Install	S.F.	1.23	1.95		3.18	Cost based on 26 gauge corrugated
Reinstall	S.F.		1.03		1.03	or ribbed steel roofing.
Minimum Charge	Job		161		161	
28 Gauge						
Demolish	S.F.		.67		.67	Includes material and labor to install
Install	S.F.	1.76	1.22		2.98	galvanized steel sheet metal roofing.
Demolish and Install	S.F.	1.76	1.89		3.65	Cost based on 28 gauge corrugated
Reinstall	S.F.		.98		.98	or ribbed steel roofing.
Minimum Charge	Job		161		161	

Roofing

Galvanized Steel

Galvanized Steel	Unit	Material	Labor	Equip.	Total	Specification
30 Gauge						
Demolish	S.F.		.67		.67	Includes material and labor to install
Install	S.F.	1.65	1.17		2.82	galvanized steel sheet metal roofing.
Demolish and Install	S.F.	1.65	1.84		3.49	Cost based on 30 gauge corrugated
Reinstall	S.F.		.93		.93	or ribbed steel roofing.
Minimum Charge	Job		161		161	
Ribbed						
24 Gauge						
Demolish	S.F.		.67		.67	Includes material and labor to install
Install	S.F.	2.86	1.35		4.21	galvanized steel sheet metal roofing.
Demolish and Install	S.F.	2.86	2.02		4.88	Cost based on 24 gauge corrugated
Reinstall	S.F.		1.08		1.08	or ribbed steel roofing.
Minimum Charge	Job		161		161	
26 Gauge						
Demolish	S.F.		.67		.67	Includes material and labor to install
Install	S.F.	1.23	1.28		2.51	galvanized steel sheet metal roofing.
Demolish and Install	S.F.	1.23	1.95		3.18	Cost based on 26 gauge corrugated
Reinstall	S.F.		1.03		1.03	or ribbed steel roofing.
Minimum Charge	Job		161		161	
28 Gauge						
Demolish	S.F.		.67		.67	Includes material and labor to install
Install	S.F.	1.76	1.22		2.98	galvanized steel sheet metal roofing.
Demolish and Install	S.F.	1.76	1.89		3.65	Cost based on 28 gauge corrugated
Reinstall	S.F.		.98		.98	or ribbed steel roofing.
Minimum Charge	Job		161		161	
30 Gauge						
Demolish	S.F.		.67		.67	Includes material and labor to install
Install	S.F.	1.65	1.17		2.82	galvanized steel sheet metal roofing.
Demolish and Install	S.F.	1.65	1.84		3.49	Cost based on 30 gauge corrugated
Reinstall	S.F.		.93		.93	or ribbed steel roofing.
Minimum Charge	Job		161		161	

Standing Seam

Standing Seam	Unit	Material	Labor	Equip.	Total	Specification
Copper						
16 Ounce						
Demolish	Sq.		67		67	Includes material and labor to install
Install	Sq.	930	298		1228	copper standing seam metal roofing.
Demolish and Install	Sq.	930	365		1295	Cost based on 16 ounce copper
Reinstall	Sq.		238.03		238.03	roofing.
Minimum Charge	Job		193		193	
18 Ounce						
Demolish	Sq.		67		67	Includes material and labor to install
Install	Sq.	1050	320		1370	copper standing seam metal roofing.
Demolish and Install	Sq.	1050	387		1437	Cost based on 18 ounce copper
Reinstall	Sq.		257.87		257.87	roofing.
Minimum Charge	Job		193		193	
20 Ounce						
Demolish	Sq.		67		67	Includes material and labor to install
Install	Sq.	1125	350		1475	copper standing seam metal roofing.
Demolish and Install	Sq.	1125	417		1542	Cost based on 20 ounce copper
Reinstall	Sq.		281.31		281.31	roofing.

Roofing

Standing Seam

Standing Seam	Unit	Material	Labor	Equip.	Total	Specification
Stainless Steel						
26 Gauge						
Demolish	Sq.		67		67	Cost includes material and labor to
Install	Sq.	710	335		1045	install 26 gauge stainless steel
Demolish and Install	Sq.	710	402		1112	standing seam metal roofing.
Reinstall	Sq.		269.08		269.08	
Minimum Charge	Job		193		193	
28 Gauge						
Demolish	Sq.		67		67	Includes material and labor to install
Install	Sq.	570	320		890	standing seam metal roofing. Cost
Demolish and Install	Sq.	570	387		957	based on 28 gauge stainless steel
Reinstall	Sq.		257.87		257.87	roofing.
Minimum Charge	Job		193		193	

Elastomeric Roofing

Elastomeric Roofing	Unit	Material	Labor	Equip.	Total	Specification
Neoprene						
1/16″						
Demolish	Sq.		23.50		23.50	Includes material and labor to install
Install	Sq.	234	154	26.50	414.50	neoprene membrane, fully adhered,
Demolish and Install	Sq.	234	177.50	26.50	438	1/16″ thick.
Minimum Charge	Job		735	84	819	
GRM Membrane						
Demolish	Sq.		23.50		23.50	Includes material and labor to install
Install	Sq.	256	245	28	529	coal tar based elastomeric roofing
Demolish and Install	Sq.	256	268.50	28	552.50	membrane, polyester fiber reinforced.
Minimum Charge	Job		735	84	819	
Cant Strips						
3″						
Demolish	L.F.		.72		.72	Includes material and labor to install
Install	L.F.	.55	.99		1.54	cants 3″ x 3″, treated timber, cut
Demolish and Install	L.F.	.55	1.71		2.26	diagonally.
Minimum Charge	Job		161		161	
4″						
Demolish	L.F.		.72		.72	Includes material and labor to install
Install	L.F.	1.52	.99		2.51	cants 4″ x 4″, treated timber, cut
Demolish and Install	L.F.	1.52	1.71		3.23	diagonally.
Minimum Charge	Job		161		161	
Polyvinyl Chloride (PVC)						
45 mil (loose laid)						
Demolish	Sq.		23.50		23.50	Cost includes material and labor to
Install	Sq.	102	29	3.30	134.30	install 45 mil PVC elastomeric roofing,
Demolish and Install	Sq.	102	52.50	3.30	157.80	loose-laid with stone.
Minimum Charge	Job		735	84	819	
48 mil (loose laid)						
Demolish	Sq.		23.50		23.50	Cost includes material and labor to
Install	Sq.	119	29	3.30	151.30	install 48 mil PVC elastomeric roofing,
Demolish and Install	Sq.	119	52.50	3.30	174.80	loose-laid with stone.
Minimum Charge	Job		735	84	819	
60 mil (loose laid)						
Demolish	Sq.		23.50		23.50	Cost includes material and labor to
Install	Sq.	121	29	3.30	153.30	install 60 mil PVC elastomeric roofing,
Demolish and Install	Sq.	121	52.50	3.30	176.80	loose-laid with stone.
Minimum Charge	Job		735	84	819	

Roofing

Elastomeric Roofing	Unit	Material	Labor	Equip.	Total	Specification
45 mil (attached)						
Demolish	Sq.		34.50		34.50	Cost includes material and labor to
Install	Sq.	88.50	56.50	6.50	151.50	install 45 mil PVC elastomeric roofing
Demolish and Install	Sq.	88.50	91	6.50	186	fully attached to roof deck.
Minimum Charge	Job		735	84	819	
48 mil (attached)						
Demolish	Sq.		34.50		34.50	Cost includes material and labor to
Install	Sq.	157	56.50	6.50	220	install 48 mil PVC elastomeric roofing
Demolish and Install	Sq.	157	91	6.50	254.50	fully attached to roof deck.
Minimum Charge	Job		735	84	819	
60 mil (attached)						
Demolish	Sq.		34.50		34.50	Cost includes material and labor to
Install	Sq.	160	56.50	6.50	223	install 60 mil PVC elastomeric roofing
Demolish and Install	Sq.	160	91	6.50	257.50	fully attached to roof deck.
Minimum Charge	Job		735	84	819	
CSPE Type						
45 mil (loose laid w / stone)						
Demolish	Sq.		23.50		23.50	Cost includes material and labor to
Install	Sq.	151	29	3.30	183.30	install 45 mil CSPE (chlorosulfonated
Demolish and Install	Sq.	151	52.50	3.30	206.80	polyethylene-hypalon) roofing material
Minimum Charge	Job		735	84	819	loose laid on deck and ballasted with
						stone.
45 mil (attached at seams)						
Demolish	Sq.		55.50		55.50	Cost includes material and labor to
Install	Sq.	145	42	4.81	191.81	install 45 mil CSPE (chlorosulfonated
Demolish and Install	Sq.	145	97.50	4.81	247.31	polyethylene-hypalon) roofing material
Minimum Charge	Job		735	84	819	attached to deck at seams with batten
						strips.
45 mil (fully attached)						
Demolish	Sq.		34.50		34.50	Cost includes material and labor to
Install	Sq.	142	56.50	6.50	205	install 45 mil CSPE (chlorosulfonated
Demolish and Install	Sq.	142	91	6.50	239.50	polyethylene-hypalon) roofing material
Minimum Charge	Job		735	84	819	fully attached to deck.
EPDM Type						
45 mil (attached at seams)						
Demolish	Sq.		55.50		55.50	Cost includes material and labor to
Install	Sq.	59.50	42	4.81	106.31	install 45 mil EPDM (ethylene
Demolish and Install	Sq.	59.50	97.50	4.81	161.81	propylene diene monomer) roofing
Minimum Charge	Job		735	84	819	material attached to deck at seams.
45 mil (fully attached)						
Demolish	Sq.		34.50		34.50	Cost includes material and labor to
Install	Sq.	86.50	56.50	6.50	149.50	install 45 mil EPDM (ethylene
Demolish and Install	Sq.	86.50	91	6.50	184	propylene diene monomer) roofing
Minimum Charge	Job		735	84	819	material fully attached to deck.
45 mil (loose laid w / stone)						
Demolish	Sq.		23.50		23.50	Cost includes material and labor to
Install	Sq.	70	29	3.30	102.30	install 45 mil EPDM (ethylene
Demolish and Install	Sq.	70	52.50	3.30	125.80	propylene diene monomer) roofing
Minimum Charge	Job		735	84	819	material loose laid on deck and
						ballasted with stones (approx. 1/2 ton
						of stone per square).
60 mil (attached at seams)						
Demolish	Sq.		55.50		55.50	Cost includes material and labor to
Install	Sq.	76.50	42	4.81	123.31	install 60 mil EPDM (ethylene
Demolish and Install	Sq.	76.50	97.50	4.81	178.81	propylene diene monomer) roofing
Minimum Charge	Job		735	84	819	material attached to deck at seams.

Roofing

Elastomeric Roofing

Elastomeric Roofing	Unit	Material	Labor	Equip.	Total	Specification
60 mil (fully attached)						Cost includes material and labor to install 60 mil EPDM (ethylene propylene diene monomer) roofing material fully attached to deck.
Demolish	Sq.		34.50		34.50	
Install	Sq.	103	56.50	6.50	166	
Demolish and Install	Sq.	103	91	6.50	200.50	
Minimum Charge	Job		735	84	819	
60 mil (loose laid w / stone)						Cost includes material and labor to install 60 mil EPDM (ethylene propylene diene monomer) roofing material loose laid on deck and ballasted with stones (approx. 1/2 ton of stone per square).
Demolish	Sq.		23.50		23.50	
Install	Sq.	88	29	3.30	120.30	
Demolish and Install	Sq.	88	52.50	3.30	143.80	
Minimum Charge	Job		735	84	819	
Silicone-urethane Foam						
Prep Roof for Foam System						Includes labor and material to removal of any gravel and prepare existing built-up roof for application of silicone-urethane foam system.
Install	Sq.	5.50	129		134.50	
Minimum Charge	Job		735	84	819	

Modified Bitumen Roofing

Modified Bitumen Roofing	Unit	Material	Labor	Equip.	Total	Specification
Fully Attached						
120 mil						Cost includes material and labor to install 120 mil modified bitumen roofing, fully attached to deck by torch.
Demolish	Sq.		55.50		55.50	
Install	Sq.	88.50	7.15	1.24	96.89	
Demolish and Install	Sq.	88.50	62.65	1.24	152.39	
Minimum Charge	Job		161		161	
150 mil						Cost includes material and labor to install 150 mil modified bitumen roofing, fully attached to deck by torch.
Demolish	Sq.		55.50		55.50	
Install	Sq.	128	17.60	3.05	148.65	
Demolish and Install	Sq.	128	73.10	3.05	204.15	
Minimum Charge	Job		1050	183	1233	
160 mil						Cost includes material and labor to install 160 mil modified bitumen roofing, fully attached to deck by torch.
Demolish	Sq.		55.50		55.50	
Install	Sq.	102	18.40	3.18	123.58	
Demolish and Install	Sq.	102	73.90	3.18	179.08	
Minimum Charge	Job		1050	183	1233	

Built-up Roofing

Built-up Roofing	Unit	Material	Labor	Equip.	Total	Specification
Emergency Repairs						Minimum charge for work requiring hot asphalt.
Install	Ea.	174	1500	261	1935	
Minimum Charge	Job		1050	183	1233	
Asphalt						
3-ply Asphalt						Cost includes material and labor to install 3-ply asphalt built-up roof including one 30 lb and two 15 lb layers of felt.
Demolish	Sq.		57.50		57.50	
Install	Sq.	49	88	15.25	152.25	
Demolish and Install	Sq.	49	145.50	15.25	209.75	
Minimum Charge	Job		1050	183	1233	

Roofing

Built-up Roofing

Built-up Roofing	Unit	Material	Labor	Equip.	Total	Specification
4-ply Asphalt						
Demolish	Sq.		68.50		68.50	Cost includes material and labor to
Install	Sq.	69	96	16.60	181.60	install 4-ply asphalt built-up roof
Demolish and Install	Sq.	69	164.50	16.60	250.10	including one 30 lb and three 15 lb
Minimum Charge	Job		1050	183	1233	layers of felt.
5-ply Asphalt						
Demolish	Sq.		90		90	Cost includes material and labor to
Install	Sq.	97	106	18.30	221.30	install 5-ply asphalt built-up roof
Demolish and Install	Sq.	97	196	18.30	311.30	including one 30 lb and four 15 lb
Minimum Charge	Job		1050	183	1233	layers of felt.

Components / Accessories

	Unit	Material	Labor	Equip.	Total	Specification
Flood Coat						
Install	Sq.	17.35	49.50		66.85	Includes labor and material to install flood coat.
R / R Gravel and Flood Coat						
Install	Sq.	23	161		184	Includes labor and material to install flood coat and gravel.
Aluminum UV Coat						
Install	Sq.	17.60	184	21	222.60	Includes labor and material to install aluminum-based UV coating.
Gravel Coat						
Demolish	Sq.		29		29	Includes labor and material to install
Install	Sq.	5.60	27		32.60	gravel coat.
Demolish and Install	Sq.	5.60	56		61.60	
Mineral Surface Cap Sheet						
Install	Sq.	24	5.55		29.55	Cost includes labor and material to install 50 # mineral surfaced cap sheet.
Fibrated Aluminum UV Coating						
Install	Sq.	20	49.50		69.50	Includes labor and material to install fibered aluminum-based UV coating.

Impregnated Walk

	Unit	Material	Labor	Equip.	Total	Specification
1"						
Install	S.F.	2.53	1.72		4.25	Cost includes labor and material to install 1" layer of asphalt coating.
1/2"						
Install	S.F.	1.76	1.36		3.12	Cost includes labor and material to install 1/2" layer of asphalt coating.
3/4"						
Install	S.F.	2.20	1.43		3.63	Cost includes labor and material to install 3/4" layer of asphalt coating.

Doors	Roof Window	Double Hung Window

Exterior Type Door		Unit	Material	Labor	Equip.	Total	Specification
Entry							
Flush (1-3/4″)							
	Demolish	Ea.		32.50		32.50	Cost includes material and labor to
	Install	Ea.	222	18.35		240.35	install 3′ x 6′ 8″ pre-hung 1-3/4″
	Demolish and Install	Ea.	222	50.85		272.85	door, including casing and stop,
	Reinstall	Ea.		18.36		18.36	hinges, jamb, aluminum sill,
	Clean	Ea.	3.11	10.60		13.71	weatherstripped.
	Paint	Ea.	9.25	43.50		52.75	
	Minimum Charge	Job		174		174	
Raised Panel							
	Demolish	Ea.		32.50		32.50	Cost includes material and labor to
	Install	Ea.	435	19.40		454.40	install 3′ x 6′ 8″ pre-hung 1-3/4″
	Demolish and Install	Ea.	435	51.90		486.90	door, jamb, including casing and
	Reinstall	Ea.		19.38		19.38	stop, hinges, aluminum sill,
	Clean	Ea.	3.11	10.60		13.71	weatherstripping.
	Paint	Ea.	9.25	43.50		52.75	
	Minimum Charge	Job		174		174	
Entry w / Lights							
Good Quality							
	Demolish	Ea.		32.50		32.50	Cost includes material and labor to
	Install	Ea.	290	43.50		333.50	install 3′ x 6′ 8″ pre-hung 1-3/4″
	Demolish and Install	Ea.	290	76		366	door, solid wood lower panel, with up
	Reinstall	Ea.		19.88		19.88	to 9-light glass panels, casing, sill,
	Clean	Ea.	3.11	10.60		13.71	weatherstripped.
	Paint	Ea.	6	20.50		26.50	
	Minimum Charge	Job		174		174	
Better Quality							
	Demolish	Ea.		32.50		32.50	Cost includes material and labor to
	Install	Ea.	305	43.50		348.50	install 3′ x 6′ 8″ pre-hung 1-3/4″ Ash
	Demolish and Install	Ea.	305	76		381	door, solid wood, raised panel door
	Reinstall	Ea.		19.88		19.88	without glass panels.
	Clean	Ea.	3.11	10.60		13.71	
	Paint	Ea.	9.25	43.50		52.75	
	Minimum Charge	Job		174		174	
Premium Quality							
	Demolish	Ea.		32.50		32.50	Cost includes material and labor to
	Install	Ea.	695	20.50		715.50	install 3′ x 6′ 8″ pre-hung 1-3/4″ Oak
	Demolish and Install	Ea.	695	53		748	door, solid wood, raised panel door
	Reinstall	Ea.		20.52		20.52	without glass panels.
	Clean	Ea.	3.11	10.60		13.71	
	Paint	Ea.	9.25	43.50		52.75	
	Minimum Charge	Job		174		174	

Exterior Doors and Windows

Exterior Type Door		Unit	Material	Labor	Equip.	Total	Specification
Custom Quality							
	Demolish	Ea.		32.50		32.50	Cost includes material and labor to
	Install	Ea.	485	46.50		531.50	install 3' x 6' 8" pre-hung 1-3/4"
	Demolish and Install	Ea.	485	79		564	hand-made, solid wood, raised panel
	Reinstall	Ea.		19.88		19.88	door with custom stained glass.
	Clean	Ea.	.21	10.60		10.81	
	Paint	Ea.	2.11	102		104.11	
	Minimum Charge	Job		174		174	
Miami Style							
	Demolish	Ea.		32.50		32.50	Cost includes material and labor to
	Install	Ea.	210	17.45		227.45	install 3' x 6' 8" pre-hung 1-3/4"
	Demolish and Install	Ea.	210	49.95		259.95	lauan door of solid wood.
	Reinstall	Ea.		17.44		17.44	
	Clean	Ea.	.21	10.60		10.81	
	Paint	Ea.	9.25	43.50		52.75	
	Minimum Charge	Job		174		174	
French Style							
1-Light							
	Demolish	Ea.		32.50		32.50	Cost includes material and labor to
	Install	Ea.	380	25		405	install 3' x 6' 8" pre-hung 1-3/4"
	Demolish and Install	Ea.	380	57.50		437.50	french style door with 1 light.
	Reinstall	Ea.		24.91		24.91	
	Clean	Ea.	1.95	15.90		17.85	
	Paint	Ea.	9.30	76.50		85.80	
	Minimum Charge	Job		174		174	
9-Light							
	Demolish	Ea.		32.50		32.50	Cost includes material and labor to
	Install	Ea.	365	25		390	install 3' x 6' 8" pre-hung 1-3/4"
	Demolish and Install	Ea.	365	57.50		422.50	french style door with 9 lights.
	Reinstall	Ea.		24.91		24.91	
	Clean	Ea.	.43	21		21.43	
	Paint	Ea.	9.30	76.50		85.80	
	Minimum Charge	Job		174		174	
10-Light							
	Demolish	Ea.		35.50		35.50	Cost includes material and labor to
	Install	Ea.	585	43.50		628.50	install 3' x 6' 8" pre-hung 1-3/4"
	Demolish and Install	Ea.	585	79		664	french style door with 10 lights.
	Reinstall	Ea.		43.60		43.60	
	Clean	Ea.	.43	21		21.43	
	Paint	Ea.	9.30	76.50		85.80	
	Minimum Charge	Job		174		174	
15-Light							
	Demolish	Ea.		35.50		35.50	Cost includes material and labor to
	Install	Ea.	655	50		705	install 3' x 6' 8" pre-hung 1-3/4"
	Demolish and Install	Ea.	655	85.50		740.50	french style door with 15 lights.
	Reinstall	Ea.		49.83		49.83	
	Clean	Ea.	.43	21		21.43	
	Paint	Ea.	9.30	76.50		85.80	
	Minimum Charge	Job		174		174	
Jamb							
Wood							
	Demolish	Ea.		19.40		19.40	Includes material and labor to install
	Install	Ea.	103	9.45		112.45	flat pine jamb with square cut heads
	Demolish and Install	Ea.	103	28.85		131.85	and rabbeted sides for 6' 8" high and
	Clean	Ea.	.10	6.35		6.45	3-9/16" door including trim sets for
	Paint	Ea.	6	20.50		26.50	both sides.
	Minimum Charge	Job		174		174	

Exterior Doors and Windows

Exterior Type Door		Unit	Material	Labor	Equip.	Total	Specification
Metal 6' 8" High							
	Demolish	Ea.		19.40		19.40	Cost includes material and labor to
	Install	Ea.	134	22		156	install 18 gauge hollow metal door
	Demolish and Install	Ea.	134	41.40		175.40	frame to fit a 4-1/2" jamb 6' 8" high
	Reinstall	Ea.		21.80		21.80	and up to 3' 6" opening.
	Clean	Ea.	.65	5.30		5.95	
	Paint	Ea.	1.10	6.10		7.20	
	Minimum Charge	Job		174		174	
Metal 7' High							
	Demolish	Ea.		19.40		19.40	Cost includes material and labor to
	Install	Ea.	140	23.50		163.50	install 18 gauge hollow metal door
	Demolish and Install	Ea.	140	42.90		182.90	frame to fit a 4-1/2" jamb 7' high and
	Reinstall	Ea.		23.25		23.25	up to 3' 6" opening.
	Clean	Ea.	.65	5.30		5.95	
	Paint	Ea.	1.10	6.10		7.20	
	Minimum Charge	Job		174		174	
Metal 8' High							
	Demolish	Ea.		19.40		19.40	Cost includes material and labor to
	Install	Ea.	171	29		200	install 18 gauge hollow metal door
	Demolish and Install	Ea.	171	48.40		219.40	frame to fit a 4-1/2" jamb 8' high and
	Reinstall	Ea.		29.07		29.07	up to 3' 6" opening.
	Clean	Ea.	.91	6.35		7.26	
	Paint	Ea.	1.22	6.80		8.02	
	Minimum Charge	Job		174		174	
Metal 9' High							
	Demolish	Ea.		19.40		19.40	Cost includes material and labor to
	Install	Ea.	188	35		223	install 18 gauge hollow metal door
	Demolish and Install	Ea.	188	54.40		242.40	frame to fit a 4-1/2" jamb 9' high and
	Reinstall	Ea.		34.88		34.88	up to 3' 6" opening.
	Clean	Ea.	.91	6.35		7.26	
	Paint	Ea.	1.22	6.80		8.02	
	Minimum Charge	Job		174		174	
Casing Trim							
Single Width							
	Demolish	Ea.		6.15		6.15	Cost includes material and labor to
	Install	Opng.	25.50	59		84.50	install 11/16" x 2-1/2" pine ranch
	Demolish and Install	Opng.	25.50	65.15		90.65	style casing for one side of a standard
	Reinstall	Opng.		47.29		47.29	door opening.
	Clean	Opng.	.44	2.54		2.98	
	Paint	Ea.	3.22	6.10		9.32	
	Minimum Charge	Job		174		174	
Double Width							
	Demolish	Ea.		7.30		7.30	Cost includes material and labor to
	Install	Opng.	29.50	70		99.50	install 11/16" x 2-1/2" pine ranch
	Demolish and Install	Opng.	29.50	77.30		106.80	style door casing for one side of a
	Reinstall	Opng.		55.81		55.81	double door opening.
	Clean	Opng.	.52	5.10		5.62	
	Paint	Ea.	3.22	6.95		10.17	
	Minimum Charge	Job		174		174	

Exterior Doors and Windows

Residential Metal Door

Residential Metal Door	Unit	Material	Labor	Equip.	Total	Specification
Pre-hung, Steel-clad						
2' 8" x 6' 8"						
Demolish	Ea.		32.50		32.50	Cost includes material and labor to
Install	Ea.	241	43.50		284.50	install 2' 6" x 6' 8" pre-hung 1-3/4"
Demolish and Install	Ea.	241	76		317	24 gauge steel clad door with 4-1/2"
Reinstall	Ea.		43.60		43.60	wood frame, trims, hinges, aluminum
Clean	Ea.	3.11	10.60		13.71	sill, weatherstripped.
Paint	Ea.	9.30	43.50		52.80	
Minimum Charge	Job		174		174	
3' x 6' 8"						
Demolish	Ea.		32.50		32.50	Cost includes material and labor to
Install	Ea.	241	46.50		287.50	install 3' x 6' 8" pre-hung 1-3/4" 24
Demolish and Install	Ea.	241	79		320	gauge steel clad door with 4-1/2"
Reinstall	Ea.		46.51		46.51	wood frame, trims, hinges, aluminum
Clean	Ea.	3.11	10.60		13.71	sill, weatherstripped.
Paint	Ea.	9.30	43.50		52.80	
Minimum Charge	Job		174		174	

Metal Door

Metal Door	Unit	Material	Labor	Equip.	Total	Specification
2' 8" x 6' 8", 18 Gauge						
Demolish	Ea.		6.90		6.90	Cost includes material and labor to
Install	Ea.	325	22		347	install 18 gauge 2' 8" x 6' 8" steel
Demolish and Install	Ea.	325	28.90		353.90	door only, 1-3/4" thick, hinges.
Reinstall	Ea.		21.80		21.80	
Clean	Ea.	1.30	7.95		9.25	
Paint	Ea.	3.97	25.50		29.47	
Minimum Charge	Job		174		174	
3' x 6' 8", 18 Gauge						
Demolish	Ea.		6.90		6.90	Cost includes material and labor to
Install	Ea.	315	23.50		338.50	install 18 gauge 3' x 6' 8" steel door
Demolish and Install	Ea.	315	30.40		345.40	only, 1-3/4" thick, hinges.
Reinstall	Ea.		23.25		23.25	
Clean	Ea.	1.30	7.95		9.25	
Paint	Ea.	3.97	25.50		29.47	
Minimum Charge	Job		174		174	
2' 8" x 7', 18 Gauge						
Demolish	Ea.		6.90		6.90	Cost includes material and labor to
Install	Ea.	415	23.50		438.50	install 18 gauge 2' 8" x 7' steel door
Demolish and Install	Ea.	415	30.40		445.40	only, 1-3/4" thick, hinges.
Reinstall	Ea.		23.25		23.25	
Clean	Ea.	1.30	7.95		9.25	
Paint	Ea.	3.97	25.50		29.47	
Minimum Charge	Job		174		174	
3' x 7', 18 Gauge						
Demolish	Ea.		6.90		6.90	Cost includes material and labor to
Install	Ea.	390	25		415	install 18 gauge 3' x 7' steel door
Demolish and Install	Ea.	390	31.90		421.90	only, 1-3/4" thick, hinges.
Reinstall	Ea.		24.91		24.91	
Clean	Ea.	1.30	7.95		9.25	
Paint	Ea.	3.97	25.50		29.47	
Minimum Charge	Job		174		174	

Exterior Doors and Windows

Metal Door		Unit	Material	Labor	Equip.	Total	Specification
3' 6" x 7', 18 Gauge							
	Demolish	Ea.		6.90		6.90	Cost includes material and labor to
	Install	Ea.	420	27		447	install 18 gauge 3' 6" x 7' steel door
	Demolish and Install	Ea.	420	33.90		453.90	only, 1-3/4" thick, hinges.
	Reinstall	Ea.		26.83		26.83	
	Clean	Ea.	1.30	7.95		9.25	
	Paint	Ea.	3.97	25.50		29.47	
	Minimum Charge	Job		174		174	
Casing Trim							
Single Width							
	Demolish	Ea.		6.15		6.15	Cost includes material and labor to
	Install	Opng.	25.50	59		84.50	install 11/16" x 2-1/2" pine ranch
	Demolish and Install	Opng.	25.50	65.15		90.65	style casing for one side of a standard
	Reinstall	Opng.		47.29		47.29	door opening.
	Clean	Opng.	.44	2.54		2.98	
	Paint	Ea.	3.22	6.10		9.32	
	Minimum Charge	Job		174		174	
Double Width							
	Demolish	Ea.		7.30		7.30	Cost includes material and labor to
	Install	Opng.	29.50	70		99.50	install 11/16" x 2-1/2" pine ranch
	Demolish and Install	Opng.	29.50	77.30		106.80	style door casing for one side of a
	Reinstall	Opng.		55.81		55.81	double door opening.
	Clean	Opng.	.52	5.10		5.62	
	Paint	Ea.	3.22	6.95		10.17	
	Minimum Charge	Job		174		174	

Solid Core Door		Unit	Material	Labor	Equip.	Total	Specification
Exterior Type							
Flush (1-3/4")							
	Demolish	Ea.		32.50		32.50	Cost includes material and labor to
	Install	Ea.	390	43.50		433.50	install 3' x 6' 8" pre-hung 1-3/4"
	Demolish and Install	Ea.	390	76		466	door, including casing and stop,
	Reinstall	Ea.		43.60		43.60	hinges, jamb, aluminum sill,
	Clean	Ea.	3.11	10.60		13.71	weatherstripped.
	Paint	Ea.	9.30	43.50		52.80	
	Minimum Charge	Job		174		174	
Raised Panel							
	Demolish	Ea.		32.50		32.50	Cost includes material and labor to
	Install	Ea.	435	19.40		454.40	install 3' x 6' 8" pre-hung 1-3/4"
	Demolish and Install	Ea.	435	51.90		486.90	door, jamb, including casing and
	Reinstall	Ea.		19.38		19.38	stop, hinges, aluminum sill,
	Clean	Ea.	3.11	10.60		13.71	weatherstripping.
	Paint	Ea.	9.30	102		111.30	
	Minimum Charge	Job		174		174	
Entry w / Lights							
	Demolish	Ea.		32.50		32.50	Cost includes material and labor to
	Install	Ea.	290	43.50		333.50	install 3' x 6' 8" pre-hung 1-3/4"
	Demolish and Install	Ea.	290	76		366	door, solid wood lower panel, with up
	Reinstall	Ea.		43.60		43.60	to 9-light glass panels, casing, sill,
	Clean	Ea.	3.11	10.60		13.71	weatherstripped.
	Paint	Ea.	2.11	102		104.11	
	Minimum Charge	Job		174		174	

Solid Core Door	Unit	Material	Labor	Equip.	Total	Specification
Custom Entry						
Demolish	Ea.		32.50		32.50	Includes material and labor to install
Install	Ea.	1450	70		1520	prehung, fir, single glazed with lead
Demolish and Install	Ea.	1450	102.50		1552.50	caming, 1-3/4" x 6' 8" x 3'.
Reinstall	Ea.		69.76		69.76	
Clean	Ea.	3.11	10.60		13.71	
Paint	Ea.	3.97	25.50		29.47	
Minimum Charge	Job		174		174	
Dutch Style						
Demolish	Ea.		32.50		32.50	Includes material and labor to install
Install	Ea.	745	58		803	dutch door, pine, 1-3/4" x 6' 8" x 2'
Demolish and Install	Ea.	745	90.50		835.50	8" wide.
Reinstall	Ea.		58.13		58.13	
Clean	Ea.	3.11	10.60		13.71	
Paint	Ea.	2.11	102		104.11	
Minimum Charge	Job		174		174	
Miami Style						
Demolish	Ea.		32.50		32.50	Cost includes material and labor to
Install	Ea.	210	17.45		227.45	install 3' x 6' 8" pre-hung 1-3/4"
Demolish and Install	Ea.	210	49.95		259.95	lauan door of solid wood.
Reinstall	Ea.		17.44		17.44	
Clean	Ea.	3.11	10.60		13.71	
Paint	Ea.	9.25	43.50		52.75	
Minimum Charge	Job		174		174	
French Style						
Demolish	Ea.		35.50		35.50	Cost includes material and labor to
Install	Ea.	655	50		705	install 3' x 6' 8" pre-hung 1-3/4"
Demolish and Install	Ea.	655	85.50		740.50	french style door with 15 lights.
Reinstall	Ea.		49.83		49.83	
Clean	Ea.	.21	10.60		10.81	
Paint	Ea.	6.95	153		159.95	
Minimum Charge	Job		174		174	
Double						
Demolish	Ea.		23.50		23.50	Includes material and labor to install a
Install	Ea.	1150	87		1237	pre-hung double entry door including
Demolish and Install	Ea.	1150	110.50		1260.50	frame and exterior trim.
Clean	Ea.	.12	36.50		36.62	
Paint	Ea.	18.60	153		171.60	
Minimum Charge	Job		87		87	
Commercial Grade						
3' x 6' 8"						
Demolish	Ea.		32.50		32.50	Includes material and labor to install
Install	Ea.	296	20.50		316.50	flush solid core 1-3/4" hardwood
Demolish and Install	Ea.	296	53		349	veneer 3' x 6' 8' door only.
Reinstall	Ea.		20.52		20.52	
Clean	Ea.	3.11	10.60		13.71	
Paint	Ea.	6.60	25.50		32.10	
Minimum Charge	Job		174		174	
3' 6" x 6' 8"						
Demolish	Ea.		32.50		32.50	Includes material and labor to install
Install	Ea.	400	20.50		420.50	flush solid core 1-3/4" hardwood
Demolish and Install	Ea.	400	53		453	veneer 3' 6" x 6' 8' door only.
Reinstall	Ea.		20.52		20.52	
Clean	Ea.	3.11	10.60		13.71	
Paint	Ea.	6.60	25.50		32.10	
Minimum Charge	Job		174		174	

Exterior Doors and Windows

Solid Core Door

	Unit	Material	Labor	Equip.	Total	Specification
3' x 7'						
Demolish	Ea.		32.50		32.50	Includes material and labor to install
Install	Ea.	315	50		365	flush solid core 1-3/4" hardwood
Demolish and Install	Ea.	315	82.50		397.50	veneer 3' x 7' door only.
Reinstall	Ea.		49.83		49.83	
Clean	Ea.	3.11	10.60		13.71	
Paint	Ea.	6.60	25.50		32.10	
Minimum Charge	Job		174		174	
3' 6" x 7'						
Demolish	Ea.		32.50		32.50	Includes material and labor to install
Install	Ea.	430	25		455	flush solid core 1-3/4" hardwood
Demolish and Install	Ea.	430	57.50		487.50	veneer 3' 6" x 7' door only.
Reinstall	Ea.		24.91		24.91	
Clean	Ea.	3.11	10.60		13.71	
Paint	Ea.	6.60	25.50		32.10	
Minimum Charge	Job		174		174	
3' x 8'						
Demolish	Ea.		32.50		32.50	Includes material and labor to install
Install	Ea.	370	43.50		413.50	flush solid core 1-3/4" hardwood
Demolish and Install	Ea.	370	76		446	veneer 3' x 8' door only.
Reinstall	Ea.		43.60		43.60	
Clean	Ea.	3.74	15.90		19.64	
Paint	Ea.	6.60	25.50		32.10	
Minimum Charge	Job		174		174	
3' 6" x 8'						
Demolish	Ea.		32.50		32.50	Includes material and labor to install
Install	Ea.	455	43.50		498.50	flush solid core 1-3/4" hardwood
Demolish and Install	Ea.	455	76		531	veneer 3' 6" x 8' door only.
Reinstall	Ea.		43.60		43.60	
Clean	Ea.	3.74	15.90		19.64	
Paint	Ea.	6.60	25.50		32.10	
Minimum Charge	Job		174		174	
3' x 9'						
Demolish	Ea.		32.50		32.50	Includes material and labor to install
Install	Ea.	480	50		530	flush solid core 1-3/4" hardwood
Demolish and Install	Ea.	480	82.50		562.50	veneer 3' x 9' door only.
Reinstall	Ea.		49.83		49.83	
Clean	Ea.	3.74	15.90		19.64	
Paint	Ea.	6.60	25.50		32.10	
Minimum Charge	Job		174		174	
Stain						
Install	Ea.	6.45	34		40.45	Includes labor and material to stain
Minimum Charge	Job		153		153	exterior type door and trim on both sides.

Door Only (SC)

	Unit	Material	Labor	Equip.	Total	Specification
2' 4" x 6' 8"						
Demolish	Ea.		6.90		6.90	Cost includes material and labor to
Install	Ea.	200	19.40		219.40	install 2' 4" x 6' 8" lauan solid core
Demolish and Install	Ea.	200	26.30		226.30	door slab.
Reinstall	Ea.		19.38		19.38	
Clean	Ea.	1.30	7.95		9.25	
Paint	Ea.	2.82	38		40.82	
Minimum Charge	Job		174		174	

Solid Core Door		Unit	Material	Labor	Equip.	Total	Specification
2' 8" x 6' 8"							
	Demolish	Ea.		6.90		6.90	Cost includes material and labor to
	Install	Ea.	205	19.40		224.40	install 2' 8" x 6' 8" lauan solid core
	Demolish and Install	Ea.	205	26.30		231.30	door slab.
	Reinstall	Ea.		19.38		19.38	
	Clean	Ea.	1.30	7.95		9.25	
	Paint	Ea.	2.82	38		40.82	
	Minimum Charge	Job		174		174	
3' x 6' 8"							
	Demolish	Ea.		6.90		6.90	Cost includes material and labor to
	Install	Ea.	212	20.50		232.50	install 3' x 6' 8" lauan solid core door
	Demolish and Install	Ea.	212	27.40		239.40	slab.
	Reinstall	Ea.		20.52		20.52	
	Clean	Ea.	1.30	7.95		9.25	
	Paint	Ea.	2.82	38		40.82	
	Minimum Charge	Job		174		174	
3' 6" x 6' 8"							
	Demolish	Ea.		6.90		6.90	Cost includes material and labor to
	Install	Ea.	241	22		263	install 3' 6" x 6' 8" lauan solid core
	Demolish and Install	Ea.	241	28.90		269.90	door slab.
	Reinstall	Ea.		21.80		21.80	
	Clean	Ea.	1.30	7.95		9.25	
	Paint	Ea.	2.82	38		40.82	
	Minimum Charge	Job		174		174	
Panel							
	Demolish	Ea.		6.90		6.90	Cost includes material and labor to
	Install	Ea.	177	41		218	install 3' x 6' 8" six (6) panel interior
	Demolish and Install	Ea.	177	47.90		224.90	door.
	Reinstall	Ea.		41.04		41.04	
	Clean	Ea.	1.56	7.95		9.51	
	Paint	Ea.	2.82	38		40.82	
	Minimum Charge	Job		174		174	
Casing Trim							
Single Width							
	Demolish	Ea.		6.15		6.15	Cost includes material and labor to
	Install	Opng.	25.50	59		84.50	install 11/16" x 2-1/2" pine ranch
	Demolish and Install	Opng.	25.50	65.15		90.65	style casing for one side of a standard
	Reinstall	Opng.		47.29		47.29	door opening.
	Clean	Opng.	.44	2.54		2.98	
	Paint	Ea.	3.22	6.10		9.32	
	Minimum Charge	Job		174		174	
Double Width							
	Demolish	Ea.		7.30		7.30	Cost includes material and labor to
	Install	Opng.	29.50	70		99.50	install 11/16" x 2-1/2" pine ranch
	Demolish and Install	Opng.	29.50	77.30		106.80	style door casing for one side of a
	Reinstall	Opng.		55.81		55.81	double door opening.
	Clean	Opng.	.52	5.10		5.62	
	Paint	Ea.	3.22	6.95		10.17	
	Minimum Charge	Job		174		174	

Exterior Doors and Windows

Components		Unit	Material	Labor	Equip.	Total	Specification
Hardware							
Doorknob							
	Demolish	Ea.		12.90		12.90	Includes material and labor to install
	Install	Ea.	18.75	22		40.75	residential passage lockset, keyless
	Demolish and Install	Ea.	18.75	34.90		53.65	bored type.
	Reinstall	Ea.		17.44		17.44	
	Clean	Ea.	.31	7.95		8.26	
	Minimum Charge	Job		174		174	
Doorknob w / Lock							
	Demolish	Ea.		12.90		12.90	Includes material and labor to install
	Install	Ea.	21.50	22		43.50	privacy lockset.
	Demolish and Install	Ea.	21.50	34.90		56.40	
	Reinstall	Ea.		17.44		17.44	
	Clean	Ea.	.31	7.95		8.26	
	Minimum Charge	Job		174		174	
Lever Handle							
	Demolish	Ea.		12.90		12.90	Includes material and labor to install
	Install	Ea.	28.50	35		63.50	residential passage lockset, keyless
	Demolish and Install	Ea.	28.50	47.90		76.40	bored type with lever handle.
	Reinstall	Ea.		27.90		27.90	
	Clean	Ea.	.31	7.95		8.26	
	Minimum Charge	Job		174		174	
Deadbolt							
	Demolish	Ea.		12.10		12.10	Includes material and labor to install a
	Install	Ea.	55	25		80	deadbolt.
	Demolish and Install	Ea.	55	37.10		92.10	
	Reinstall	Ea.		19.93		19.93	
	Clean	Ea.	.31	7.95		8.26	
	Minimum Charge	Job		174		174	
Grip Handle Entry							
	Demolish	Ea.		11.05		11.05	Includes material and labor to install
	Install	Ea.	131	39		170	residential keyed entry lock set with
	Demolish and Install	Ea.	131	50.05		181.05	grip type handle.
	Reinstall	Ea.		31		31	
	Clean	Ea.	.31	7.95		8.26	
	Minimum Charge	Job		174		174	
Panic Device							
	Demolish	Ea.		28		28	Includes material and labor to install
	Install	Ea.	370	58		428	touch bar, low profile, exit only.
	Demolish and Install	Ea.	370	86		456	
	Reinstall	Ea.		46.51		46.51	
	Clean	Ea.	.15	14.10		14.25	
	Minimum Charge	Job		174		174	
Closer							
	Demolish	Ea.		7.25		7.25	Includes material and labor to install
	Install	Ea.	177	58		235	pneumatic heavy duty closer for
	Demolish and Install	Ea.	177	65.25		242.25	exterior type doors.
	Reinstall	Ea.		46.51		46.51	
	Clean	Ea.	.15	10.60		10.75	
	Minimum Charge	Job		174		174	
Peep-hole							
	Demolish	Ea.		6.25		6.25	Includes material and labor to install
	Install	Ea.	15.75	10.90		26.65	peep-hole for door.
	Demolish and Install	Ea.	15.75	17.15		32.90	
	Reinstall	Ea.		8.72		8.72	
	Clean	Ea.	.02	2.81		2.83	
	Minimum Charge	Job		174		174	

Components		Unit	Material	Labor	Equip.	Total	Specification
Exit Sign							
	Demolish	Ea.		19.35		19.35	Includes material and labor to install a
	Install	Ea.	53	48		101	wall mounted interior electric exit sign.
	Demolish and Install	Ea.	53	67.35		120.35	
	Reinstall	Ea.		38.32		38.32	
	Clean	Ea.	.02	5.70		5.72	
	Minimum Charge	Job		174		174	
Kickplate							
	Demolish	Ea.		10.75		10.75	Includes material and labor to install
	Install	Ea.	22	23.50		45.50	aluminum kickplate, 10" x 28".
	Demolish and Install	Ea.	22	34.25		56.25	
	Reinstall	Ea.		18.60		18.60	
	Clean	Ea.	.31	7.95		8.26	
	Minimum Charge	Job		174		174	
Threshold							
	Install	L.F.	4.13	3.49		7.62	Includes labor and material to install
	Minimum Charge	Job		87		87	oak threshold.
Weatherstripping							
	Install	L.F.	.30	.87		1.17	Includes labor and material to install
	Minimum Charge	Job		87		87	rubber weatherstripping for door.
Metal Jamb							
6' 8" High							
	Demolish	Ea.		19.40		19.40	Cost includes material and labor to
	Install	Ea.	134	22		156	install 18 gauge hollow metal door
	Demolish and Install	Ea.	134	41.40		175.40	frame to fit a 4-1/2" jamb 6' 8" high
	Reinstall	Ea.		21.80		21.80	and up to 3' 6" opening.
	Clean	Ea.	.65	5.30		5.95	
	Paint	Ea.	1.10	6.10		7.20	
	Minimum Charge	Job		174		174	
7' High							
	Demolish	Ea.		19.40		19.40	Cost includes material and labor to
	Install	Ea.	140	23.50		163.50	install 18 gauge hollow metal door
	Demolish and Install	Ea.	140	42.90		182.90	frame to fit a 4-1/2" jamb 7' high and
	Reinstall	Ea.		23.25		23.25	up to 3' 6" opening.
	Clean	Ea.	.65	5.30		5.95	
	Paint	Ea.	1.10	6.10		7.20	
	Minimum Charge	Job		174		174	
8' High							
	Demolish	Ea.		19.40		19.40	Cost includes material and labor to
	Install	Ea.	171	29		200	install 18 gauge hollow metal door
	Demolish and Install	Ea.	171	48.40		219.40	frame to fit a 4-1/2" jamb 8' high and
	Reinstall	Ea.		29.07		29.07	up to 3' 6" opening.
	Clean	Ea.	.91	6.35		7.26	
	Paint	Ea.	1.22	6.80		8.02	
	Minimum Charge	Job		174		174	
9' High							
	Demolish	Ea.		19.40		19.40	Cost includes material and labor to
	Install	Ea.	188	35		223	install 18 gauge hollow metal door
	Demolish and Install	Ea.	188	54.40		242.40	frame to fit a 4-1/2" jamb 9' high and
	Reinstall	Ea.		34.88		34.88	up to 3' 6" opening.
	Clean	Ea.	.91	6.35		7.26	
	Paint	Ea.	1.22	6.80		8.02	
	Minimum Charge	Job		174		174	

Exterior Doors and Windows

Components		Unit	Material	Labor	Equip.	Total	Specification
Wood Jamb							
	Demolish	Ea.		24		24	Includes material and labor to install
	Install	Ea.	103	9.45		112.45	flat pine jamb with square cut heads
	Demolish and Install	Ea.	103	33.45		136.45	and rabbeted sides for 6' 8" high and
	Clean	Ea.	.10	6.35		6.45	3-9/16" door including trim sets for
	Paint	Ea.	6	20.50		26.50	both sides.
	Minimum Charge	Job		87		87	
Stain							
	Install	Ea.	4.44	25.50		29.94	Includes labor and material to stain
	Minimum Charge	Job		153		153	single door and trim on both sides.
Shave & Refit							
	Install	Ea.		29		29	Includes labor to shave and rework
	Minimum Charge	Job		87		87	door to fit opening at the job site.

Sliding Patio Door		Unit	Material	Labor	Equip.	Total	Specification
Aluminum							
6' Single Pane							
	Demolish	Ea.		48.50		48.50	Cost includes material and labor to
	Install	Ea.	430	174		604	install 6' x 6' 8" sliding anodized
	Demolish and Install	Ea.	430	222.50		652.50	aluminum patio door with single pane
	Reinstall	Ea.		139.52		139.52	tempered safety glass, anodized
	Clean	Ea.	1.95	15.90		17.85	frame, screen, weatherstripped.
	Minimum Charge	Job		174		174	
8' Single Pane							
	Demolish	Ea.		48.50		48.50	Cost includes material and labor to
	Install	Ea.	490	233		723	install 8' x 6' 8" sliding aluminum
	Demolish and Install	Ea.	490	281.50		771.50	patio door with single pane tempered
	Reinstall	Ea.		186.03		186.03	safety glass, anodized frame, screen,
	Clean	Ea.	1.95	15.90		17.85	weatherstripped.
	Minimum Charge	Job		174		174	
10' Single Pane							
	Demolish	Ea.		48.50		48.50	Cost includes material and labor to
	Install	Ea.	640	350		990	install 10' x 6' 8" sliding aluminum
	Demolish and Install	Ea.	640	398.50		1038.50	patio door with single pane tempered
	Reinstall	Ea.		279.04		279.04	safety glass, anodized frame, screen,
	Clean	Ea.	1.95	15.90		17.85	weatherstripped.
	Minimum Charge	Job		174		174	
6' Insulated							
	Demolish	Ea.		48.50		48.50	Cost includes material and labor to
	Install	Ea.	595	174		769	install 6' x 6' 8" sliding aluminum
	Demolish and Install	Ea.	595	222.50		817.50	patio door with double insulated
	Reinstall	Ea.		139.52		139.52	tempered glass, baked-on enamel
	Clean	Ea.	1.95	15.90		17.85	finish, screen, weatherstripped.
	Minimum Charge	Job		174		174	
8' Insulated							
	Demolish	Ea.		48.50		48.50	Cost includes material and labor to
	Install	Ea.	695	233		928	install 8' x 6' 8" sliding aluminum
	Demolish and Install	Ea.	695	281.50		976.50	patio door with insulated tempered
	Reinstall	Ea.		186.03		186.03	glass, baked-on enamel finish, screen,
	Clean	Ea.	1.95	15.90		17.85	weatherstripped.
	Minimum Charge	Job		174		174	

Exterior Doors and Windows

Sliding Patio Door	Unit	Material	Labor	Equip.	Total	Specification
10' Insulated						
Demolish	Ea.		48.50		48.50	Cost includes material and labor to
Install	Ea.	770	350		1120	install 10' x 6' 8" sliding aluminum
Demolish and Install	Ea.	770	398.50		1168.50	patio door with double insulated
Reinstall	Ea.		279.04		279.04	tempered glass, baked-on enamel
Clean	Ea.	1.95	15.90		17.85	finish, screen, weatherstripped.
Minimum Charge	Job		174		174	
Vinyl						
6' Single Pane						
Demolish	Ea.		48.50		48.50	Cost includes material and labor to
Install	Ea.	420	174		594	install 6' x 6' 8" sliding vinyl patio
Demolish and Install	Ea.	420	222.50		642.50	door with single pane tempered safety
Reinstall	Ea.		139.52		139.52	glass, anodized frame, screen,
Clean	Ea.	1.95	15.90		17.85	weatherstripped.
Minimum Charge	Job		174		174	
8' Single Pane						
Demolish	Ea.		48.50		48.50	Cost includes material and labor to
Install	Ea.	470	233		703	install 8' x 6' 8" sliding vinyl patio
Demolish and Install	Ea.	470	281.50		751.50	door with single pane tempered safety
Reinstall	Ea.		186.03		186.03	glass, anodized frame, screen,
Clean	Ea.	1.95	15.90		17.85	weatherstripped.
Minimum Charge	Job		174		174	
10' Single Pane						
Demolish	Ea.		48.50		48.50	Cost includes material and labor to
Install	Ea.	610	350		960	install 10' x 6' 8" sliding vinyl patio
Demolish and Install	Ea.	610	398.50		1008.50	door with single pane tempered safety
Reinstall	Ea.		279.04		279.04	glass, anodized frame, screen,
Clean	Ea.	1.95	15.90		17.85	weatherstripped.
Minimum Charge	Job		174		174	
6' Insulated						
Demolish	Ea.		48.50		48.50	Cost includes material and labor to
Install	Ea.	730	174		904	install 6' x 6' 8" sliding vinyl patio
Demolish and Install	Ea.	730	222.50		952.50	door with double insulated tempered
Reinstall	Ea.		139.52		139.52	glass, anodized frame, screen,
Clean	Ea.	1.95	15.90		17.85	weatherstripped.
Minimum Charge	Job		174		174	
8' Insulated						
Demolish	Ea.		48.50		48.50	Cost includes material and labor to
Install	Ea.	940	233		1173	install 8' x 6' 8" sliding vinyl patio
Demolish and Install	Ea.	940	281.50		1221.50	door with double insulated tempered
Reinstall	Ea.		186.03		186.03	glass, anodized frame, screen,
Clean	Ea.	1.95	15.90		17.85	weatherstripped.
Minimum Charge	Job		174		174	
10' Insulated						
Demolish	Ea.		48.50		48.50	Cost includes material and labor to
Install	Ea.	1000	350		1350	install 10' x 6' 8" sliding vinyl patio
Demolish and Install	Ea.	1000	398.50		1398.50	door with double insulated tempered
Reinstall	Ea.		279.04		279.04	glass, anodized frame, screen,
Clean	Ea.	1.95	15.90		17.85	weatherstripped.
Minimum Charge	Job		174		174	

Sliding Patio Door

Sliding Patio Door	Unit	Material	Labor	Equip.	Total	Specification
Wood Framed						
6' Premium Insulated						
Demolish	Ea.		48.50		48.50	Cost includes material and labor to
Install	Ea.	1275	174		1449	install 6' x 6' 8" sliding wood patio
Demolish and Install	Ea.	1275	222.50		1497.50	door with pine frame, double insulated
Reinstall	Ea.		139.52		139.52	glass, screen, lock and dead bolt,
Clean	Ea.	1.95	15.90		17.85	weatherstripped. Add for casing and
Minimum Charge	Job		174		174	finishing.
8' Premium Insulated						
Demolish	Ea.		48.50		48.50	Cost includes material and labor to
Install	Ea.	1475	233		1708	install 8' x 6' 8" sliding wood patio
Demolish and Install	Ea.	1475	281.50		1756.50	door with pine frame, double insulated
Reinstall	Ea.		186.03		186.03	glass, screen, lock and dead bolt,
Clean	Ea.	1.95	15.90		17.85	weatherstripped. Add for casing and
Minimum Charge	Job		174		174	finishing.
9' Premium Insulated						
Demolish	Ea.		48.50		48.50	Cost includes material and labor to
Install	Ea.	1475	233		1708	install 9' x 6' 8" sliding wood patio
Demolish and Install	Ea.	1475	281.50		1756.50	door with pine frame, double insulated
Reinstall	Ea.		186.03		186.03	glass, screen, lock and dead bolt,
Clean	Ea.	1.95	15.90		17.85	weatherstripped. Add for casing and
Minimum Charge	Job		174		174	finishing.
Stain						
Install	Ea.	6.45	34		40.45	Includes labor and material to stain
Minimum Charge	Job		153		153	exterior type door and trim on both sides.
Casing Trim						
Demolish	Ea.		7.30		7.30	Cost includes material and labor to
Install	Opng.	29.50	70		99.50	install 11/16" x 2-1/2" pine ranch
Demolish and Install	Opng.	29.50	77.30		106.80	style door casing for one side of a
Reinstall	Opng.		55.81		55.81	double door opening.
Clean	Opng.	.52	5.10		5.62	
Paint	Ea.	3.22	6.95		10.17	
Minimum Charge	Job		174		174	
Sliding Screen						
Demolish	Ea.		6.90		6.90	Includes material and labor to install
Install	Ea.	118	17.45		135.45	sliding screen door, 6' 8" x 3'.
Demolish and Install	Ea.	118	24.35		142.35	
Reinstall	Ea.		13.95		13.95	
Clean	Ea.	.65	5.30		5.95	
Minimum Charge	Job		174		174	

Screen Door

Screen Door	Unit	Material	Labor	Equip.	Total	Specification
Aluminum						
Demolish	Ea.		32.50		32.50	Includes material and labor to install
Install	Ea.	108	50		158	aluminum frame screen door with
Demolish and Install	Ea.	108	82.50		190.50	fiberglass screen cloth, closer, hinges,
Reinstall	Ea.		39.86		39.86	latch.
Clean	Ea.	3.11	10.60		13.71	
Paint	Ea.	3.10	12.20		15.30	
Minimum Charge	Job		174		174	

Exterior Doors and Windows

Screen Door

Screen Door	Unit	Material	Labor	Equip.	Total	Specification
Wood						Includes material and labor to install wood screen door with aluminum cloth screen. Frame and hardware not included.
Demolish	Ea.		32.50		32.50	
Install	Ea.	144	58		202	
Demolish and Install	Ea.	144	90.50		234.50	
Reinstall	Ea.		46.51		46.51	
Clean	Ea.	3.11	10.60		13.71	
Paint	Ea.	3.10	12.20		15.30	
Minimum Charge	Job		174		174	
Shave & Refit						Includes labor to shave and rework door to fit opening at the job site.
Install	Ea.		29		29	
Minimum Charge	Job		174		174	

Storm Door

Storm Door	Unit	Material	Labor	Equip.	Total	Specification
Good Grade						Includes material and labor to install aluminum frame storm door with fiberglass screen cloth, upper glass panel, closer, hinges, latch.
Demolish	Ea.		32.50		32.50	
Install	Ea.	168	50		218	
Demolish and Install	Ea.	168	82.50		250.50	
Reinstall	Ea.		39.86		39.86	
Clean	Ea.	3.11	10.60		13.71	
Paint	Ea.	3.10	12.20		15.30	
Minimum Charge	Job		174		174	
Better Grade						Includes material and labor to install aluminum frame storm door with fiberglass screen cloth, upper and lower glass panel, closer, latch.
Demolish	Ea.		32.50		32.50	
Install	Ea.	223	50		273	
Demolish and Install	Ea.	223	82.50		305.50	
Reinstall	Ea.		39.86		39.86	
Clean	Ea.	3.11	10.60		13.71	
Paint	Ea.	3.10	12.20		15.30	
Minimum Charge	Job		174		174	
Premium Grade						Includes material and labor to install aluminum frame storm door with fiberglass screen cloth, upper and lower glass panel, closer, latch.
Demolish	Ea.		32.50		32.50	
Install	Ea.	272	50		322	
Demolish and Install	Ea.	272	82.50		354.50	
Reinstall	Ea.		39.86		39.86	
Clean	Ea.	3.11	10.60		13.71	
Paint	Ea.	3.10	12.20		15.30	
Minimum Charge	Job		174		174	
Insulated						Includes material and labor to install aluminum frame storm door with fiberglass screen cloth, upper and lower glass panel, closer, latch.
Demolish	Ea.		32.50		32.50	
Install	Ea.	251	25		276	
Demolish and Install	Ea.	251	57.50		308.50	
Reinstall	Ea.		19.93		19.93	
Clean	Ea.	3.11	10.60		13.71	
Paint	Ea.	3.10	12.20		15.30	
Minimum Charge	Job		174		174	

Garage Door		Unit	Material	Labor	Equip.	Total	Specification
9' One Piece Metal							
	Demolish	Ea.		81.50		81.50	Includes material and labor to install
	Install	Ea.	400	87		487	one piece steel garage door 9' x 7'
	Demolish and Install	Ea.	400	168.50		568.50	primed including hardware.
	Reinstall	Ea.		69.76		69.76	
	Clean	Ea.	3.89	32		35.89	
	Paint	Ea.	8.15	76.50		84.65	
	Minimum Charge	Job		174		174	
16' One Piece Metal							
	Demolish	Ea.		97		97	Includes material and labor to install
	Install	Ea.	625	116		741	one piece steel garage door 16' x 7'
	Demolish and Install	Ea.	625	213		838	primed including hardware.
	Reinstall	Ea.		93.01		93.01	
	Clean	Ea.	7.80	42.50		50.30	
	Paint	Ea.	16.25	87		103.25	
	Minimum Charge	Job		174		174	
9' Metal Sectional							
	Demolish	Ea.		81.50		81.50	Includes material and labor to install
	Install	Ea.	545	132		677	sectional metal garage door 9' x 7'
	Demolish and Install	Ea.	545	213.50		758.50	primed including hardware.
	Reinstall	Ea.		105.70		105.70	
	Clean	Ea.	3.98	36.50		40.48	
	Paint	Ea.	16.25	87		103.25	
	Minimum Charge	Job		174		174	
16' Metal Sectional							
	Demolish	Ea.		97		97	Includes material and labor to install
	Install	Ea.	695	233		928	sectional metal garage door 16' x 7'
	Demolish and Install	Ea.	695	330		1025	primed including hardware.
	Reinstall	Ea.		186.03		186.03	
	Clean	Ea.	7.80	42.50		50.30	
	Paint	Ea.	16.25	87		103.25	
	Minimum Charge	Job		174		174	
Metal / Steel Roll Up							
8' x 8' Steel							
	Demolish	Ea.		111		111	Cost includes material and labor to
	Install	Ea.	940	540		1480	install 8' x 8' steel overhead roll-up,
	Demolish and Install	Ea.	940	651		1591	chain hoist operated service door.
	Reinstall	Ea.		432.40		432.40	
	Clean	Ea.	3.98	36.50		40.48	
	Paint	Ea.	17.35	102		119.35	
	Minimum Charge	Job		865		865	
10' x 10' Steel							
	Demolish	Ea.		129		129	Cost includes material and labor to
	Install	Ea.	1625	620		2245	install 10' x 10' steel overhead roll-up,
	Demolish and Install	Ea.	1625	749		2374	chain hoist operated service door.
	Reinstall	Ea.		494.17		494.17	
	Clean	Ea.	6.20	51		57.20	
	Paint	Ea.	27	102		129	
	Minimum Charge	Job		865		865	
12' x 12' Steel							
	Demolish	Ea.		155		155	Cost includes material and labor to
	Install	Ea.	1625	720		2345	install 12' x 12' steel overhead roll-up,
	Demolish and Install	Ea.	1625	875		2500	chain hoist operated service door.
	Reinstall	Ea.		576.53		576.53	
	Clean	Ea.	8.95	56.50		65.45	
	Paint	Ea.	39	122		161	
	Minimum Charge	Job		865		865	

Garage Door	Unit	Material	Labor	Equip.	Total	Specification
14' x 14' Steel						
Demolish	Ea.		194		194	Cost includes material and labor to
Install	Ea.	3025	1075		4100	install 14' x 14' steel overhead roll-up,
Demolish and Install	Ea.	3025	1269		4294	chain hoist operated service door.
Reinstall	Ea.		864.80		864.80	
Clean	Ea.	12.20	72.50		84.70	
Paint	Ea.	45.50	153		198.50	
Minimum Charge	Job		865		865	
18' x 18' Steel						
Demolish	Ea.		215		215	Cost includes material and labor to
Install	Opng.	2850	1450		4300	install 18' x 18' steel overhead roll-up,
Demolish and Install	Opng.	2850	1665		4515	chain hoist operated service door.
Reinstall	Opng.		1153.07		1153.07	
Clean	Ea.	12.20	42.50		54.70	
Paint	Ea.	88	305		393	
Minimum Charge	Job		865		865	
Automatic Motor						
Install	Ea.	420	35		455	Includes labor and material to install
Minimum Charge	Job		174		174	electric opener motor, 1/3 HP.
9' One Piece Wood						
Demolish	Ea.		81.50		81.50	Includes material and labor to install
Install	Ea.	435	87		522	one piece wood garage door 9' x 7'
Demolish and Install	Ea.	435	168.50		603.50	primed including hardware.
Reinstall	Ea.		69.76		69.76	
Clean	Ea.	3.89	32		35.89	
Paint	Ea.	8.15	76.50		84.65	
Minimum Charge	Job		174		174	
9' Wood Sectional						
Demolish	Ea.		81.50		81.50	Includes material and labor to install
Install	Ea.	580	87		667	sectional wood garage door 9' x 7'
Demolish and Install	Ea.	580	168.50		748.50	unfinished including hardware.
Reinstall	Ea.		69.76		69.76	
Clean	Ea.	3.89	32		35.89	
Paint	Ea.	8.15	76.50		84.65	
Minimum Charge	Job		174		174	
16' Wood Sectional						
Demolish	Ea.		97		97	Includes material and labor to install
Install	Ea.	1275	116		1391	sectional fiberglass garage door 16' x
Demolish and Install	Ea.	1275	213		1488	7' unfinished including hardware.
Reinstall	Ea.		93.01		93.01	
Clean	Ea.	7.80	42.50		50.30	
Paint	Ea.	16.25	87		103.25	
Minimum Charge	Job		174		174	
Electric Opener						
Demolish	Ea.		56.50		56.50	Includes material and labor to install
Install	Ea.	278	65.50		343.50	electric opener, 1/3 HP economy.
Demolish and Install	Ea.	278	122		400	
Reinstall	Ea.		52.35		52.35	
Clean	Ea.	.15	14.10		14.25	
Minimum Charge	Job		174		174	
Springs						
Install	Pr.	42.50	87		129.50	Includes labor and material to replace
Minimum Charge	Job		174		174	springs for garage door.

Commercial Metal Door	Unit	Material	Labor	Equip.	Total	Specification
Pre-hung Steel Door						
2' 8" x 6' 8"						
Demolish	Ea.		48.50		48.50	Cost includes material and labor to
Install	Ea.	555	22		577	install 2' 8" x 6' 8" pre-hung 1-3/4"
Demolish and Install	Ea.	555	70.50		625.50	18 gauge steel door with steel frame,
Reinstall	Ea.		21.80		21.80	hinges, aluminum sill,
Clean	Ea.	3.11	10.60		13.71	weather-stripped.
Paint	Ea.	3.97	25.50		29.47	
Minimum Charge	Job		174		174	
3' x 6' 8"						
Demolish	Ea.		48.50		48.50	Cost includes material and labor to
Install	Ea.	525	23.50		548.50	install 3' x 6' 8" pre-hung 1-3/4" 18
Demolish and Install	Ea.	525	72		597	gauge steel door with steel frame,
Reinstall	Ea.		23.25		23.25	hinges, aluminum sill,
Clean	Ea.	3.11	10.60		13.71	weather-stripped.
Paint	Ea.	3.97	25.50		29.47	
Minimum Charge	Job		174		174	
3' 6" x 6' 8"						
Demolish	Ea.		48.50		48.50	Cost includes material and labor to
Install	Ea.	600	27		627	install 3' 6" x 6' 8" pre-hung 1-3/4"
Demolish and Install	Ea.	600	75.50		675.50	18 gauge steel door with steel frame,
Reinstall	Ea.		26.83		26.83	hinges, aluminum sill,
Clean	Ea.	3.11	10.60		13.71	weather-stripped.
Paint	Ea.	3.97	25.50		29.47	
Minimum Charge	Job		174		174	
4' x 6' 8"						
Demolish	Ea.		48.50		48.50	Cost includes material and labor to
Install	Ea.	655	35		690	install 4' x 6' 8" pre-hung 1-3/4" 18
Demolish and Install	Ea.	655	83.50		738.50	gauge steel door with steel frame,
Reinstall	Ea.		34.88		34.88	hinges, aluminum sill,
Clean	Ea.	3.11	10.60		13.71	weather-stripped.
Paint	Ea.	3.97	25.50		29.47	
Minimum Charge	Job		174		174	
Metal Jamb						
6' 8" High						
Demolish	Ea.		19.40		19.40	Cost includes material and labor to
Install	Ea.	134	22		156	install 18 gauge hollow metal door
Demolish and Install	Ea.	134	41.40		175.40	frame to fit a 4-1/2" jamb 6' 8" high
Reinstall	Ea.		21.80		21.80	and up to 3' 6" opening.
Clean	Ea.	.65	5.30		5.95	
Paint	Ea.	1.10	6.10		7.20	
Minimum Charge	Job		174		174	
7' High						
Demolish	Ea.		19.40		19.40	Cost includes material and labor to
Install	Ea.	140	23.50		163.50	install 18 gauge hollow metal door
Demolish and Install	Ea.	140	42.90		182.90	frame to fit a 4-1/2" jamb 7' high and
Reinstall	Ea.		23.25		23.25	up to 3' 6" opening.
Clean	Ea.	.65	5.30		5.95	
Paint	Ea.	1.10	6.10		7.20	
Minimum Charge	Job		174		174	
8' High						
Demolish	Ea.		19.40		19.40	Cost includes material and labor to
Install	Ea.	171	29		200	install 18 gauge hollow metal door
Demolish and Install	Ea.	171	48.40		219.40	frame to fit a 4-1/2" jamb 8' high and
Reinstall	Ea.		29.07		29.07	up to 3' 6" opening.
Clean	Ea.	.91	6.35		7.26	
Paint	Ea.	1.22	6.80		8.02	
Minimum Charge	Job		174		174	

Commercial Metal Door	Unit	Material	Labor	Equip.	Total	Specification
9' High						
Demolish	Ea.		19.40		19.40	Cost includes material and labor to
Install	Ea.	188	35		223	install 18 gauge hollow metal door
Demolish and Install	Ea.	188	54.40		242.40	frame to fit a 4-1/2" jamb 9' high and
Reinstall	Ea.		34.88		34.88	up to 3' 6" opening.
Clean	Ea.	.91	6.35		7.26	
Paint	Ea.	1.22	6.80		8.02	
Minimum Charge	Job		174		174	
Hardware						
Doorknob w / Lock						
Demolish	Ea.		12.90		12.90	Includes material and labor to install
Install	Ea.	110	29		139	keyed entry lock set, bored type
Demolish and Install	Ea.	110	41.90		151.90	including knob trim, strike and strike
Reinstall	Ea.		23.25		23.25	box for standard commercial
Clean	Ea.	.31	7.95		8.26	application.
Minimum Charge	Job		174		174	
Lever Handle						
Demolish	Ea.		12.90		12.90	Includes material and labor to install
Install	Ea.	58.50	35		93.50	passage lockset, keyless bored type
Demolish and Install	Ea.	58.50	47.90		106.40	with lever handle, non-locking
Reinstall	Ea.		27.90		27.90	standard commercial latchset.
Clean	Ea.	.31	7.95		8.26	
Minimum Charge	Job		174		174	
Deadbolt						
Demolish	Ea.		12.10		12.10	Includes material and labor to install a
Install	Ea.	55	25		80	deadbolt.
Demolish and Install	Ea.	55	37.10		92.10	
Reinstall	Ea.		19.93		19.93	
Clean	Ea.	.31	7.95		8.26	
Minimum Charge	Job		174		174	
Panic Device						
Demolish	Ea.		28		28	Includes material and labor to install
Install	Ea.	370	58		428	touch bar, low profile, exit only.
Demolish and Install	Ea.	370	86		456	
Reinstall	Ea.		46.51		46.51	
Clean	Ea.	.15	14.10		14.25	
Minimum Charge	Job		174		174	
Closer						
Demolish	Ea.		7.25		7.25	Includes material and labor to install
Install	Ea.	188	58		246	pneumatic heavy duty model for
Demolish and Install	Ea.	188	65.25		253.25	exterior type doors.
Reinstall	Ea.		46.51		46.51	
Clean	Ea.	.15	10.60		10.75	
Minimum Charge	Job		174		174	
Peep-Hole						
Demolish	Ea.		6.25		6.25	Includes material and labor to install
Install	Ea.	15.75	10.90		26.65	peep-hole for door.
Demolish and Install	Ea.	15.75	17.15		32.90	
Reinstall	Ea.		8.72		8.72	
Clean	Ea.	.02	2.81		2.83	
Minimum Charge	Job		174		174	
Exit Sign						
Demolish	Ea.		19.35		19.35	Includes material and labor to install a
Install	Ea.	53	48		101	wall mounted interior electric exit sign.
Demolish and Install	Ea.	53	67.35		120.35	
Reinstall	Ea.		38.32		38.32	
Clean	Ea.	.02	5.70		5.72	
Minimum Charge	Job		174		174	

Exterior Doors and Windows

Commercial Metal Door	Unit	Material	Labor	Equip.	Total	Specification
Kickplate						
Demolish	Ea.		10.75		10.75	Includes material and labor to install
Install	Ea.	49.50	23.50		73	stainless steel kickplate, 10" x 36".
Demolish and Install	Ea.	49.50	34.25		83.75	
Reinstall	Ea.		18.60		18.60	
Clean	Ea.	.31	7.95		8.26	
Minimum Charge	Job		174		174	

Aluminum Window	Unit	Material	Labor	Equip.	Total	Specification
Single Hung Two Light						
2' x 2'						
Demolish	Ea.		17.65		17.65	Cost includes material and labor to
Install	Ea.	66	78.50		144.50	install 2' x 2' single hung, two light,
Demolish and Install	Ea.	66	96.15		162.15	double glazed aluminum window
Reinstall	Ea.		62.89		62.89	including screen.
Clean	Ea.	.03	6.50		6.53	
Minimum Charge	Job		216		216	
2' x 2' 6"						
Demolish	Ea.		17.65		17.65	Cost includes material and labor to
Install	Ea.	69.50	78.50		148	install 2' x 2' 6" single hung, two
Demolish and Install	Ea.	69.50	96.15		165.65	light, double glazed aluminum
Reinstall	Ea.		62.89		62.89	window including screen.
Clean	Ea.	.03	6.50		6.53	
Minimum Charge	Job		216		216	
3' x 1' 6"						
Demolish	Ea.		17.65		17.65	Cost includes material and labor to
Install	Ea.	65	86.50		151.50	install 3' x 1' 6" single hung, two
Demolish and Install	Ea.	65	104.15		169.15	light, double glazed aluminum
Reinstall	Ea.		69.18		69.18	window including screen.
Clean	Ea.	.02	5.70		5.72	
Minimum Charge	Job		216		216	
3' x 2'						
Demolish	Ea.		17.65		17.65	Cost includes material and labor to
Install	Ea.	78.50	86.50		165	install 3' x 2' single hung, two light,
Demolish and Install	Ea.	78.50	104.15		182.65	double glazed aluminum window
Reinstall	Ea.		69.18		69.18	including screen.
Clean	Ea.	.02	5.70		5.72	
Minimum Charge	Job		216		216	
3' x 2' 6"						
Demolish	Ea.		17.65		17.65	Cost includes material and labor to
Install	Ea.	82.50	86.50		169	install 3' x 2' 6" single hung, two
Demolish and Install	Ea.	82.50	104.15		186.65	light, double glazed aluminum
Reinstall	Ea.		69.18		69.18	window including screen.
Clean	Ea.	.03	6.50		6.53	
Minimum Charge	Job		216		216	
3' x 3'						
Demolish	Ea.		17.65		17.65	Cost includes material and labor to
Install	Ea.	92	86.50		178.50	install 3' x 3' single hung, two light,
Demolish and Install	Ea.	92	104.15		196.15	double glazed aluminum window
Reinstall	Ea.		69.18		69.18	including screen.
Clean	Ea.	.03	6.50		6.53	
Minimum Charge	Job		216		216	

Exterior Doors and Windows

Aluminum Window	Unit	Material	Labor	Equip.	Total	Specification
3' x 3' 6"						
Demolish	Ea.		17.65		17.65	Cost includes material and labor to
Install	Ea.	100	86.50		186.50	install 3' x 3' 6" single hung, two
Demolish and Install	Ea.	100	104.15		204.15	light, double glazed aluminum
Reinstall	Ea.		69.18		69.18	window including screen.
Clean	Ea.	.04	7.60		7.64	
Minimum Charge	Job		216		216	
3' x 4'						
Demolish	Ea.		17.65		17.65	Cost includes material and labor to
Install	Ea.	109	86.50		195.50	install 3' x 4' single hung, two light,
Demolish and Install	Ea.	109	104.15		213.15	double glazed aluminum window
Reinstall	Ea.		69.18		69.18	including screen.
Clean	Ea.	.04	7.60		7.64	
Minimum Charge	Job		216		216	
3' x 5'						
Demolish	Ea.		25.50		25.50	Cost includes material and labor to
Install	Ea.	117	86.50		203.50	install 3' x 5' single hung, two light,
Demolish and Install	Ea.	117	112		229	double glazed aluminum window
Reinstall	Ea.		69.18		69.18	including screen.
Clean	Ea.	.06	9.10		9.16	
Minimum Charge	Job		216		216	
3' x 6'						
Demolish	Ea.		25.50		25.50	Cost includes material and labor to
Install	Ea.	131	96		227	install 3' x 6' single hung, two light,
Demolish and Install	Ea.	131	121.50		252.50	double glazed aluminum window
Reinstall	Ea.		76.87		76.87	including screen.
Clean	Ea.	.07	10.75		10.82	
Minimum Charge	Job		216		216	

Double Hung Two Light	Unit	Material	Labor	Equip.	Total	Specification
2' x 2'						
Demolish	Ea.		17.65		17.65	Cost includes material and labor to
Install	Ea.	110	78.50		188.50	install 2' x 2' double hung, two light,
Demolish and Install	Ea.	110	96.15		206.15	double glazed aluminum window
Reinstall	Ea.		62.89		62.89	including screen.
Clean	Ea.	.02	5.70		5.72	
Minimum Charge	Job		216		216	
2' x 2' 6"						
Demolish	Ea.		17.65		17.65	Cost includes material and labor to
Install	Ea.	120	78.50		198.50	install 2' x 2' 6" double hung, two
Demolish and Install	Ea.	120	96.15		216.15	light, double glazed aluminum
Reinstall	Ea.		62.89		62.89	window including screen.
Clean	Ea.	.02	5.70		5.72	
Minimum Charge	Job		216		216	
3' x 1' 6"						
Demolish	Ea.		17.65		17.65	Cost includes material and labor to
Install	Ea.	284	86.50		370.50	install 3' x 1' 6" double hung, two
Demolish and Install	Ea.	284	104.15		388.15	light, double glazed aluminum
Reinstall	Ea.		69.18		69.18	window including screen.
Clean	Ea.	.02	5.70		5.72	
Minimum Charge	Job		216		216	
3' x 2'						
Demolish	Ea.		17.65		17.65	Cost includes material and labor to
Install	Ea.	310	86.50		396.50	install 3' x 2' double hung, two light,
Demolish and Install	Ea.	310	104.15		414.15	double glazed aluminum window
Reinstall	Ea.		69.18		69.18	including screen.
Clean	Ea.	.02	5.70		5.72	
Minimum Charge	Job		216		216	

Exterior Doors and Windows

Aluminum Window		Unit	Material	Labor	Equip.	Total	Specification
3' x 2' 6"							
	Demolish	Ea.		17.65		17.65	Cost includes material and labor to
	Install	Ea.	305	86.50		391.50	install 3' x 2' 6" double hung, two
	Demolish and Install	Ea.	305	104.15		409.15	light, double glazed aluminum
	Reinstall	Ea.		69.18		69.18	window including screen.
	Clean	Ea.	.03	6.50		6.53	
	Minimum Charge	Job		216		216	
3' x 3'							
	Demolish	Ea.		17.65		17.65	Cost includes material and labor to
	Install	Ea.	330	86.50		416.50	install 3' x 3' double hung, two light,
	Demolish and Install	Ea.	330	104.15		434.15	double glazed aluminum window
	Reinstall	Ea.		69.18		69.18	including screen.
	Clean	Ea.	.03	6.50		6.53	
	Minimum Charge	Job		216		216	
3' x 3' 6"							
	Demolish	Ea.		17.65		17.65	Cost includes material and labor to
	Install	Ea.	365	86.50		451.50	install 3' x 3' 6" double hung, two
	Demolish and Install	Ea.	365	104.15		469.15	light, double glazed aluminum
	Reinstall	Ea.		69.18		69.18	window including screen.
	Clean	Ea.	.04	7.60		7.64	
	Minimum Charge	Job		216		216	
3' x 4'							
	Demolish	Ea.		17.65		17.65	Cost includes material and labor to
	Install	Ea.	375	86.50		461.50	install 3' x 4' double hung, two light,
	Demolish and Install	Ea.	375	104.15		479.15	double glazed aluminum window
	Reinstall	Ea.		69.18		69.18	including screen.
	Clean	Ea.	.04	7.60		7.64	
	Minimum Charge	Job		216		216	
3' x 5'							
	Demolish	Ea.		25.50		25.50	Cost includes material and labor to
	Install	Ea.	425	86.50		511.50	install 3' x 5' double hung, two light,
	Demolish and Install	Ea.	425	112		537	double glazed aluminum window
	Reinstall	Ea.		69.18		69.18	including screen.
	Clean	Ea.	.06	9.10		9.16	
	Minimum Charge	Job		216		216	
3' x 6'							
	Demolish	Ea.		25.50		25.50	Cost includes material and labor to
	Install	Ea.	550	96		646	install 3' x 6' double hung, two light,
	Demolish and Install	Ea.	550	121.50		671.50	double glazed aluminum window
	Reinstall	Ea.		76.87		76.87	including screen.
	Clean	Ea.	.07	10.75		10.82	
	Minimum Charge	Job		216		216	
Sliding Sash							
1' 6" x 3'							
	Demolish	Ea.		17.65		17.65	Cost includes material and labor to
	Install	Ea.	63.50	78.50		142	install 1' 6" x 3' sliding sash, two
	Demolish and Install	Ea.	63.50	96.15		159.65	light, double glazed aluminum
	Reinstall	Ea.		62.89		62.89	window including screen.
	Clean	Ea.	.02	5.70		5.72	
	Minimum Charge	Job		216		216	
2' x 2'							
	Demolish	Ea.		17.65		17.65	Cost includes material and labor to
	Install	Ea.	56	86.50		142.50	install 2' x 2' sliding sash, two light,
	Demolish and Install	Ea.	56	104.15		160.15	double glazed aluminum window
	Reinstall	Ea.		69.18		69.18	including screen.
	Clean	Ea.	.02	5.70		5.72	
	Minimum Charge	Job		216		216	

Exterior Doors and Windows

Aluminum Window

	Unit	Material	Labor	Equip.	Total	Specification
3' x 2'						
Demolish	Ea.		17.65		17.65	Cost includes material and labor to
Install	Ea.	73.50	86.50		160	install 3' x 2' sliding sash, two light,
Demolish and Install	Ea.	73.50	104.15		177.65	double glazed aluminum window
Reinstall	Ea.		69.18		69.18	including screen.
Clean	Ea.	.02	5.70		5.72	
Minimum Charge	Job		216		216	
3' x 2' 6"						
Demolish	Ea.		17.65		17.65	Cost includes material and labor to
Install	Ea.	76	86.50		162.50	install 3' x 2' 6" sliding sash, two
Demolish and Install	Ea.	76	104.15		180.15	light, double glazed aluminum
Reinstall	Ea.		69.18		69.18	window including screen.
Clean	Ea.	.03	6.50		6.53	
Minimum Charge	Job		216		216	
3' x 3'						
Demolish	Ea.		17.65		17.65	Cost includes material and labor to
Install	Ea.	78	86.50		164.50	install 3' x 3' sliding sash, two light,
Demolish and Install	Ea.	78	104.15		182.15	double glazed aluminum window
Reinstall	Ea.		69.18		69.18	including screen.
Clean	Ea.	.03	6.50		6.53	
Minimum Charge	Job		216		216	
3' x 3' 6"						
Demolish	Ea.		17.65		17.65	Cost includes material and labor to
Install	Ea.	90	86.50		176.50	install 3' x 3' 6" sliding sash, two
Demolish and Install	Ea.	90	104.15		194.15	light, double glazed aluminum
Reinstall	Ea.		69.18		69.18	window including screen.
Clean	Ea.	.04	7.60		7.64	
Minimum Charge	Job		216		216	
3' x 4'						
Demolish	Ea.		17.65		17.65	Cost includes material and labor to
Install	Ea.	98.50	96		194.50	install 3' x 4' sliding sash, two light,
Demolish and Install	Ea.	98.50	113.65		212.15	double glazed aluminum window
Reinstall	Ea.		76.87		76.87	including screen.
Clean	Ea.	.04	7.60		7.64	
Minimum Charge	Job		216		216	
3' x 5'						
Demolish	Ea.		25.50		25.50	Cost includes material and labor to
Install	Ea.	125	96		221	install 3' x 5' sliding sash, two light,
Demolish and Install	Ea.	125	121.50		246.50	double glazed aluminum window
Reinstall	Ea.		76.87		76.87	including screen.
Clean	Ea.	.06	9.10		9.16	
Minimum Charge	Job		216		216	
4' x 5'						
Demolish	Ea.		25.50		25.50	Cost includes material and labor to
Install	Ea.	127	124		251	install 4' x 5' sliding sash, two light,
Demolish and Install	Ea.	127	149.50		276.50	double glazed aluminum window
Reinstall	Ea.		98.83		98.83	including screen.
Clean	Ea.	.10	13.05		13.15	
Minimum Charge	Job		216		216	
5' x 5'						
Demolish	Ea.		25.50		25.50	Cost includes material and labor to
Install	Ea.	178	124		302	install 5' x 5' sliding sash, two light,
Demolish and Install	Ea.	178	149.50		327.50	double glazed aluminum window
Reinstall	Ea.		98.83		98.83	including screen.
Clean	Ea.	.10	13.05		13.15	
Minimum Charge	Job		216		216	

Exterior Doors and Windows

Aluminum Window	Unit	Material	Labor	Equip.	Total	Specification
Awning Sash						
26" x 26", 2 Light						
Demolish	Ea.		17.65		17.65	Cost includes material and labor to
Install	Ea.	278	96		374	install 26" x 26" two light awning,
Demolish and Install	Ea.	278	113.65		391.65	double glazed aluminum window
Reinstall	Ea.		76.87		76.87	including screen.
Clean	Ea.	.03	6.50		6.53	
Minimum Charge	Job		216		216	
26" x 37", 2 Light						
Demolish	Ea.		17.65		17.65	Cost includes material and labor to
Install	Ea.	315	96		411	install 26" x 37" two light awning,
Demolish and Install	Ea.	315	113.65		428.65	double glazed aluminum window
Reinstall	Ea.		76.87		76.87	including screen.
Clean	Ea.	.03	6.50		6.53	
Minimum Charge	Job		216		216	
38" x 26", 3 Light						
Demolish	Ea.		17.65		17.65	Cost includes material and labor to
Install	Ea.	350	96		446	install 38" x 26" three light awning,
Demolish and Install	Ea.	350	113.65		463.65	double glazed aluminum window
Reinstall	Ea.		76.87		76.87	including screen.
Clean	Ea.	.04	7.60		7.64	
Minimum Charge	Job		216		216	
38" x 37", 3 Light						
Demolish	Ea.		17.65		17.65	Cost includes material and labor to
Install	Ea.	365	108		473	install 38" x 37" three light awning,
Demolish and Install	Ea.	365	125.65		490.65	double glazed aluminum window
Reinstall	Ea.		86.48		86.48	including screen.
Clean	Ea.	.04	7.60		7.64	
Minimum Charge	Job		216		216	
38" x 53", 3 Light						
Demolish	Ea.		25.50		25.50	Cost includes material and labor to
Install	Ea.	595	108		703	install 38" x 53" three light awning,
Demolish and Install	Ea.	595	133.50		728.50	double glazed aluminum window
Reinstall	Ea.		86.48		86.48	including screen.
Clean	Ea.	.06	9.10		9.16	
Minimum Charge	Job		216		216	
50" x 37", 4 Light						
Demolish	Ea.		25.50		25.50	Cost includes material and labor to
Install	Ea.	595	108		703	install 50" x 37" four light awning,
Demolish and Install	Ea.	595	133.50		728.50	double glazed aluminum window
Reinstall	Ea.		86.48		86.48	including screen.
Clean	Ea.	.06	9.10		9.16	
Minimum Charge	Job		216		216	
50" x 53", 4 Light						
Demolish	Ea.		25.50		25.50	Cost includes material and labor to
Install	Ea.	985	124		1109	install 50" x 53" four light awning,
Demolish and Install	Ea.	985	149.50		1134.50	double glazed aluminum window
Reinstall	Ea.		98.83		98.83	including screen.
Clean	Ea.	.07	10.75		10.82	
Minimum Charge	Job		216		216	
63" x 37", 5 Light						
Demolish	Ea.		25.50		25.50	Cost includes material and labor to
Install	Ea.	770	124		894	install 63" x 37" five light awning,
Demolish and Install	Ea.	770	149.50		919.50	double glazed aluminum window
Reinstall	Ea.		98.83		98.83	including screen.
Clean	Ea.	.07	10.75		10.82	
Minimum Charge	Job		216		216	

Aluminum Window	Unit	Material	Labor	Equip.	Total	Specification
63″ x 53″, 5 Light						
Demolish	Ea.		25.50		25.50	Cost includes material and labor to
Install	Ea.	1075	124		1199	install 63″ x 53″ five light awning,
Demolish and Install	Ea.	1075	149.50		1224.50	double glazed aluminum window
Reinstall	Ea.		98.83		98.83	including screen.
Clean	Ea.	.10	13.05		13.15	
Minimum Charge	Job		216		216	
Mullion Bars						
Install	Ea.	10.45	34.50		44.95	Includes labor and material to install
Minimum Charge	Job		216		216	mullions for sliding aluminum window.
Glass Louvered						
2′ x 2′						
Demolish	Ea.		17.65		17.65	Cost includes material and labor to
Install	Ea.	51	78.50		129.50	install 2′ x 2′ aluminum window
Demolish and Install	Ea.	51	96.15		147.15	including 4″ glass louvers, hardware,
Reinstall	Ea.		62.89		62.89	screen.
Clean	Ea.	.02	5.70		5.72	
Minimum Charge	Job		216		216	
2′ x 2′ 6″						
Demolish	Ea.		17.65		17.65	Cost includes material and labor to
Install	Ea.	62	78.50		140.50	install 2′ x 2′ 6″ aluminum window
Demolish and Install	Ea.	62	96.15		158.15	including 4″ glass louvers, hardware,
Reinstall	Ea.		62.89		62.89	screen.
Clean	Ea.	.02	5.70		5.72	
Minimum Charge	Job		216		216	
3′ x 1′ 6″						
Demolish	Ea.		17.65		17.65	Cost includes material and labor to
Install	Ea.	65.50	86.50		152	install 3′ x 1′ 6″ aluminum window
Demolish and Install	Ea.	65.50	104.15		169.65	including 4″ glass louvers, hardware,
Reinstall	Ea.		69.18		69.18	screen.
Clean	Ea.	.02	5.70		5.72	
Minimum Charge	Job		216		216	
3′ x 2′						
Demolish	Ea.		17.65		17.65	Cost includes material and labor to
Install	Ea.	75.50	86.50		162	install 3′ x 2′ aluminum window
Demolish and Install	Ea.	75.50	104.15		179.65	including 4″ glass louvers, hardware,
Reinstall	Ea.		69.18		69.18	screen.
Clean	Ea.	.02	5.70		5.72	
Minimum Charge	Job		216		216	
3′ x 2′ 6″						
Demolish	Ea.		17.65		17.65	Cost includes material and labor to
Install	Ea.	82.50	86.50		169	install 3′ x 2′ 6″ aluminum window
Demolish and Install	Ea.	82.50	104.15		186.65	including 4″ glass louvers, hardware,
Reinstall	Ea.		69.18		69.18	screen.
Clean	Ea.	.03	6.50		6.53	
Minimum Charge	Job		216		216	
3′ x 3′						
Demolish	Ea.		17.65		17.65	Cost includes material and labor to
Install	Ea.	95.50	96		191.50	install 3′ x 3′ aluminum window
Demolish and Install	Ea.	95.50	113.65		209.15	including 4″ glass louvers, hardware,
Reinstall	Ea.		76.87		76.87	screen.
Clean	Ea.	.03	6.50		6.53	
Minimum Charge	Job		216		216	

Exterior Doors and Windows

Aluminum Window		Unit	Material	Labor	Equip.	Total	Specification
3' x 3' 6"							
	Demolish	Ea.		17.65		17.65	Cost includes material and labor to
	Install	Ea.	108	96		204	install 3' x 3' 6" aluminum window
	Demolish and Install	Ea.	108	113.65		221.65	including 4" glass louvers, hardware,
	Reinstall	Ea.		76.87		76.87	screen.
	Clean	Ea.	.04	7.60		7.64	
	Minimum Charge	Job		216		216	
3' x 4'							
	Demolish	Ea.		17.65		17.65	Cost includes material and labor to
	Install	Ea.	121	96		217	install 3' x 4' aluminum window
	Demolish and Install	Ea.	121	113.65		234.65	including 4" glass louvers, hardware,
	Reinstall	Ea.		76.87		76.87	screen.
	Clean	Ea.	.04	7.60		7.64	
	Minimum Charge	Job		216		216	
3' x 5'							
	Demolish	Ea.		25.50		25.50	Cost includes material and labor to
	Install	Ea.	154	108		262	install 3' x 5' aluminum window with
	Demolish and Install	Ea.	154	133.50		287.50	4" glass louvers, hardware, screen.
	Reinstall	Ea.		86.48		86.48	
	Clean	Ea.	.06	9.10		9.16	
	Minimum Charge	Job		216		216	
3' x 6'							
	Demolish	Ea.		25.50		25.50	Cost includes material and labor to
	Install	Ea.	177	108		285	install 3' x 6' aluminum window with
	Demolish and Install	Ea.	177	133.50		310.50	4" glass louvers, hardware, screen.
	Reinstall	Ea.		86.48		86.48	
	Clean	Ea.	.07	10.75		10.82	
	Minimum Charge	Job		216		216	
Bay							
4' 8" x 6' 4"							
	Demolish	Ea.		18.80		18.80	Cost includes material and labor to
	Install	Ea.	1250	87		1337	install 4' 8" x 6' 4" angle bay double
	Demolish and Install	Ea.	1250	105.80		1355.80	hung unit, pine frame with aluminum
	Reinstall	Ea.		69.76		69.76	cladding, picture center sash,
	Clean	Ea.	.13	15.20		15.33	hardware, screens.
	Minimum Charge	Job		216		216	
4' 8" x 7' 2"							
	Demolish	Ea.		18.80		18.80	Cost includes material and labor to
	Install	Ea.	1275	87		1362	install 4' 8" x 7' 2" angle bay double
	Demolish and Install	Ea.	1275	105.80		1380.80	hung unit, pine frame with aluminum
	Reinstall	Ea.		69.76		69.76	cladding, picture center sash,
	Clean	Ea.	.13	15.20		15.33	hardware, screens.
	Minimum Charge	Job		216		216	
4' 8" x 7' 10"							
	Demolish	Ea.		18.80		18.80	Cost includes material and labor to
	Install	Ea.	1300	87		1387	install 4' 8" x 7' 10" angle bay
	Demolish and Install	Ea.	1300	105.80		1405.80	double hung unit, pine frame with
	Reinstall	Ea.		69.76		69.76	aluminum cladding, picture center
	Clean	Ea.	.13	15.20		15.33	sash, hardware, screens.
	Minimum Charge	Job		216		216	
4' 8" x 8' 6"							
	Demolish	Ea.		21.50		21.50	Cost includes material and labor to
	Install	Ea.	1375	99.50		1474.50	install 4' 8" x 8' 6" angle bay double
	Demolish and Install	Ea.	1375	121		1496	hung unit, pine frame with aluminum
	Reinstall	Ea.		79.73		79.73	cladding, picture center sash,
	Clean	Ea.	.02	20.50		20.52	hardware, screens.
	Minimum Charge	Job		216		216	

Exterior Doors and Windows

Aluminum Window		Unit	Material	Labor	Equip.	Total	Specification
4' 8" x 8' 10"							
	Demolish	Ea.		21.50		21.50	Cost includes material and labor to
	Install	Ea.	1400	99.50		1499.50	install 4' 8" x 8' 10" angle bay
	Demolish and Install	Ea.	1400	121		1521	double hung unit, pine frame with
	Reinstall	Ea.		79.73		79.73	aluminum cladding, picture center
	Clean	Ea.	.02	20.50		20.52	sash, hardware, screens.
	Minimum Charge	Job		216		216	
4' 8" x 9' 2"							
	Demolish	Ea.		21.50		21.50	Cost includes material and labor to
	Install	Ea.	1475	174		1649	install 4' 8" x 9' 2" angle bay double
	Demolish and Install	Ea.	1475	195.50		1670.50	hung unit, pine frame with aluminum
	Reinstall	Ea.		139.52		139.52	cladding, picture center sash,
	Clean	Ea.	.02	20.50		20.52	hardware, screens.
	Minimum Charge	Job		216		216	
4' 8" x 9' 10"							
	Demolish	Ea.		21.50		21.50	Cost includes material and labor to
	Install	Ea.	1575	174		1749	install 4' 8" x 9' 10" angle bay
	Demolish and Install	Ea.	1575	195.50		1770.50	double hung unit, pine frame with
	Reinstall	Ea.		139.52		139.52	aluminum cladding, picture center
	Clean	Ea.	.02	20.50		20.52	sash, hardware, screens.
	Minimum Charge	Job		216		216	
Casing							
	Demolish	Ea.		8.05		8.05	Cost includes material and labor to
	Install	Opng.	33.50	27		60.50	install 11/16" x 2-1/2" pine ranch
	Demolish and Install	Opng.	33.50	35.05		68.55	style window casing.
	Reinstall	Opng.		21.46		21.46	
	Clean	Opng.	.44	2.54		2.98	
	Paint	Ea.	3.22	6.10		9.32	
	Minimum Charge	Job		216		216	
Storm Window							
2' x 3' 6"							
	Demolish	Ea.		10.45		10.45	Includes material and labor to install
	Install	Ea.	76.50	23.50		100	residential aluminum storm window
	Demolish and Install	Ea.	76.50	33.95		110.45	with mill finish, 2' x 3' 5" high.
	Reinstall	Ea.		18.60		18.60	
	Clean	Ea.	.02	5.70		5.72	
	Minimum Charge	Job		216		216	
2' 6" x 5'							
	Demolish	Ea.		10.45		10.45	Includes material and labor to install
	Install	Ea.	85.50	25		110.50	residential aluminum storm window
	Demolish and Install	Ea.	85.50	35.45		120.95	with mill finish, 2' 6" x 5' high.
	Reinstall	Ea.		19.93		19.93	
	Clean	Ea.	.04	7.60		7.64	
	Minimum Charge	Job		216		216	
4' x 6'							
	Demolish	Ea.		20		20	Includes material and labor to install
	Install	Ea.	102	28		130	residential aluminum storm window, 4'
	Demolish and Install	Ea.	102	48		150	x 6' high.
	Reinstall	Ea.		22.32		22.32	
	Clean	Ea.	.09	11.40		11.49	
	Minimum Charge	Job		216		216	
Hardware							
Weatherstripping							
	Install	L.F.	.29	2.91		3.20	Includes labor and material to install
	Minimum Charge	Job		116		116	vinyl V strip weatherstripping for
							double-hung window.

Exterior Doors and Windows

Aluminum Window

	Unit	Material	Labor	Equip.	Total	Specification
Trim Set						
Demolish	Ea.		8.05		8.05	Cost includes material and labor to
Install	Opng.	33.50	27		60.50	install 11/16" x 2-1/2" pine ranch
Demolish and Install	Opng.	33.50	35.05		68.55	style window casing.
Reinstall	Opng.		21.46		21.46	
Clean	Opng.	.44	2.54		2.98	
Paint	Ea.	3.22	6.10		9.32	
Minimum Charge	Job		174		174	

Vinyl Window

	Unit	Material	Labor	Equip.	Total	Specification
Double Hung						
2' x 2' 6"						
Demolish	Ea.		12.85		12.85	Cost includes material and labor to
Install	Ea.	180	46.50		226.50	install 2' x 2' 6" double hung, two
Demolish and Install	Ea.	180	59.35		239.35	light, vinyl window with screen.
Reinstall	Ea.		37.21		37.21	
Clean	Ea.	.02	5.70		5.72	
Minimum Charge	Job		350		350	
2' x 3' 6"						
Demolish	Ea.		12.85		12.85	Cost includes material and labor to
Install	Ea.	188	50		238	install 2' x 3' 6" double hung, two
Demolish and Install	Ea.	188	62.85		250.85	light, vinyl window with screen.
Reinstall	Ea.		39.86		39.86	
Clean	Ea.	.03	6.50		6.53	
Minimum Charge	Job		350		350	
2' 6" x 4' 6"						
Demolish	Ea.		12.85		12.85	Cost includes material and labor to
Install	Ea.	227	53.50		280.50	install 2' 6" x 4' 6" double hung, two
Demolish and Install	Ea.	227	66.35		293.35	light, vinyl window with screen.
Reinstall	Ea.		42.93		42.93	
Clean	Ea.	.04	7.60		7.64	
Minimum Charge	Job		350		350	
3' x 4'						
Demolish	Ea.		12.85		12.85	Includes material and labor to install
Install	Ea.	230	70		300	medium size double hung, 3' x 4',
Demolish and Install	Ea.	230	82.85		312.85	two light, vinyl window with screen.
Reinstall	Ea.		55.81		55.81	
Clean	Ea.	.04	7.60		7.64	
Minimum Charge	Job		350		350	
3' x 4' 6"						
Demolish	Ea.		15.70		15.70	Cost includes material and labor to
Install	Ea.	254	77.50		331.50	install 3' x 4' 6" double hung, two
Demolish and Install	Ea.	254	93.20		347.20	light, vinyl window with screen.
Reinstall	Ea.		62.01		62.01	
Clean	Ea.	.06	9.10		9.16	
Minimum Charge	Job		350		350	
4' x 4' 6"						
Demolish	Ea.		15.70		15.70	Cost includes material and labor to
Install	Ea.	281	87		368	install 4' x 4' 6" double hung, two
Demolish and Install	Ea.	281	102.70		383.70	light, vinyl window with screen.
Reinstall	Ea.		69.76		69.76	
Clean	Ea.	.07	10.75		10.82	
Minimum Charge	Job		350		350	
4' x 6'						
Demolish	Ea.		15.70		15.70	Includes material and labor to install
Install	Ea.	325	99.50		424.50	large double hung, 4' x 6', two light,
Demolish and Install	Ea.	325	115.20		440.20	vinyl window with screen.
Reinstall	Ea.		79.73		79.73	
Clean	Ea.	.10	13.05		13.15	
Minimum Charge	Job		350		350	

Exterior Doors and Windows

Vinyl Window

Vinyl Window	Unit	Material	Labor	Equip.	Total	Specification
Trim Set						
Demolish	Ea.		8.05		8.05	Cost includes material and labor to
Install	Opng.	33.50	27		60.50	install 11/16" x 2-1/2" pine ranch
Demolish and Install	Opng.	33.50	35.05		68.55	style window casing.
Reinstall	Opng.		21.46		21.46	
Clean	Opng.	.44	2.54		2.98	
Paint	Ea.	3.22	6.10		9.32	
Minimum Charge	Job		216		216	

Wood Window

Wood Window	Unit	Material	Labor	Equip.	Total	Specification
2' 2" x 3' 4"						
Demolish	Ea.		12.85		12.85	Cost includes material and labor to
Install	Ea.	198	46.50		244.50	install 2' 2" x 3' 4" double hung, two
Demolish and Install	Ea.	198	59.35		257.35	light, double glazed wood window
Reinstall	Ea.		37.21		37.21	with screen.
Clean	Ea.	.03	6.50		6.53	
Paint	Ea.	2.01	12.70		14.71	
Minimum Charge	Job		174		174	
2' 2" x 4' 4"						
Demolish	Ea.		12.85		12.85	Cost includes material and labor to
Install	Ea.	222	50		272	install 2' 2" x 4' 4" double hung, two
Demolish and Install	Ea.	222	62.85		284.85	light, double glazed wood window
Reinstall	Ea.		39.86		39.86	with screen.
Clean	Ea.	.03	6.50		6.53	
Paint	Ea.	2.01	12.70		14.71	
Minimum Charge	Job		174		174	
2' 6" x 3' 4"						
Demolish	Ea.		12.85		12.85	Cost includes material and labor to
Install	Ea.	209	53.50		262.50	install 2' 6" x 3' 4" double hung, two
Demolish and Install	Ea.	209	66.35		275.35	light, double glazed wood window
Reinstall	Ea.		42.93		42.93	with screen.
Clean	Ea.	.04	7.60		7.64	
Paint	Ea.	2.59	15.25		17.84	
Minimum Charge	Job		174		174	
2' 6" x 4'						
Demolish	Ea.		12.85		12.85	Cost includes material and labor to
Install	Ea.	226	58		284	install 2' 6" x 4' double hung, two
Demolish and Install	Ea.	226	70.85		296.85	light, double glazed wood window
Reinstall	Ea.		46.51		46.51	with screen.
Clean	Ea.	.04	7.60		7.64	
Paint	Ea.	2.59	15.25		17.84	
Minimum Charge	Job		174		174	
2' 6" x 4' 8"						
Demolish	Ea.		12.85		12.85	Cost includes material and labor to
Install	Ea.	248	58		306	install 2' 6" x 4' 8" double hung, two
Demolish and Install	Ea.	248	70.85		318.85	light, double glazed wood window
Reinstall	Ea.		46.51		46.51	with screen.
Clean	Ea.	.04	7.60		7.64	
Paint	Ea.	2.59	15.25		17.84	
Minimum Charge	Job		174		174	
2' 10" x 3' 4"						
Demolish	Ea.		12.85		12.85	Cost includes material and labor to
Install	Ea.	220	70		290	install 2' 10" x 3' 4" double hung, two
Demolish and Install	Ea.	220	82.85		302.85	light, double glazed wood window
Reinstall	Ea.		55.81		55.81	with screen.
Clean	Ea.	.04	7.60		7.64	
Paint	Ea.	2.59	15.25		17.84	
Minimum Charge	Job		174		174	

Exterior Doors and Windows

Wood Window		Unit	Material	Labor	Equip.	Total	Specification
2' 10" x 4'							
	Demolish	Ea.		12.85		12.85	Cost includes material and labor to
	Install	Ea.	241	70		311	install 2' 10" x 4' double hung, two
	Demolish and Install	Ea.	241	82.85		323.85	light, double glazed wood window
	Reinstall	Ea.		55.81		55.81	with screen.
	Clean	Ea.	.04	7.60		7.64	
	Paint	Ea.	2.59	15.25		17.84	
	Minimum Charge	Job		174		174	
3' 7" x 3' 4"							
	Demolish	Ea.		12.85		12.85	Cost includes material and labor to
	Install	Ea.	248	77.50		325.50	install 3' 7" x 3' 4" double hung, two
	Demolish and Install	Ea.	248	90.35		338.35	light, double glazed wood window
	Reinstall	Ea.		62.01		62.01	with screen.
	Clean	Ea.	.04	7.60		7.64	
	Paint	Ea.	2.59	15.25		17.84	
	Minimum Charge	Job		174		174	
3' 7" x 5' 4"							
	Demolish	Ea.		15.70		15.70	Cost includes material and labor to
	Install	Ea.	320	77.50		397.50	install 3' 7" x 5' 4" double hung, two
	Demolish and Install	Ea.	320	93.20		413.20	light, double glazed wood window
	Reinstall	Ea.		62.01		62.01	with screen.
	Clean	Ea.	.10	13.05		13.15	
	Paint	Ea.	4.30	25.50		29.80	
	Minimum Charge	Job		174		174	
4' 3" x 5' 4"							
	Demolish	Ea.		15.70		15.70	Cost includes material and labor to
	Install	Ea.	500	87		587	install 4' 3" x 5' 4" double hung, two
	Demolish and Install	Ea.	500	102.70		602.70	light, double glazed wood window
	Reinstall	Ea.		69.76		69.76	with screen.
	Clean	Ea.	.10	13.05		13.15	
	Paint	Ea.	2.56	30.50		33.06	
	Minimum Charge	Job		174		174	
Casement							
2' x 3' 4"							
	Demolish	Ea.		12.85		12.85	Cost includes material and labor to
	Install	Ea.	238	35		273	install 2' x 3' 4" casement window
	Demolish and Install	Ea.	238	47.85		285.85	with 3/4" insulated glass, screens,
	Reinstall	Ea.		27.90		27.90	weatherstripping, hardware.
	Clean	Ea.	.03	6.50		6.53	
	Paint	Ea.	2.01	12.70		14.71	
	Minimum Charge	Job		350		350	
2' x 4'							
	Demolish	Ea.		12.85		12.85	Cost includes material and labor to
	Install	Ea.	261	39		300	install 2' x 4' casement window with
	Demolish and Install	Ea.	261	51.85		312.85	3/4" insulated glass, screens,
	Reinstall	Ea.		31		31	weatherstripping, hardware.
	Clean	Ea.	.03	6.50		6.53	
	Paint	Ea.	2.01	12.70		14.71	
	Minimum Charge	Job		350		350	
2' x 5'							
	Demolish	Ea.		12.85		12.85	Cost includes material and labor to
	Install	Ea.	298	41		339	install 2' x 5' casement window with
	Demolish and Install	Ea.	298	53.85		351.85	3/4" insulated glass, screens,
	Reinstall	Ea.		32.83		32.83	weatherstripping, hardware.
	Clean	Ea.	.04	7.60		7.64	
	Paint	Ea.	2.59	15.25		17.84	
	Minimum Charge	Job		350		350	

Wood Window	Unit	Material	Labor	Equip.	Total	Specification
2' x 6'						
Demolish	Ea.		12.85		12.85	Cost includes material and labor to
Install	Ea.	295	43.50		338.50	install 2' x 6' casement window with
Demolish and Install	Ea.	295	56.35		351.35	3/4" insulated glass, screens,
Reinstall	Ea.		34.88		34.88	weatherstripping, hardware.
Clean	Ea.	.04	7.60		7.64	
Paint	Ea.	2.59	15.25		17.84	
Minimum Charge	Job		350		350	
4' x 3' 4"						
Demolish	Ea.		15.70		15.70	Cost includes material and labor to
Install	Ea.	600	46.50		646.50	install 4' x 3' 4 " casement windows
Demolish and Install	Ea.	600	62.20		662.20	with 3/4" insulated glass, screens,
Reinstall	Ea.		37.21		37.21	weatherstripping, hardware.
Clean	Ea.	.04	7.60		7.64	
Paint	Ea.	1.29	15.25		16.54	
Minimum Charge	Job		350		350	
4' x 4'						
Demolish	Ea.		15.70		15.70	Cost includes material and labor to
Install	Ea.	660	46.50		706.50	install 4' x 4' casement windows with
Demolish and Install	Ea.	660	62.20		722.20	3/4" insulated glass, screens,
Reinstall	Ea.		37.21		37.21	weatherstripping, hardware.
Clean	Ea.	.07	10.75		10.82	
Paint	Ea.	1.52	19.10		20.62	
Minimum Charge	Job		350		350	
4' x 5'						
Demolish	Ea.		15.70		15.70	Cost includes material and labor to
Install	Ea.	750	50		800	install 4' x 5' casement windows with
Demolish and Install	Ea.	750	65.70		815.70	3/4" insulated glass, screens,
Reinstall	Ea.		39.86		39.86	weatherstripping, hardware.
Clean	Ea.	.10	13.05		13.15	
Paint	Ea.	1.98	25.50		27.48	
Minimum Charge	Job		350		350	
4' x 6'						
Demolish	Ea.		15.70		15.70	Cost includes material and labor to
Install	Ea.	840	58		898	install 4' x 6' casement windows with
Demolish and Install	Ea.	840	73.70		913.70	3/4" insulated glass, screens,
Reinstall	Ea.		46.51		46.51	weatherstripping, hardware.
Clean	Ea.	.10	13.05		13.15	
Paint	Ea.	2.33	30.50		32.83	
Minimum Charge	Job		350		350	
Casement Bow						
4' 8" x 8' 1"						
Demolish	Ea.		21.50		21.50	Includes material and labor to install
Install	Ea.	1600	131		1731	casement bow window unit with 4
Demolish and Install	Ea.	1600	152.50		1752.50	lights, pine frame, 1/2" insulated
Reinstall	Ea.		104.64		104.64	glass, weatherstripping, hardware,
Clean	Ea.	.13	15.20		15.33	screens.
Paint	Ea.	4.31	61		65.31	
Minimum Charge	Job		350		350	
4' 8" x 10'						
Demolish	Ea.		21.50		21.50	Includes material and labor to install
Install	Ea.	1700	174		1874	casement bow window unit with 5
Demolish and Install	Ea.	1700	195.50		1895.50	lights, pine frame, 1/2" insulated
Reinstall	Ea.		139.52		139.52	glass, weatherstripping, hardware,
Clean	Ea.	.02	20.50		20.52	screens.
Paint	Ea.	5.35	76.50		81.85	
Minimum Charge	Job		350		350	

Exterior Doors and Windows

Wood Window		Unit	Material	Labor	Equip.	Total	Specification
5′ 4″ x 8′ 1″							
	Demolish	Ea.		21.50		21.50	Includes material and labor to install
	Install	Ea.	1750	131		1881	casement bow window unit with 4
	Demolish and Install	Ea.	1750	152.50		1902.50	lights, pine frame, 1/2″ insulated
	Reinstall	Ea.		104.64		104.64	glass, weatherstripping, hardware,
	Clean	Ea.	.02	20.50		20.52	screens.
	Paint	Ea.	4.90	61		65.90	
	Minimum Charge	Job		350		350	
5′ 4″ x 10′							
	Demolish	Ea.		21.50		21.50	Includes material and labor to install
	Install	Ea.	1925	174		2099	casement bow window unit with 5
	Demolish and Install	Ea.	1925	195.50		2120.50	lights, pine frame, 1/2″ insulated
	Reinstall	Ea.		139.52		139.52	glass, weatherstripping, hardware,
	Clean	Ea.	.02	20.50		20.52	screens.
	Paint	Ea.	6.15	76.50		82.65	
	Minimum Charge	Job		350		350	
6′ x 8′ 1″							
	Demolish	Ea.		21.50		21.50	Includes material and labor to install
	Install	Ea.	1700	174		1874	casement bow window unit with 5
	Demolish and Install	Ea.	1700	195.50		1895.50	lights, pine frame, 1/2″ insulated
	Reinstall	Ea.		139.52		139.52	glass, weatherstripping, hardware,
	Clean	Ea.	.02	20.50		20.52	screens.
	Paint	Ea.	5.60	76.50		82.10	
	Minimum Charge	Job		350		350	
6′ x 10′							
	Demolish	Ea.		21.50		21.50	Includes material and labor to install
	Install	Ea.	2000	174		2174	casement bow window unit with 5
	Demolish and Install	Ea.	2000	195.50		2195.50	lights, pine frame, 1/2″ insulated
	Reinstall	Ea.		139.52		139.52	glass, weatherstripping, hardware,
	Clean	Ea.	.02	20.50		20.52	screens.
	Paint	Ea.	7	102		109	
	Minimum Charge	Job		350		350	
Casing							
	Demolish	Ea.		8.05		8.05	Cost includes material and labor to
	Install	Opng.	33.50	27		60.50	install 11/16″ x 2-1/2″ pine ranch
	Demolish and Install	Opng.	33.50	35.05		68.55	style window casing.
	Reinstall	Opng.		21.46		21.46	
	Clean	Opng.	.44	2.54		2.98	
	Paint	Ea.	3.22	6.10		9.32	
	Minimum Charge	Job		216		216	
Trim Set							
	Demolish	Ea.		6.15		6.15	Cost includes material and labor to
	Install	Opng.	33.50	27		60.50	install 11/16″ x 2-1/2″ pine ranch
	Demolish and Install	Opng.	33.50	33.15		66.65	style window casing.
	Reinstall	Opng.		21.46		21.46	
	Clean	Opng.	.44	2.54		2.98	
	Paint	Ea.	3.22	6.10		9.32	
	Minimum Charge	Job		174		174	
Wood Picture Window							
3′ 6″ x 4′							
	Demolish	Ea.		15.70		15.70	Includes material and labor to install a
	Install	Ea.	425	58		483	wood framed picture window
	Demolish and Install	Ea.	425	73.70		498.70	including exterior trim.
	Clean	Ea.	.10	13.05		13.15	
	Paint	Ea.	2.01	23.50		25.51	
	Minimum Charge	Job		350		350	

Exterior Doors and Windows

Wood Window

	Unit	Material	Labor	Equip.	Total	Specification
4' x 4' 6"						
Demolish	Ea.		15.70		15.70	Includes material and labor to install a
Install	Ea.	495	63.50		558.50	wood framed picture window
Demolish and Install	Ea.	495	79.20		574.20	including exterior trim.
Clean	Ea.	.10	13.05		13.15	
Paint	Ea.	2.01	23.50		25.51	
Minimum Charge	Job		350		350	
5' x 4'						
Demolish	Ea.		15.70		15.70	Includes material and labor to install a
Install	Ea.	580	63.50		643.50	wood framed picture window
Demolish and Install	Ea.	580	79.20		659.20	including exterior trim.
Clean	Ea.	.10	13.05		13.15	
Paint	Ea.	4.29	38		42.29	
Minimum Charge	Job		350		350	
6' x 4' 6"						
Demolish	Ea.		15.70		15.70	Includes material and labor to install a
Install	Ea.	620	70		690	wood framed picture window
Demolish and Install	Ea.	620	85.70		705.70	including exterior trim.
Clean	Ea.	.10	13.05		13.15	
Paint	Ea.	4.29	38		42.29	
Minimum Charge	Job		350		350	

Window Screens

	Unit	Material	Labor	Equip.	Total	Specification
Small						
Demolish	Ea.		2.94		2.94	Includes material and labor to install
Install	Ea.	17.25	10.55		27.80	small window screen, 4 S.F.
Demolish and Install	Ea.	17.25	13.49		30.74	
Reinstall	Ea.		8.46		8.46	
Clean	Ea.	.31	5.30		5.61	
Minimum Charge	Job		87		87	
Standard Size						
Demolish	Ea.		2.94		2.94	Includes material and labor to install
Install	Ea.	42	10.55		52.55	medium window screen, 12 S.F.
Demolish and Install	Ea.	42	13.49		55.49	
Reinstall	Ea.		8.46		8.46	
Clean	Ea.	.31	5.30		5.61	
Minimum Charge	Job		87		87	
Large						
Demolish	Ea.		2.94		2.94	Includes material and labor to install
Install	Ea.	56	10.55		66.55	large window screen, 16 S.F.
Demolish and Install	Ea.	56	13.49		69.49	
Reinstall	Ea.		8.46		8.46	
Clean	Ea.	.31	5.30		5.61	
Minimum Charge	Job		87		87	

Window Glass

	Unit	Material	Labor	Equip.	Total	Specification
Plate, Clear, 1/4"						
Demolish	S.F.		1.41		1.41	Includes material and labor to install
Install	S.F.	7.80	5.65		13.45	clear glass.
Demolish and Install	S.F.	7.80	7.06		14.86	
Reinstall	S.F.		4.52		4.52	
Clean	S.F.	.02	.15		.17	
Minimum Charge	Job		169		169	

Exterior Doors and Windows

Window Glass		Unit	Material	Labor	Equip.	Total	Specification
Insulated							
	Demolish	S.F.		1.41		1.41	Includes material and labor to install
	Install	S.F.	12.85	9.70		22.55	insulated glass.
	Demolish and Install	S.F.	12.85	11.11		23.96	
	Reinstall	S.F.		7.74		7.74	
	Clean	S.F.	.02	.15		.17	
	Minimum Charge	Job		169		169	
Tinted							
	Demolish	S.F.		1.41		1.41	Includes material and labor to install
	Install	S.F.	8	5.20		13.20	tinted glass.
	Demolish and Install	S.F.	8	6.61		14.61	
	Reinstall	S.F.		4.17		4.17	
	Clean	S.F.	.02	.15		.17	
	Minimum Charge	Job		169		169	
Obscure / Florentine							
	Demolish	S.F.		1.41		1.41	Includes material and labor to install
	Install	S.F.	8.70	4.84		13.54	obscure glass.
	Demolish and Install	S.F.	8.70	6.25		14.95	
	Reinstall	S.F.		3.87		3.87	
	Clean	S.F.	.02	.15		.17	
	Minimum Charge	Job		169		169	
Laminated Safety							
	Demolish	S.F.		1.41		1.41	Includes material and labor to install
	Install	S.F.	11	7.55		18.55	clear laminated glass.
	Demolish and Install	S.F.	11	8.96		19.96	
	Reinstall	S.F.		6.02		6.02	
	Clean	S.F.	.02	.15		.17	
	Minimum Charge	Job		169		169	
Wired							
	Demolish	S.F.		1.41		1.41	Includes material and labor to install
	Install	S.F.	14.30	5		19.30	rough obscure wire glass.
	Demolish and Install	S.F.	14.30	6.41		20.71	
	Reinstall	S.F.		4.02		4.02	
	Clean	S.F.	.02	.15		.17	
	Minimum Charge	Job		169		169	
Tempered							
	Demolish	S.F.		1.41		1.41	Includes material and labor to install
	Install	S.F.	7.80	5.65		13.45	clear glass.
	Demolish and Install	S.F.	7.80	7.06		14.86	
	Reinstall	S.F.		4.52		4.52	
	Clean	S.F.	.02	.15		.17	
	Minimum Charge	Job		169		169	
Stained							
	Demolish	S.F.		1.41		1.41	Includes material and labor to install
	Install	S.F.	58.50	5.65		64.15	stained glass.
	Demolish and Install	S.F.	58.50	7.06		65.56	
	Reinstall	S.F.		4.52		4.52	
	Clean	S.F.	.02	.15		.17	
	Minimum Charge	Job		169		169	
Etched							
	Demolish	S.F.		1.41		1.41	Includes material and labor to install
	Install	S.F.	48	5.65		53.65	etched glass.
	Demolish and Install	S.F.	48	7.06		55.06	
	Reinstall	S.F.		4.52		4.52	
	Clean	S.F.	.02	.15		.17	
	Minimum Charge	Job		169		169	

Exterior Doors and Windows

Window Glass

Window Glass	Unit	Material	Labor	Equip.	Total	Specification
Wall Mirror						
Demolish	S.F.		1.41		1.41	Includes material and labor to install
Install	S.F.	7.60	5.40		13	distortion-free float glass mirror, 1/4"
Demolish and Install	S.F.	7.60	6.81		14.41	thick, cut and polished edges.
Reinstall	S.F.		4.34		4.34	
Clean	S.F.	.02	.15		.17	
Minimum Charge	Job		169		169	
Bullet Resistant						
Demolish	S.F.		1.41		1.41	Includes material and labor to install
Install	S.F.	103	68		171	bullet resistant clear laminated glass.
Demolish and Install	S.F.	103	69.41		172.41	
Reinstall	S.F.		54.21		54.21	
Clean	S.F.	.02	.15		.17	
Minimum Charge	Job		169		169	

Storefront System

Storefront System	Unit	Material	Labor	Equip.	Total	Specification
Clear Glass						
Stub Wall to 8'						
Demolish	S.F.		2.20		2.20	Includes material and labor to install
Install	S.F.	17.95	9.30		27.25	commercial grade 4-1/2" section,
Demolish and Install	S.F.	17.95	11.50		29.45	anodized aluminum with clear glass in
Reinstall	S.F.		7.45		7.45	upper section and safety glass in
Clean	S.F.		.30		.30	lower section.
Minimum Charge	Job		216		216	
Floor to 10'						
Demolish	S.F.		2.20		2.20	Includes material and labor to install
Install	S.F.	19.65	9.30		28.95	commercial grade 4-1/2" section,
Demolish and Install	S.F.	19.65	11.50		31.15	anodized aluminum with clear glass in
Reinstall	S.F.		7.45		7.45	upper section and safety glass in
Clean	S.F.		.30		.30	lower section.
Minimum Charge	Job		216		216	
Polished Plate Glass						
Stub Wall to 9'						
Demolish	S.F.		2.20		2.20	Cost includes material and labor to
Install	S.F.	13.60	9.30		22.90	install 4' x 9' polished plate glass
Demolish and Install	S.F.	13.60	11.50		25.10	window built on 8" stub wall.
Reinstall	S.F.		7.45		7.45	
Clean	S.F.		.30		.30	
Minimum Charge	Job		216		216	
Stub Wall to 13'						
Demolish	S.F.		2.20		2.20	Cost includes material and labor to
Install	S.F.	14.20	9.30		23.50	install 4' x 13' polished plate glass
Demolish and Install	S.F.	14.20	11.50		25.70	window built on 8" stub wall.
Reinstall	S.F.		7.45		7.45	
Clean	S.F.		.30		.30	
Minimum Charge	Job		216		216	
Floor to 9'						
Demolish	S.F.		2.20		2.20	Cost includes material and labor to
Install	S.F.	14.70	9.30		24	install 4' x 9' polished plate glass
Demolish and Install	S.F.	14.70	11.50		26.20	window built on floor.
Reinstall	S.F.		7.45		7.45	
Clean	S.F.		.30		.30	
Minimum Charge	Job		216		216	

Exterior Doors and Windows

Storefront System		Unit	Material	Labor	Equip.	Total	Specification
Floor to 13'							
	Demolish	S.F.		2.20		2.20	Cost includes material and labor to
	Install	S.F.	15.85	9.30		25.15	install 4' x 13' polished plate glass
	Demolish and Install	S.F.	15.85	11.50		27.35	window built on floor.
	Reinstall	S.F.		7.45		7.45	
	Clean	S.F.		.30		.30	
	Minimum Charge	Job		216		216	
Clear Plate Glass							
Floor to 9'							
	Demolish	S.F.		2.20		2.20	Cost includes material and labor to
	Install	S.F.	21.50	9.30		30.80	install 4' x 9' clear plate glass window
	Demolish and Install	S.F.	21.50	11.50		33	built on floor.
	Reinstall	S.F.		7.45		7.45	
	Clean	S.F.		.30		.30	
	Minimum Charge	Job		216		216	
Floor to 11'							
	Demolish	S.F.		2.20		2.20	Cost includes material and labor to
	Install	S.F.	20.50	9.30		29.80	install 4' x 11' clear plate glass
	Demolish and Install	S.F.	20.50	11.50		32	window built on floor.
	Reinstall	S.F.		7.45		7.45	
	Clean	S.F.		.30		.30	
	Minimum Charge	Job		216		216	
Floor to 13'							
	Demolish	S.F.		2.20		2.20	Cost includes material and labor to
	Install	S.F.	21	9.30		30.30	install 4' x 13' clear plate glass
	Demolish and Install	S.F.	21	11.50		32.50	window built on floor.
	Reinstall	S.F.		7.45		7.45	
	Clean	S.F.		.30		.30	
	Minimum Charge	Job		216		216	
Decorative Type							
	Demolish	S.F.		2.20		2.20	Includes material and labor to install
	Install	S.F.	16.25	6.80		23.05	decorative storefront system.
	Demolish and Install	S.F.	16.25	9		25.25	
	Reinstall	S.F.		5.42		5.42	
	Clean	S.F.		.30		.30	
	Minimum Charge	Job		169		169	
Decorative Curved							
	Demolish	S.F.		2.20		2.20	Includes material and labor to install
	Install	S.F.	31	13.55		44.55	decorative curved storefront system.
	Demolish and Install	S.F.	31	15.75		46.75	
	Reinstall	S.F.		10.84		10.84	
	Clean	S.F.		.30		.30	
	Minimum Charge	Job		169		169	
Entrance							
Narrow Stile 3' x 7"							
	Demolish	Ea.		19.40		19.40	Cost includes material and labor to
	Install	Ea.	800	233		1033	install 3' x 7' narrow stile door, center
	Demolish and Install	Ea.	800	252.40		1052.40	pivot concealed closer.
	Reinstall	Ea.		186.03		186.03	
	Clean	Ea.	3.89	32		35.89	
	Minimum Charge	Job		216		216	
Tempered Glass							
	Demolish	Ea.		19.40		19.40	Cost includes material and labor to
	Install	Ea.	2725	465		3190	install 3' x 7' tempered glass door,
	Demolish and Install	Ea.	2725	484.40		3209.40	center pivot concealed closer,.
	Reinstall	Ea.		373.97		373.97	
	Clean	Ea.	3.89	32		35.89	
	Minimum Charge	Job		216		216	

Exterior Doors and Windows

Storefront System

Storefront System	Unit	Material	Labor	Equip.	Total	Specification
Heavy Section						
Demolish	Ea.		19.40		19.40	Includes material and labor to install
Install	Ea.	1000	430		1430	wide stile door with 3' 6" x 7'
Demolish and Install	Ea.	1000	449.40		1449.40	opening.
Reinstall	Ea.		345.92		345.92	
Clean	Ea.	.78	10.60		11.38	
Minimum Charge	Job		680		680	

Curtain Wall System

Curtain Wall System	Unit	Material	Labor	Equip.	Total	Specification
Regular Weight						
Bronze Anodized						
Demolish	S.F.		2.20		2.20	Includes material and labor to install
Install	S.F.	33.50	9.30		42.80	structural curtain wall system with
Demolish and Install	S.F.	33.50	11.50		45	bronze anodized frame of heavy
Clean	S.F.		.30		.30	weight, float glass.
Minimum Charge	Job		174		174	
Black Anodized						
Demolish	S.F.		2.20		2.20	Includes material and labor to install
Install	S.F.	36	9.30		45.30	structural curtain wall system with
Demolish and Install	S.F.	36	11.50		47.50	black anodized frame of regular
Clean	S.F.		.30		.30	weight, float glass.
Minimum Charge	Job		174		174	
Heavy Weight						
Bronze Anodized						
Demolish	S.F.		2.20		2.20	Includes material and labor to install
Install	S.F.	33.50	9.30		42.80	structural curtain wall system with
Demolish and Install	S.F.	33.50	11.50		45	bronze anodized frame of heavy
Clean	S.F.		.30		.30	weight, float glass.
Minimum Charge	Job		174		174	
Black Anodized						
Demolish	S.F.		2.20		2.20	Includes material and labor to install
Install	S.F.	34	9.30		43.30	structural curtain wall system with
Demolish and Install	S.F.	34	11.50		45.50	black anodized frame of heavy
Clean	S.F.		.30		.30	weight, float glass.
Minimum Charge	Job		174		174	
Glass						
1" Insulated						
Install	S.F.	15.90	9.05		24.95	Cost includes labor and material to
Minimum Charge	Job		169		169	install 1" insulated glass.
Heat Absorbing						
Install	S.F.	8.85	9.05		17.90	Includes labor an material to install
Minimum Charge	Job		169		169	heat absorbing glass.
Low Transmission						
Install	S.F.	5.85	9.05		14.90	Includes labor an material to install
Minimum Charge	Job		169		169	low transmission glass.

Skylight	Unit	Material	Labor	Equip.	Total	Specification
Dome						
22" x 22"						
Demolish	Ea.		67.50		67.50	Cost includes material and labor to
Install	Ea.	49	107		156	install 22" x 22" fixed dome skylight.
Demolish and Install	Ea.	49	174.50		223.50	
Reinstall	Ea.		85.44		85.44	
Clean	Ea.	.35	9.05		9.40	
Minimum Charge	Job		640		640	
22" x 46"						
Demolish	Ea.		67.50		67.50	Cost includes material and labor to
Install	Ea.	110	128		238	install 22" x 46" fixed dome skylight.
Demolish and Install	Ea.	110	195.50		305.50	
Reinstall	Ea.		102.53		102.53	
Clean	Ea.		14.10		14.10	
Minimum Charge	Job		640		640	
30" x 30"						
Demolish	Ea.		67.50		67.50	Cost includes material and labor to
Install	Ea.	92	107		199	install 30" x 30" fixed dome skylight.
Demolish and Install	Ea.	92	174.50		266.50	
Reinstall	Ea.		85.44		85.44	
Clean	Ea.		14.10		14.10	
Minimum Charge	Job		640		640	
30" x 46"						
Demolish	Ea.		67.50		67.50	Cost includes material and labor to
Install	Ea.	160	128		288	install 30" x 46" fixed dome skylight.
Demolish and Install	Ea.	160	195.50		355.50	
Reinstall	Ea.		102.53		102.53	
Clean	Ea.		14.10		14.10	
Minimum Charge	Job		640		640	
Fixed						
22" x 27"						
Demolish	Ea.		67.50		67.50	Includes material and labor to install
Install	Ea.	234	107		341	double glazed skylight, 22" x 27",
Demolish and Install	Ea.	234	174.50		408.50	fixed.
Reinstall	Ea.		85.44		85.44	
Clean	Ea.	.35	9.05		9.40	
Minimum Charge	Job		640		640	
22" x 46"						
Demolish	Ea.		67.50		67.50	Includes material and labor to install
Install	Ea.	295	128		423	double glazed skylight, 22" x 46",
Demolish and Install	Ea.	295	195.50		490.50	fixed.
Reinstall	Ea.		102.53		102.53	
Clean	Ea.		14.10		14.10	
Minimum Charge	Job		640		640	
44" x 46"						
Demolish	Ea.		67.50		67.50	Includes material and labor to install
Install	Ea.	420	128		548	double glazed skylight, 44" x 46",
Demolish and Install	Ea.	420	195.50		615.50	fixed.
Reinstall	Ea.		102.53		102.53	
Clean	Ea.		15.90		15.90	
Minimum Charge	Job		640		640	
Operable						
22" x 27"						
Demolish	Ea.		67.50		67.50	Includes material and labor to install
Install	Ea.	345	107		452	double glazed skylight, 22" x 27",
Demolish and Install	Ea.	345	174.50		519.50	operable.
Reinstall	Ea.		85.44		85.44	
Clean	Ea.	.35	9.05		9.40	
Minimum Charge	Job		640		640	

121

Exterior Doors and Windows

Skylight		Unit	Material	Labor	Equip.	Total	Specification
22″ x 46″							
	Demolish	Ea.		67.50		67.50	Includes material and labor to install
	Install	Ea.	405	128		533	double glazed skylight, 22″ x 46″,
	Demolish and Install	Ea.	405	195.50		600.50	operable.
	Reinstall	Ea.		102.53		102.53	
	Clean	Ea.		14.10		14.10	
	Minimum Charge	Job		640		640	
44″ x 46″							
	Demolish	Ea.		67.50		67.50	Includes material and labor to install
	Install	Ea.	665	128		793	double glazed skylight, 44″ x 46″,
	Demolish and Install	Ea.	665	195.50		860.50	operable.
	Reinstall	Ea.		102.53		102.53	
	Clean	Ea.		15.90		15.90	
	Minimum Charge	Job		640		640	

Industrial Steel Window		Unit	Material	Labor	Equip.	Total	Specification
100% Fixed							
	Demolish	S.F.		1.81		1.81	Includes material and labor to install
	Install	S.F.	29	4.32		33.32	commercial grade steel windows,
	Demolish and Install	S.F.	29	6.13		35.13	including glazing.
	Reinstall	S.F.		3.46		3.46	
	Clean	Ea.	.14	.95		1.09	
	Paint	S.F.	.11	.68		.79	
	Minimum Charge	Job		216		216	
50% Vented							
	Demolish	S.F.		1.81		1.81	Includes material and labor to install
	Install	S.F.	43	4.32		47.32	commercial grade steel windows,
	Demolish and Install	S.F.	43	6.13		49.13	including glazing.
	Reinstall	S.F.		3.46		3.46	
	Clean	Ea.	.14	.95		1.09	
	Paint	S.F.	.11	.68		.79	
	Minimum Charge	Job		216		216	

Masonry

Brick Veneer

Concrete Masonry Units

Masonry Wall	Unit	Material	Labor	Equip.	Total	Specification
Unreinforced CMU						
4″ x 8″ x 16″						
Demolish	S.F.		1.57		1.57	Includes material and labor to install
Install	S.F.	1.42	3.47		4.89	normal weight 4″ x 8″ x 16″ block.
Demolish and Install	S.F.	1.42	5.04		6.46	
Clean	S.F.	.04	.39		.43	
Paint	S.F.	.14	.69		.83	
Minimum Charge	Job		177		177	
6″ x 8″ x 16″						
Demolish	S.F.		1.66		1.66	Includes material and labor to install
Install	S.F.	2.05	3.63		5.68	normal weight 6″ x 8″ x 16″ block.
Demolish and Install	S.F.	2.05	5.29		7.34	
Clean	S.F.	.04	.39		.43	
Paint	S.F.	.14	.69		.83	
Minimum Charge	Job		177		177	
8″ x 8″ x 16″						
Demolish	S.F.		1.88		1.88	Includes material and labor to install
Install	S.F.	2.22	3.99		6.21	normal weight 8″ x 8″ x 16″ block.
Demolish and Install	S.F.	2.22	5.87		8.09	
Clean	S.F.	.04	.39		.43	
Paint	S.F.	.14	.69		.83	
Minimum Charge	Job		177		177	
12″ x 8″ x 16″						
Demolish	S.F.		1.88		1.88	Includes material and labor to install
Install	S.F.	3.27	6		9.27	normal weight 12″ x 8″ x 16″ block.
Demolish and Install	S.F.	3.27	7.88		11.15	
Clean	S.F.	.04	.39		.43	
Paint	S.F.	.14	.69		.83	
Minimum Charge	Job		177		177	
Reinforced CMU						
4″ x 8″ x 16″						
Demolish	S.F.		1.57	.80	2.37	Includes material and labor to install
Install	S.F.	1.55	3.55		5.10	normal weight 4″ x 8″ x 16″ block,
Demolish and Install	S.F.	1.55	5.12	.80	7.47	mortar and 1/2″ re-bar installed
Clean	S.F.	.04	.39		.43	vertically every 24″ O.C.
Paint	S.F.	.14	.69		.83	
Minimum Charge	Job		310		310	
6″ x 8″ x 16″						
Demolish	S.F.		1.65	.39	2.04	Includes material and labor to install
Install	S.F.	2.18	3.71		5.89	normal weight 6″ x 8″ x 16″ block,
Demolish and Install	S.F.	2.18	5.36	.39	7.93	mortar and 1/2″ re-bar installed
Clean	S.F.	.04	.39		.43	vertically every 24″ O.C.
Paint	S.F.	.14	.69		.83	
Minimum Charge	Job		310		310	

Masonry

Masonry Wall		Unit	Material	Labor	Equip.	Total	Specification
8" x 8" x 16"							
	Demolish	S.F.		1.74	.41	2.15	Includes material and labor to install
	Install	S.F.	2.37	4.04		6.41	normal weight 8" x 8" x 16" block,
	Demolish and Install	S.F.	2.37	5.78	.41	8.56	mortar and 1/2" re-bar installed
	Clean	S.F.	.04	.39		.43	vertically every 24" O.C.
	Paint	S.F.	.14	.69		.83	
	Minimum Charge	Job		310		310	
12" x 8" x 16"							
	Demolish	S.F.		1.80	.42	2.22	Includes material and labor to install
	Install	S.F.	3.42	6.20		9.62	normal weight 12" x 8" x 16" block,
	Demolish and Install	S.F.	3.42	8	.42	11.84	mortar and 1/2" re-bar installed
	Clean	S.F.	.04	.39		.43	vertically every 24" O.C.
	Paint	S.F.	.14	.69		.83	
	Minimum Charge	Job		310		310	
Structural Brick							
4" Single Row							
	Demolish	S.F.		2.26		2.26	Cost includes material and labor to
	Install	S.F.	4.17	7.40		11.57	install 4" red brick wall, running bond
	Demolish and Install	S.F.	4.17	9.66		13.83	with 3/8" concave joints, mortar and
	Clean	S.F.	.04	.33		.37	ties.
	Paint	S.F.	.14	.69		.83	
	Minimum Charge	Job		310		310	
8" Double Row							
	Demolish	S.F.		2.26		2.26	Cost includes material and labor to
	Install	S.F.	6.45	11.80		18.25	install 8" red brick wall, running bond,
	Demolish and Install	S.F.	6.45	14.06		20.51	with 3/8" concave joints, mortar and
	Clean	S.F.	.04	.33		.37	wall ties.
	Paint	S.F.	.14	.69		.83	
	Minimum Charge	Job		310		310	
12" Triple Row							
	Demolish	S.F.		3.14		3.14	Cost includes material and labor to
	Install	S.F.	9.70	16.80		26.50	install 12" red brick wall, running
	Demolish and Install	S.F.	9.70	19.94		29.64	bond, with 3/8" concave joints,
	Clean	S.F.	.04	.33		.37	mortar and wall ties.
	Paint	S.F.	.14	.69		.83	
	Minimum Charge	Job		310		310	
Brick Veneer							
4" Red Common							
	Demolish	S.F.		2.26		2.26	Cost includes material and labor to
	Install	S.F.	3.03	6.95		9.98	install 4" common red face brick,
	Demolish and Install	S.F.	3.03	9.21		12.24	running bond with 3/8" concave
	Clean	S.F.	.04	.33		.37	joints, mortar and wall ties.
	Paint	S.F.	.14	.69		.83	
	Minimum Charge	Job		177		177	
4" Used							
	Demolish	S.F.		2.26		2.26	Cost includes material and labor to
	Install	S.F.	4.17	7.40		11.57	install 4" red brick wall, running bond
	Demolish and Install	S.F.	4.17	9.66		13.83	with 3/8" concave joints, mortar and
	Clean	S.F.	.04	.33		.37	ties.
	Paint	S.F.	.14	.69		.83	
	Minimum Charge	Job		177		177	

Masonry

Masonry Wall	Unit	Material	Labor	Equip.	Total	Specification
6" Jumbo						
Demolish	S.F.		2.26		2.26	Cost includes material and labor to
Install	S.F.	4.71	3.67		8.38	install 6" red jumbo brick veneer,
Demolish and Install	S.F.	4.71	5.93		10.64	running bond with 3/8" concave
Clean	S.F.	.04	.33		.37	joints, mortar and ties.
Paint	S.F.	.14	.69		.83	
Minimum Charge	Job		177		177	
Norman (11-1/2")						
Demolish	S.F.		2.26		2.26	Cost includes material and labor to
Install	S.F.	5.10	4.99		10.09	install 4" norman red brick wall,
Demolish and Install	S.F.	5.10	7.25		12.35	running bond with 3/8" concave
Clean	S.F.	.04	.33		.37	joints, mortar and ties.
Paint	S.F.	.14	.69		.83	
Minimum Charge	Job		177		177	
Glazed						
Demolish	S.F.		2.26		2.26	Includes material, labor and
Install	S.F.	9.10	7.60		16.70	equipment to install glazed brick.
Demolish and Install	S.F.	9.10	9.86		18.96	
Clean	S.F.	.04	.33		.37	
Paint	S.F.	.14	.69		.83	
Minimum Charge	Job		177		177	
Slumpstone						
6" x 4"						
Demolish	S.F.		1.57		1.57	Cost includes material and labor to
Install	S.F.	5.60	3.88		9.48	install 6" x 4" x 16" concrete slump
Demolish and Install	S.F.	5.60	5.45		11.05	block.
Clean	S.F.	.04	.39		.43	
Paint	S.F.	.14	.69		.83	
Minimum Charge	Job		177		177	
6" x 6"						
Demolish	S.F.		1.66		1.66	Cost includes material and labor to
Install	S.F.	4.94	4.01		8.95	install 6" x 6" x 16" concrete slump
Demolish and Install	S.F.	4.94	5.67		10.61	block.
Clean	S.F.	.04	.39		.43	
Paint	S.F.	.14	.69		.83	
Minimum Charge	Job		177		177	
8" x 6"						
Demolish	S.F.		1.66		1.66	Cost includes material and labor to
Install	S.F.	6.80	4.14		10.94	install 8" x 6" x 16" concrete slump
Demolish and Install	S.F.	6.80	5.80		12.60	block.
Clean	S.F.	.04	.39		.43	
Paint	S.F.	.14	.69		.83	
Minimum Charge	Job		177		177	
12" x 4"						
Demolish	S.F.		1.57		1.57	Cost includes material and labor to
Install	S.F.	11.85	4.78		16.63	install 12" x 4" x 16" concrete slump
Demolish and Install	S.F.	11.85	6.35		18.20	block.
Clean	S.F.	.04	.39		.43	
Paint	S.F.	.14	.69		.83	
Minimum Charge	Job		177		177	
12" x 6"						
Demolish	S.F.		1.66		1.66	Cost includes material and labor to
Install	S.F.	10.85	5.15		16	install 12" x 6" x 16" concrete slump
Demolish and Install	S.F.	10.85	6.81		17.66	block.
Clean	S.F.	.04	.39		.43	
Paint	S.F.	.14	.69		.83	
Minimum Charge	Job		177		177	

Masonry

Masonry Wall	Unit	Material	Labor	Equip.	Total	Specification
Stucco						
Three Coat						
Demolish	S.Y.		1.95		1.95	Includes material and labor to install
Install	S.Y.	3.85	7.35	.62	11.82	stucco on exterior masonry walls.
Demolish and Install	S.Y.	3.85	9.30	.62	13.77	
Clean	S.Y.		3.53		3.53	
Paint	S.F.	.12	.50		.62	
Minimum Charge	Job		155		155	
Three Coat w / Wire Mesh						
Demolish	S.Y.		3.53		3.53	Includes material and labor to install
Install	S.Y.	6.20	28	1.98	36.18	stucco on wire mesh over wood
Demolish and Install	S.Y.	6.20	31.53	1.98	39.71	framing.
Clean	S.Y.		3.53		3.53	
Paint	S.F.	.12	.50		.62	
Minimum Charge	Job		155		155	
Fill Cracks						
Install	S.F.	.43	2.49		2.92	Includes labor and material to fill-in
Minimum Charge	Job		155		155	hairline cracks up to 1/8" with filler, including sand and fill per S.F.
Furring						
1" x 2"						
Demolish	L.F.		.11		.11	Cost includes material and labor to
Install	L.F.	.13	.70		.83	install 1" x 2" furring on masonry wall.
Demolish and Install	L.F.	.13	.81		.94	
Minimum Charge	Job		174		174	
2" x 2"						
Demolish	L.F.		.12		.12	Cost includes material and labor to
Install	L.F.	.22	.78		1	install 2" x 2" furring on masonry wall.
Demolish and Install	L.F.	.22	.90		1.12	
Minimum Charge	Job		174		174	
2" x 4"						
Demolish	L.F.		.13		.13	Cost includes material and labor to
Install	L.F.	.41	.88		1.29	install 2" x 4" furring on masonry wall.
Demolish and Install	L.F.	.41	1.01		1.42	
Reinstall	L.F.		.71		.71	
Minimum Charge	Job		174		174	
Maintenance / Cleaning						
Repoint						
Install	S.F.	.52	4.43		4.95	Includes labor and material to cut and
Minimum Charge	Hr.		44.50		44.50	repoint brick, hard mortar, running bond.
Acid Etch						
Install	S.F.	.03	.63		.66	Includes labor and material to acid
Minimum Charge	Job		127	29.50	156.50	wash smooth brick.
Pressure Wash						
Clean	S.F.		.86	.13	.99	Includes labor and material to high
Minimum Charge	Job		127	29.50	156.50	pressure water wash masonry, average.
Waterproof						
Install	S.F.	.26	.64		.90	Includes labor and material to install
Minimum Charge	Job		161		161	masory waterproofing.

Masonry

Chimney

Chimney		Unit	Material	Labor	Equip.	Total	Specification
Demo & Haul Exterior							
	Demolish	C.F.		4.62		4.62	Includes minimum labor and
	Minimum Charge	Job		141		141	equipment to remove masonry and haul debris to truck or dumpster.
Demo & Haul Interior							
	Demolish	C.F.		6.35		6.35	Includes minimum labor and
	Minimum Charge	Job		141		141	equipment to remove masonry and haul debris to truck or dumpster.
Install New (w / flue)							
Brick							
	Install	V.L.F.	32	39		71	Includes material and labor to install
	Minimum Charge	Job		177		177	brick chimney with flue from ground level up. Foundation not included.
Natural Stone							
	Install	L.F.	94	62		156	Includes labor and material to install
	Minimum Charge	Job		177		177	natural stone chimney.
Install From Roof Up							
Brick							
	Install	V.L.F.	32	39		71	Includes material and labor to install
	Minimum Charge	Job		177		177	brick chimney with flue from ground level up. Foundation not included.
Natural Stone							
	Install	L.F.	94	51.50		145.50	Includes labor and material to install
	Minimum Charge	Job		177		177	natural stone chimney.
Install Base Pad							
	Install	Ea.	264	191	.98	455.98	Includes labor and material to replace
	Minimum Charge	Job		810	23.50	833.50	spread footing, 5' square x 16" deep.
Metal Cap							
	Install	Ea.	46	22		68	Includes labor and materials to install
	Minimum Charge	Job		174		174	bird and squirrel screens in a chimney flue.

Fireplace

Fireplace		Unit	Material	Labor	Equip.	Total	Specification
Common Brick							
30"							
	Demolish	Ea.		1000		1000	Includes material and labor to install a
	Install	Ea.	1675	3100		4775	complete unit which includes common
	Demolish and Install	Ea.	1675	4100		5775	brick, foundation, damper, flue lining
	Minimum Charge	Job		177		177	stack to 15' high and a 30" fire box.
36"							
	Demolish	Ea.		1000		1000	Includes material and labor to install a
	Install	Ea.	1800	3450		5250	complete unit which includes common
	Demolish and Install	Ea.	1800	4450		6250	brick, foundation, damper, flue lining
	Minimum Charge	Job		177		177	stack to 15' high and a 36" fire box.
42"							
	Demolish	Ea.		1000		1000	Includes material and labor to install a
	Install	Ea.	1950	3875		5825	complete unit which includes common
	Demolish and Install	Ea.	1950	4875		6825	brick, foundation, damper, flue lining
	Minimum Charge	Job		177		177	stack to 15' high and a 42" fire box.

Fireplace		Unit	Material	Labor	Equip.	Total	Specification
48"							
	Demolish	Ea.		1000		1000	Includes material and labor to install
	Install	Ea.	2075	4425		6500	a complete unit which includes common
	Demolish and Install	Ea.	2075	5425		7500	brick, foundation, damper, flue lining
	Minimum Charge	Job		177		177	stack to 15' high and a 48" fire box.
Raised Hearth							
Brick Hearth							
	Demolish	S.F.		4.41		4.41	Includes material and labor to install a
	Install	S.F.	16.25	26		42.25	brick fireplace hearth based on 30" x
	Demolish and Install	S.F.	16.25	30.41		46.66	29" opening.
	Clean	S.F.	.04	.33		.37	
	Minimum Charge	Job		177		177	
Marble Hearth / Facing							
	Demolish	Ea.		28		28	Includes material and labor to install
	Install	Ea.	600	105		705	marble fireplace hearth.
	Demolish and Install	Ea.	600	133		733	
	Reinstall	Ea.		84.17		84.17	
	Clean	Ea.	23.50	5.10		28.60	
	Minimum Charge	Job		177		177	
Breast (face) Brick							
	Demolish	S.F.		4.41		4.41	Includes material and labor to install
	Install	S.F.	4.24	7.25		11.49	standard red brick, running bond.
	Demolish and Install	S.F.	4.24	11.66		15.90	
	Clean	S.F.		.86	.13	.99	
	Minimum Charge	Hr.		44.50		44.50	
Fire Box							
	Demolish	Ea.		226		226	Includes material and labor to install
	Install	Ea.	174	310		484	fireplace box only (110 brick).
	Demolish and Install	Ea.	174	536		710	
	Clean	Ea.	.12	11.40		11.52	
	Minimum Charge	Hr.		44.50		44.50	
Natural Stone							
	Demolish	S.F.		4.41		4.41	Includes material and labor to install
	Install	S.F.	22.50	13.30		35.80	high priced stone, split or rock face.
	Demolish and Install	S.F.	22.50	17.71		40.21	
	Clean	S.F.	.09	.78		.87	
	Minimum Charge	Job		310		310	
Slumpstone Veneer							
	Demolish	S.F.		1.66		1.66	Cost includes material and labor to
	Install	S.F.	6.80	4.14		10.94	install 8" x 6" x 16" concrete slump
	Demolish and Install	S.F.	6.80	5.80		12.60	block.
	Clean	S.F.	.04	.39		.43	
	Paint	S.F.	.14	.69		.83	
	Minimum Charge	Job		177		177	
Prefabricated							
Firebox							
	Demolish	Ea.		40.50		40.50	Includes material and labor to install
	Install	Ea.	1725	315		2040	up to 43" radiant heat, zero
	Demolish and Install	Ea.	1725	355.50		2080.50	clearance, prefabricated fireplace.
	Reinstall	Ea.		253.67		253.67	Flues, doors or blowers not included.
	Clean	Ea.	1.56	15.90		17.46	
	Minimum Charge	Job		174		174	

Reed Construction Data®

Since its founding in 1975, Reed Construction Data has developed an online and print portfolio of innovative products and services for the construction, design and manufacturing community. Our products and services are designed specifically to help construction industry professionals advance their businesses with timely, accurate and actionable project, product and cost data. Reed Construction Data is your all-inclusive source of construction information encompassing all phases of the construction process.

Reed Bulletin and Reed Connect™ deliver the most comprehensive, timely and reliable project information to support contractors, distributors and building product manufacturers in identifying, bidding and tracking projects – private and public, general building and civil. Reed Construction Data also offers in-depth construction activity statistics and forecasts covering major project categories, many at the county and metropolitan level.

Reed Bulletin

www.reedbulletin.com

- Project leads targeted by geographic region and formatted by construction stage – available online or in print.

- Locate those hard-to-find jobs that are more profitable to your business.

- Optional automatic e-mail updates sent whenever project details change.

- Download plans and specs online or order print copies.

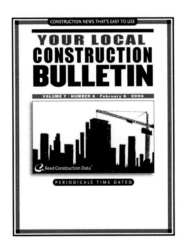

Reed Connect

www.reedconnect.com

- Customized web-based project lead delivery service featuring advanced search capabilities.

- Manage and track actionable leads from planning to quote through winning the job.

- Competitive analysis tool to analyze lost sales opportunities.

- Potential integration with your CRM application.

Reed Research & Analytics

www.buildingteamforecast.com

Reed Construction Forecast

- Delivers timely construction industry activity combining historical data, current year projections and forecasts.

- Modeled at the individual MSA-level to capture changing local market conditions.

- Covers 21 major project categories.

Reed Construction Starts

- Available in a monthly report or as an interactive database.

- Data provided in square footage and dollar value.

- Highly effective and efficient business planning tool.

Market Fundamentals

- Metropolitan area-specific reporting.

- Five-year forecast of industry performance including major projects in development and underway.

- Property types include office, retail, hotel, warehouse and apartment.

For more information about Reed Construction Data, please call 877-REED411, visit our website at www.reedconstructiondata.com, or E-mail: marketing@reedbusiness.com

Reed Construction Data
CONNECT • FIRST SOURCE • BULLETIN
RSMEANS • ACP • RESEARCH • PLANSDIRECT

Masonry

Fireplace

	Unit	Material	Labor	Equip.	Total	Specification
Flue						
Demolish	V.L.F.		4.70		4.70	Includes material and labor to install
Install	V.L.F.	12.05	13.40		25.45	prefabricated chimney vent, 8"
Demolish and Install	V.L.F.	12.05	18.10		30.15	exposed metal flue.
Reinstall	V.L.F.		10.72		10.72	
Clean	L.F.	.63	3.18		3.81	
Minimum Charge	Job		350		350	
Blower						
Demolish	Ea.		3.76		3.76	Includes material and labor to install
Install	Ea.	122	24		146	blower for metal fireplace flue.
Demolish and Install	Ea.	122	27.76		149.76	
Reinstall	Ea.		19.34		19.34	
Clean	Ea.	3.11	32		35.11	
Minimum Charge	Job		350		350	

Accessories

	Unit	Material	Labor	Equip.	Total	Specification
Wood Mantle						
Demolish	L.F.		4.70		4.70	Includes material and labor to install
Install	L.F.	7.20	9.95		17.15	wood fireplace mantel.
Demolish and Install	L.F.	7.20	14.65		21.85	
Reinstall	L.F.		9.97		9.97	
Clean	Ea.	.02	.70		.72	
Paint	L.F.	.08	8.30		8.38	
Minimum Charge	Job		174		174	
Colonial Style						
Demolish	Ea.		43		43	Includes material and labor to install
Install	Opng.	2525	174		2699	wood fireplace mantel.
Demolish and Install	Opng.	2525	217		2742	
Reinstall	Opng.		174.40		174.40	
Clean	Ea.		.51		.51	
Paint	Ea.	12.30	114		126.30	
Minimum Charge	Job		174		174	
Modern Design						
Demolish	Ea.		43		43	Includes material and labor to install
Install	Opng.	201	70		271	wood fireplace mantel.
Demolish and Install	Opng.	201	113		314	
Reinstall	Opng.		69.76		69.76	
Clean	Ea.		.51		.51	
Paint	Ea.	12.30	114		126.30	
Minimum Charge	Job		174		174	
Glass Door Unit						
Demolish	Ea.		5.90		5.90	Includes material and labor to install a
Install	Ea.	143	7.40		150.40	fireplace glass door unit.
Demolish and Install	Ea.	143	13.30		156.30	
Reinstall	Ea.		7.38		7.38	
Clean	Ea.	.37	9.75		10.12	
Minimum Charge	Job		174		174	
Fire Screen						
Demolish	Ea.		2.94		2.94	Includes material and labor to install a
Install	Ea.	44	3.69		47.69	fireplace fire screen.
Demolish and Install	Ea.	44	6.63		50.63	
Reinstall	Ea.		3.69		3.69	
Clean	Ea.	.19	8.45		8.64	
Minimum Charge	Job		174		174	

Siding

Shingles

Board & Batten

Texture 1-11

Exterior Siding		Unit	Material	Labor	Equip.	Total	Specification
Trim							
1″ x 2″							
	Demolish	L.F.		.24		.24	Cost includes material and labor to
	Install	L.F.	.35	1.06		1.41	install 1″ x 2″ pine trim.
	Demolish and Install	L.F.	.35	1.30		1.65	
	Clean	L.F.	.01	.17		.18	
	Paint	L.F.	.02	.48		.50	
	Minimum Charge	Job		174		174	
1″ x 3″							
	Demolish	L.F.		.24		.24	Cost includes material and labor to
	Install	L.F.	.24	1.16		1.40	install 1″ x 3″ pine trim.
	Demolish and Install	L.F.	.24	1.40		1.64	
	Clean	L.F.	.01	.17		.18	
	Paint	L.F.	.02	.48		.50	
	Minimum Charge	Job		174		174	
1″ x 4″							
	Demolish	L.F.		.24		.24	Cost includes material and labor to
	Install	L.F.	.68	1.40		2.08	install 1″ x 4″ pine trim.
	Demolish and Install	L.F.	.68	1.64		2.32	
	Clean	L.F.	.01	.17		.18	
	Paint	L.F.	.02	.48		.50	
	Minimum Charge	Job		174		174	
1″ x 6″							
	Demolish	L.F.		.24		.24	Cost includes material and labor to
	Install	L.F.	1.18	1.40		2.58	install 1″ x 6″ pine trim.
	Demolish and Install	L.F.	1.18	1.64		2.82	
	Clean	L.F.	.01	.17		.18	
	Paint	L.F.	.07	.48		.55	
	Minimum Charge	Job		174		174	
1″ x 8″							
	Demolish	L.F.		.26		.26	Cost includes material and labor to
	Install	L.F.	1.53	1.74		3.27	install 1″ x 8″ pine trim.
	Demolish and Install	L.F.	1.53	2		3.53	
	Clean	L.F.	.02	.18		.20	
	Paint	L.F.	.07	.48		.55	
	Minimum Charge	Job		174		174	
Aluminum							
8″ Non-Insulated							
	Demolish	S.F.		.64		.64	Includes material and labor to install
	Install	S.F.	1.43	1.38		2.81	horizontal, colored clapboard, 8″ to
	Demolish and Install	S.F.	1.43	2.02		3.45	10″ wide, plain.
	Reinstall	S.F.		1.38		1.38	
	Clean	S.F.		.25		.25	
	Minimum Charge	Job		174		174	

Siding

Exterior Siding		Unit	Material	Labor	Equip.	Total	Specification
8″ Insulated							
	Demolish	S.F.		.64		.64	Includes material and labor to install
	Install	S.F.	1.41	1.38		2.79	horizontal, colored insulated
	Demolish and Install	S.F.	1.41	2.02		3.43	clapboard, 8″ wide, plain.
	Reinstall	S.F.		1.38		1.38	
	Clean	S.F.		.25		.25	
	Minimum Charge	Job		174		174	
12″ Non-Insulated							
	Demolish	S.F.		.64		.64	Includes material and labor to install
	Install	S.F.	1.44	1.18		2.62	horizontal, colored clapboard, plain.
	Demolish and Install	S.F.	1.44	1.82		3.26	
	Reinstall	S.F.		1.18		1.18	
	Clean	S.F.		.25		.25	
	Minimum Charge	Job		174		174	
12″ Insulated							
	Demolish	S.F.		.64		.64	Includes material and labor to install
	Install	S.F.	1.67	1.18		2.85	horizontal, colored insulated
	Demolish and Install	S.F.	1.67	1.82		3.49	clapboard, plain.
	Reinstall	S.F.		1.18		1.18	
	Clean	S.F.		.25		.25	
	Minimum Charge	Job		174		174	
Clapboard 8″ to 10″							
	Demolish	S.F.		.64		.64	Includes material and labor to install
	Install	S.F.	1.43	1.38		2.81	horizontal, colored clapboard, 8″ to
	Demolish and Install	S.F.	1.43	2.02		3.45	10″ wide, plain.
	Reinstall	S.F.		1.38		1.38	
	Clean	S.F.		.25		.25	
	Minimum Charge	Job		174		174	
Corner Strips							
	Demolish	L.F.		.33		.33	Includes material and labor to install
	Install	V.L.F.	2.15	1.38		3.53	corners for horizontal lapped
	Demolish and Install	V.L.F.	2.15	1.71		3.86	aluminum siding.
	Reinstall	V.L.F.		1.38		1.38	
	Clean	L.F.		.17		.17	
	Minimum Charge	Job		174		174	
Vinyl							
Non-Insulated							
	Demolish	S.F.		.50		.50	Cost includes material and labor to
	Install	S.F.	.86	1.37		2.23	install 10″ non-insulated solid vinyl
	Demolish and Install	S.F.	.86	1.87		2.73	siding.
	Reinstall	S.F.		1.09		1.09	
	Clean	S.F.		.25		.25	
	Minimum Charge	Job		174		174	
Insulated							
	Demolish	S.F.		.50		.50	Includes material and labor to install
	Install	S.F.	.98	1.37		2.35	insulated solid vinyl siding panels.
	Demolish and Install	S.F.	.98	1.87		2.85	
	Reinstall	S.F.		1.09		1.09	
	Clean	S.F.		.25		.25	
	Minimum Charge	Job		174		174	
Corner Strips							
	Demolish	L.F.		.31		.31	Includes material and labor to install
	Install	L.F.	1.68	1.70		3.38	corner posts for vinyl siding panels.
	Demolish and Install	L.F.	1.68	2.01		3.69	
	Reinstall	L.F.		1.36		1.36	
	Clean	L.F.		.17		.17	
	Minimum Charge	Job		174		174	

Siding

Exterior Siding	Unit	Material	Labor	Equip.	Total	Specification
Hardboard						
4' x 8' Panel						
Demolish	S.F.		.52		.52	Cost includes material and labor to
Install	S.F.	1.10	.86		1.96	install 4' x 8' masonite siding 7/16"
Demolish and Install	S.F.	1.10	1.38		2.48	thick.
Clean	S.F.		.25		.25	
Paint	S.F.	.28	1.03		1.31	
Minimum Charge	Job		174		174	
Lap						
Demolish	S.F.		.52		.52	Cost includes material and labor to
Install	S.F.	1.01	.90		1.91	install 8" or 12" masonite lap siding
Demolish and Install	S.F.	1.01	1.42		2.43	7/16" thick with wood grain finish.
Clean	S.F.		.25		.25	
Paint	S.F.	.28	1.03		1.31	
Minimum Charge	Job		174		174	
Plywood Panel						
Douglas Fir (T-1-11)						
Demolish	S.F.		.39		.39	Includes material and labor to install
Install	S.F.	1.22	1.05		2.27	Douglas fir 4' x 8' plywood exterior
Demolish and Install	S.F.	1.22	1.44		2.66	siding plain or patterns.
Reinstall	S.F.		1.05		1.05	
Clean	S.F.		.25		.25	
Paint	S.F.	.28	1.03		1.31	
Minimum Charge	Job		174		174	
Redwood						
Demolish	S.F.		.39		.39	Includes material and labor to install
Install	S.F.	2.09	1.05		3.14	redwood 4' x 8' plywood exterior
Demolish and Install	S.F.	2.09	1.44		3.53	siding plain or patterns.
Reinstall	S.F.		1.05		1.05	
Clean	S.F.		.25		.25	
Paint	S.F.	.28	1.03		1.31	
Minimum Charge	Job		174		174	
Cedar						
Demolish	S.F.		.39		.39	Cost includes material and labor to
Install	S.F.	2.65	1.05		3.70	install 4' x 8' cedar plywood exterior
Demolish and Install	S.F.	2.65	1.44		4.09	siding plain or patterns.
Reinstall	S.F.		1.05		1.05	
Clean	S.F.		.25		.25	
Paint	S.F.	.28	1.03		1.31	
Minimum Charge	Job		174		174	
Tongue and Groove						
Fir						
Demolish	S.F.		.71		.71	Includes material and labor to install
Install	S.F.	1.65	.93		2.58	fir tongue and groove exterior siding
Demolish and Install	S.F.	1.65	1.64		3.29	1" x 8".
Reinstall	S.F.		.74		.74	
Clean	S.F.		.25		.25	
Paint	S.F.	.28	1.03		1.31	
Minimum Charge	Job		174		174	
Pine						
Demolish	S.F.		.71		.71	Includes material and labor to install
Install	S.F.	1.18	1.89		3.07	pine tongue and groove exterior
Demolish and Install	S.F.	1.18	2.60		3.78	siding 1" x 8".
Reinstall	S.F.		1.89		1.89	
Clean	S.F.		.25		.25	
Paint	S.F.	.28	1.03		1.31	
Minimum Charge	Job		174		174	

Siding

Exterior Siding	Unit	Material	Labor	Equip.	Total	Specification
Cedar						
Demolish	S.F.		.71		.71	Includes material and labor to install
Install	S.F.	1.65	.93		2.58	cedar tongue and groove exterior
Demolish and Install	S.F.	1.65	1.64		3.29	siding 1″ x 8″.
Reinstall	S.F.		.74		.74	
Clean	S.F.		.25		.25	
Paint	S.F.	.28	1.03		1.31	
Minimum Charge	Job		174		174	
Redwood						
Demolish	S.F.		.71		.71	Includes material and labor to install
Install	S.F.	3.19	2.37		5.56	redwood tongue and groove exterior
Demolish and Install	S.F.	3.19	3.08		6.27	siding.
Reinstall	S.F.		2.37		2.37	
Clean	S.F.		.25		.25	
Paint	S.F.	.28	1.03		1.31	
Minimum Charge	Job		174		174	
Board and Batten						
Fir						
Demolish	S.F.		.71		.71	Cost includes material and labor to
Install	S.F.	1.65	.87		2.52	install 1″ x 8″ fir boards with 1″ x 3″
Demolish and Install	S.F.	1.65	1.58		3.23	battens.
Reinstall	S.F.		.70		.70	
Clean	S.F.		.25		.25	
Paint	S.F.	.28	1.03		1.31	
Minimum Charge	Job		174		174	
Pine						
Demolish	S.F.		.71		.71	Cost includes material and labor to
Install	S.F.	.87	.87		1.74	install 1″ x 10″ pine boards with 1″ x
Demolish and Install	S.F.	.87	1.58		2.45	3″ battens.
Reinstall	S.F.		.70		.70	
Clean	S.F.		.25		.25	
Paint	S.F.	.28	1.03		1.31	
Minimum Charge	Job		174		174	
Cedar						
Demolish	S.F.		.71		.71	Cost includes material and labor to
Install	S.F.	2.18	.87		3.05	install 1″ x 10″ cedar boards with 1″ x
Demolish and Install	S.F.	2.18	1.58		3.76	3″ battens.
Reinstall	S.F.		.70		.70	
Clean	S.F.		.25		.25	
Paint	S.F.	.28	1.03		1.31	
Minimum Charge	Job		174		174	
Redwood						
Demolish	S.F.		.71		.71	Cost includes material and labor to
Install	S.F.	4.02	.87		4.89	install 1″ x 10″ redwood boards with
Demolish and Install	S.F.	4.02	1.58		5.60	1″ x 3″ battens.
Reinstall	S.F.		.70		.70	
Clean	S.F.		.25		.25	
Paint	S.F.	.28	1.03		1.31	
Minimum Charge	Job		174		174	
Wood Shingle						
Demolish	Sq.		65.50		65.50	Includes material and labor to install
Install	Sq.	135	155		290	wood shingles/shakes.
Demolish and Install	Sq.	135	220.50		355.50	
Clean	Sq.		26.50		26.50	
Paint	Sq.	18.45	91.50		109.95	
Minimum Charge	Job		174		174	

Siding

Exterior Siding	Unit	Material	Labor	Equip.	Total	Specification
Synthetic Stucco System						
Demolish	S.F.		2.35		2.35	Cost includes material and labor to
Install	S.F.	3.16	6.65	.57	10.38	install 2 coats of adhesive mixture,
Demolish and Install	S.F.	3.16	9	.57	12.73	glass fiber mesh, 1" insulation board,
Clean	S.F.		.23		.23	and textured finish coat.
Paint	S.F.	.12	.50		.62	
Minimum Charge	Job		735	62.50	797.50	

Front-end Loader

Bulldozer

Grading

Grading		Unit	Material	Labor	Equip.	Total	Specification
Hand (fine)							
	Install	S.F.		.51		.51	Includes labor and material to install
	Minimum Charge	Job		127		127	fine grade for slab on grade, hand grade.
Machine (fine)							
	Install	S.Y.		.44	.41	.85	Includes labor and material to install
	Minimum Charge	Job		300	278	578	fine grade for roadway, base or leveling course.

Asphalt Paving

Asphalt Paving		Unit	Material	Labor	Equip.	Total	Specification
Seal Coat							
	Install	S.F.	.08	.09		.17	Includes labor and material to install
	Minimum Charge	Job		127		127	sealcoat for a small area.
Wearing Course							
	Install	S.F.	.10	.27	.20	.57	Cost includes labor and material to
	Minimum Charge	Job		127		127	install 1″ thick wearing course.

Masonry Pavers

Masonry Pavers		Unit	Material	Labor	Equip.	Total	Specification
Natural Concrete							
Sand Base							
	Demolish	S.F.		2.82		2.82	Includes material and labor to install
	Install	S.F.	2.34	5.90		8.24	concrete pavers on sand base.
	Demolish and Install	S.F.	2.34	8.72		11.06	
	Reinstall	S.F.		4.73		4.73	
	Clean	S.F.		1.02		1.02	
	Minimum Charge	Job		177		177	
Mortar Base							
	Demolish	S.F.		4.68	1.49	6.17	Includes material and labor to install
	Install	S.F.	3.04	4.92		7.96	concrete pavers on mortar base.
	Demolish and Install	S.F.	3.04	9.60	1.49	14.13	
	Reinstall	S.F.		3.94		3.94	
	Clean	S.F.		.92		.92	
	Minimum Charge	Job		177		177	

Masonry Pavers		Unit	Material	Labor	Equip.	Total	Specification
Adobe							
Sand Base							
	Demolish	S.F.		2.82		2.82	Includes material and labor to install
	Install	S.F.	1.95	3.94		5.89	adobe pavers on sand base.
	Demolish and Install	S.F.	1.95	6.76		8.71	
	Reinstall	S.F.		3.15		3.15	
	Clean	S.F.		1.02		1.02	
	Minimum Charge	Job		177		177	
Mortar Base							
	Demolish	S.F.		4.68	1.49	6.17	Includes material and labor to install
	Install	S.F.	2.71	4.92		7.63	adobe pavers on mortar base.
	Demolish and Install	S.F.	2.71	9.60	1.49	13.80	
	Reinstall	S.F.		3.94		3.94	
	Clean	S.F.		.92		.92	
	Minimum Charge	Job		177		177	
Standard Brick							
Sand Base							
	Demolish	S.F.		2.82		2.82	Includes material and labor to install
	Install	S.F.	2.94	6.20		9.14	standard brick pavers on sand base.
	Demolish and Install	S.F.	2.94	9.02		11.96	
	Reinstall	S.F.		4.97		4.97	
	Clean	S.F.		1.02		1.02	
	Minimum Charge	Job		177		177	
Mortar Base							
	Demolish	S.F.		4.68	1.49	6.17	Includes material and labor to install
	Install	S.F.	5.35	8.85		14.20	standard brick pavers on mortar base.
	Demolish and Install	S.F.	5.35	13.53	1.49	20.37	
	Reinstall	S.F.		7.09		7.09	
	Clean	S.F.		.92		.92	
	Minimum Charge	Job		177		177	
Deluxe Brick							
Sand Base							
	Demolish	S.F.		2.82		2.82	Includes material and labor to install
	Install	S.F.	5.35	4.92		10.27	deluxe brick pavers on sand base.
	Demolish and Install	S.F.	5.35	7.74		13.09	
	Reinstall	S.F.		3.94		3.94	
	Clean	S.F.		1.02		1.02	
	Minimum Charge	Job		177		177	
Mortar Base							
	Demolish	S.F.		4.68	1.49	6.17	Includes material and labor to install
	Install	S.F.	5.80	8.85		14.65	deluxe brick pavers on mortar base.
	Demolish and Install	S.F.	5.80	13.53	1.49	20.82	
	Reinstall	S.F.		7.09		7.09	
	Clean	S.F.		.92		.92	
	Minimum Charge	Job		177		177	

Fencing		Unit	Material	Labor	Equip.	Total	Specification
Block w / Footing							
4"							
	Demolish	S.F.		1.57		1.57	Cost includes material and labor to
	Install	S.F.	1.34	3.71		5.05	install 4" non-reinforced concrete
	Demolish and Install	S.F.	1.34	5.28		6.62	block.
	Clean	S.F.	.14	1.08	.16	1.38	
	Paint	S.F.	.18	.40		.58	
	Minimum Charge	Job		177		177	

Site Work

Fencing

	Unit	Material	Labor	Equip.	Total	Specification
4″ Reinforced						
Demolish	S.F.		1.57	.80	2.37	Cost includes material and labor to
Install	S.F.	1.46	3.76		5.22	install 4″ reinforced concrete block.
Demolish and Install	S.F.	1.46	5.33	.80	7.59	
Clean	S.F.	.14	1.08	.16	1.38	
Paint	S.F.	.18	.40		.58	
Minimum Charge	Job		177		177	
6″						
Demolish	S.F.		1.66		1.66	Cost includes material and labor to
Install	S.F.	1.96	3.99		5.95	install 6″ non-reinforced concrete
Demolish and Install	S.F.	1.96	5.65		7.61	block.
Clean	S.F.	.14	1.08	.16	1.38	
Paint	S.F.	.18	.40		.58	
Minimum Charge	Job		177		177	
6″ Reinforced						
Demolish	S.F.		1.65	.39	2.04	Cost includes material and labor to
Install	S.F.	2.09	4.04		6.13	install 6″ reinforced concrete block.
Demolish and Install	S.F.	2.09	5.69	.39	8.17	
Clean	S.F.	.14	1.08	.16	1.38	
Paint	S.F.	.18	.40		.58	
Minimum Charge	Job		177		177	
Cap for Block Wall						
Demolish	L.F.		1.76		1.76	Includes labor and equipment to install
Install	L.F.	18.80	8.30		27.10	precast concrete coping.
Demolish and Install	L.F.	18.80	10.06		28.86	
Clean	L.F.	.01	2.65		2.66	
Paint	S.F.	.18	.40		.58	
Minimum Charge	Job		177		177	
Block Pilaster						
Demolish	Ea.		121		121	Includes material and labor to install
Install	Ea.	73.50	124		197.50	block pilaster for fence.
Demolish and Install	Ea.	73.50	245		318.50	
Reinstall	Ea.		124.16		124.16	
Paint	Ea.	4.13	13.55		17.68	
Minimum Charge	Job		177		177	
Block w / o Footing						
4″						
Demolish	S.F.		1.57		1.57	Cost includes material and labor to
Install	S.F.	1.64	2.82		4.46	install 4″ non-reinforced hollow
Demolish and Install	S.F.	1.64	4.39		6.03	lightweight block.
Clean	S.F.	.14	1.08	.16	1.38	
Paint	S.F.	.18	.40		.58	
Minimum Charge	Job		177		177	
4″ Reinforced						
Demolish	S.F.		1.57	.80	2.37	Cost includes material and labor to
Install	S.F.	2.32	3.67		5.99	install 4″ reinforced hollow lightweight
Demolish and Install	S.F.	2.32	5.24	.80	8.36	block.
Clean	S.F.	.14	1.08	.16	1.38	
Paint	S.F.	.18	.40		.58	
Minimum Charge	Job		177		177	
6″						
Demolish	S.F.		1.66		1.66	Cost includes material and labor to
Install	S.F.	2.23	3.10		5.33	install 6″ non-reinforced hollow
Demolish and Install	S.F.	2.23	4.76		6.99	lightweight block.
Clean	S.F.	.14	1.08	.16	1.38	
Paint	S.F.	.18	.40		.58	
Minimum Charge	Job		177		177	

Fencing	Unit	Material	Labor	Equip.	Total	Specification
6" Reinforced						
Demolish	S.F.		1.65	.39	2.04	Cost includes material and labor to
Install	S.F.	2.33	3.94		6.27	install 6" reinforced hollow lightweight
Demolish and Install	S.F.	2.33	5.59	.39	8.31	block.
Clean	S.F.	.14	1.08	.16	1.38	
Paint	S.F.	.18	.40		.58	
Minimum Charge	Job		177		177	
Cap for Block Wall						
Demolish	L.F.		1.76		1.76	Includes labor and equipment to install
Install	L.F.	18.80	8.30		27.10	precast concrete coping.
Demolish and Install	L.F.	18.80	10.06		28.86	
Clean	L.F.	.01	2.65		2.66	
Paint	S.F.	.18	.40		.58	
Minimum Charge	Job		177		177	
Block Pilaster						
Demolish	Ea.		121		121	Includes material and labor to install
Install	Ea.	73.50	124		197.50	block pilaster for fence.
Demolish and Install	Ea.	73.50	245		318.50	
Reinstall	Ea.		124.16		124.16	
Paint	Ea.	4.13	13.55		17.68	
Minimum Charge	Job		177		177	
Chain Link						
3' High w / o Top Rail						
Demolish	L.F.		2.20	.70	2.90	Includes material and labor to install
Install	L.F.	3.75	1.49	.31	5.55	residential galvanized fence 3' high
Demolish and Install	L.F.	3.75	3.69	1.01	8.45	with 1-3/8" steel post 10' O.C. set
Reinstall	L.F.		1.19	.25	1.44	without concrete.
Clean	L.F.	.10	.92		1.02	
Minimum Charge	Job		395		395	
4' High w / o Top Rail						
Demolish	L.F.		2.20	.70	2.90	Includes material and labor to install
Install	L.F.	3.98	1.84	.38	6.20	residential galvanized fence 4' high
Demolish and Install	L.F.	3.98	4.04	1.08	9.10	with 1-3/8" steel post 10' O.C. set
Reinstall	L.F.		1.47	.31	1.78	without concrete.
Clean	L.F.	.12	1.02		1.14	
Minimum Charge	Job		395		395	
5' High w / o Top Rail						
Demolish	L.F.		2.20	.70	2.90	Includes material and labor to install
Install	L.F.	4.68	2.40	.50	7.58	residential galvanized fence 5' high
Demolish and Install	L.F.	4.68	4.60	1.20	10.48	with 1-3/8" steel post 10' O.C. set
Reinstall	L.F.		1.92	.40	2.32	without concrete.
Clean	L.F.	.15	1.13		1.28	
Minimum Charge	Job		395		395	
3' High w / Top Rail						
Demolish	L.F.		2.34	.74	3.08	Includes material and labor to install
Install	L.F.	4.63	1.56	.32	6.51	residential galvanized fence 3' high
Demolish and Install	L.F.	4.63	3.90	1.06	9.59	with 1-3/8" steel post 10' O.C. set
Reinstall	L.F.		1.25	.26	1.51	without concrete.
Clean	L.F.	.10	.94		1.04	
Minimum Charge	Job		395		395	
4' High w / Top Rail						
Demolish	L.F.		2.34	.74	3.08	Includes material and labor to install
Install	L.F.	5.25	1.95	.41	7.61	residential galvanized fence 4' high
Demolish and Install	L.F.	5.25	4.29	1.15	10.69	with 1-5/8" steel post 10' O.C. set
Reinstall	L.F.		1.56	.32	1.88	without concrete and 1-5/8" top rail
Clean	L.F.	.12	1.04		1.16	with sleeves.
Minimum Charge	Job		395		395	

Fencing	Unit	Material	Labor	Equip.	Total	Specification
5' High w / Top Rail						Includes material and labor to install
Demolish	L.F.		2.34	.74	3.08	residential galvanized fence 5' high
Install	L.F.	5.95	2.60	.54	9.09	with 1-5/8" steel post 10' O.C. set
Demolish and Install	L.F.	5.95	4.94	1.28	12.17	without concrete and 1-5/8" top rail
Reinstall	L.F.		2.08	.43	2.51	with sleeves.
Clean	L.F.	.15	1.15		1.30	
Minimum Charge	Job		395		395	
6' High w / Top Rail						Includes material and labor to install
Demolish	L.F.		2.34	.74	3.08	residential galvanized fence 6' high
Install	L.F.	6.40	3.90	.81	11.11	with 2" steel post 10' O.C. set without
Demolish and Install	L.F.	6.40	6.24	1.55	14.19	concrete and 1-3/8" top rail with
Reinstall	L.F.		3.12	.65	3.77	sleeves.
Clean	L.F.	.19	1.30		1.49	
Minimum Charge	Job		395		395	
Barbed Wire						Cost includes material and labor to
Demolish	L.F.		1.31		1.31	install 3 strand galvanized barbed
Install	L.F.	.40	.51	.14	1.05	wire fence.
Demolish and Install	L.F.	.40	1.82	.14	2.36	
Reinstall	L.F.		.41	.11	.52	
Minimum Charge	Job		395	310	705	
6' Redwood						Cost includes material and labor to
Demolish	L.F.		8.75		8.75	install 6' high redwood fencing.
Install	L.F.	11.40	6.25	1.30	18.95	
Demolish and Install	L.F.	11.40	15	1.30	27.70	
Reinstall	L.F.		5	1.04	6.04	
Clean	L.F.	.19	1.18		1.37	
Paint	L.F.	.76	3.21		3.97	
Minimum Charge	Job		395		395	
6' Red Cedar						Cost includes material and labor to
Demolish	L.F.		8.75		8.75	install 6' cedar fence.
Install	L.F.	17.40	6.25	1.30	24.95	
Demolish and Install	L.F.	17.40	15	1.30	33.70	
Reinstall	L.F.		5	1.04	6.04	
Clean	L.F.	.19	1.18		1.37	
Paint	L.F.	.76	3.21		3.97	
Minimum Charge	Job		395		395	
6' Shadowbox						Cost includes material and labor to
Demolish	L.F.		8.75		8.75	install 6' pressure treated pine fence.
Install	L.F.	13.30	5.20	1.08	19.58	
Demolish and Install	L.F.	13.30	13.95	1.08	28.33	
Reinstall	L.F.		4.16	.87	5.03	
Clean	L.F.	.19	1.18		1.37	
Paint	L.F.	.76	3.21		3.97	
Minimum Charge	Job		395		395	
3' Picket						Includes material, labor and
Demolish	L.F.		8.75		8.75	equipment to install pressure treated
Install	L.F.	5.05	5.60	1.16	11.81	pine picket fence.
Demolish and Install	L.F.	5.05	14.35	1.16	20.56	
Reinstall	L.F.		4.46	.93	5.39	
Clean	L.F.	.10	1.02		1.12	
Paint	L.F.	.37	1.74		2.11	
Minimum Charge	Job		395		395	

Fencing	Unit	Material	Labor	Equip.	Total	Specification
Wrought Iron						
Demolish	L.F.		18.80		18.80	Cost includes material and labor to
Install	L.F.	26.50	10.50		37	install 5' high wrought iron fence with
Demolish and Install	L.F.	26.50	29.30		55.80	pickets at 4-1/2" O.C.
Reinstall	L.F.		8.42		8.42	
Clean	L.F.		1.02		1.02	
Paint	L.F.	.59	3.59		4.18	
Minimum Charge	Job		395		395	
Post						
Dirt Based Wood						
Demolish	Ea.		22		22	Includes material and labor to install
Install	Ea.	12.30	98.50		110.80	wood post set in earth.
Demolish and Install	Ea.	12.30	120.50		132.80	
Reinstall	Ea.		78.91		78.91	
Clean	Ea.		3.18		3.18	
Paint	Ea.	.83	1.22		2.05	
Minimum Charge	Job		395		395	
Concrete Based Wood						
Demolish	Ea.		31	9.90	40.90	Includes material and labor to install
Install	Ea.	12.30	132		144.30	wood post set in concrete.
Demolish and Install	Ea.	12.30	163	9.90	185.20	
Clean	Ea.		3.18		3.18	
Paint	Ea.	.83	1.22		2.05	
Minimum Charge	Job		395		395	
Block Pilaster						
Demolish	Ea.		121		121	Includes material and labor to install
Install	Ea.	73.50	124		197.50	block pilaster for fence.
Demolish and Install	Ea.	73.50	245		318.50	
Reinstall	Ea.		124.16		124.16	
Paint	Ea.	4.13	13.55		17.68	
Minimum Charge	Job		177		177	
Gate						
Chain Link						
Demolish	Ea.		29		29	Cost includes material and labor to
Install	Ea.	114	44		158	install 3' wide chain link pass gate
Demolish and Install	Ea.	114	73		187	and hardware.
Reinstall	Ea.		35.07		35.07	
Clean	Ea.		3.53		3.53	
Minimum Charge	Job		395		395	
Redwood						
Demolish	Ea.		25		25	Includes material, labor and
Install	Ea.	146	44		190	equipment to install redwood pass
Demolish and Install	Ea.	146	69		215	gate and hardware.
Reinstall	Ea.		35.07		35.07	
Clean	Ea.		3.53		3.53	
Paint	Ea.	1.24	6.10		7.34	
Minimum Charge	Job		395		395	
Red Cedar						
Demolish	Ea.		25		25	Cost includes material and labor to
Install	Ea.	78	44		122	install 3' wide red cedar pass gate
Demolish and Install	Ea.	78	69		147	and hardware.
Reinstall	Ea.		35.07		35.07	
Clean	Ea.		3.53		3.53	
Paint	Ea.	1.24	6.10		7.34	
Minimum Charge	Job		395		395	

Site Work

Fencing

Fencing	Unit	Material	Labor	Equip.	Total	Specification
Wrought Iron						
Demolish	Ea.		18.80		18.80	Includes material, labor and
Install	Ea.	138	98.50		236.50	equipment to install wrought iron gate
Demolish and Install	Ea.	138	117.30		255.30	with hardware.
Reinstall	Ea.		78.91		78.91	
Clean	Ea.		3.53		3.53	
Paint	Ea.	1.19	6.10		7.29	
Minimum Charge	Job		395		395	

Fire-extinguishing System   **Space Heater**   **Hot-air Furnace**

Plumbing		Unit	Material	Labor	Equip.	Total	Specification
Components							
3/4″ Supply							
	Demolish	L.F.		2.16		2.16	Includes material and labor to install
	Install	L.F.	5.70	5.10		10.80	type L copper tubing including
	Demolish and Install	L.F.	5.70	7.26		12.96	couplings and hangers at 10′ O.C.
	Reinstall	L.F.		5.12		5.12	
	Minimum Charge	Job		194		194	
1/2″ Fixture Supply							
	Install	Ea.	55	130		185	Includes labor and materials to install
							supply pipe only to a plumbing fixture.
2″ Single Vent Stack							
	Demolish	Ea.		36		36	Includes material and labor to install
	Install	Ea.	22.50	194		216.50	PVC vent stack pipe with couplings at
	Demolish and Install	Ea.	22.50	230		252.50	10′ O.C.
	Reinstall	Ea.		194.40		194.40	
	Minimum Charge	Job		194		194	
4″ Drain							
	Demolish	L.F.		2.88		2.88	Includes material and labor to install
	Install	L.F.	6.25	6.35		12.60	DVW type PVC pipe with couplings at
	Demolish and Install	L.F.	6.25	9.23		15.48	10′ O.C. and 3 strap hangers per 10′
	Reinstall	L.F.		6.36		6.36	run.
	Minimum Charge	Job		194		194	
1-1/2″ Drain							
	Demolish	L.F.		2.16		2.16	Includes material and labor to install
	Install	L.F.	1.95	4.14		6.09	DVW type PVC pipe with couplings at
	Demolish and Install	L.F.	1.95	6.30		8.25	10′ O.C. and 3 strap hangers per 10′
	Reinstall	L.F.		4.14		4.14	run.
	Minimum Charge	Job		194		194	
Sewer / Septic							
	Install	L.F.	2.48	3.59	1.14	7.21	Includes labor and material to install
Maintenance							sewerage and drainage piping, 6″
Flush Out / Unclog							diameter.
	Clean	Job		194		194	Minimum labor and equipment charge
	Minimum Charge	Job		194		194	to install water cooler.

Rough Mechanical

Plumbing	Unit	Material	Labor	Equip.	Total	Specification
Vent Cap						
Large						
Demolish	Ea.		9.80		9.80	Includes labor and material to install a
Install	Ea.	35	17.65		52.65	PVC plumbing pipe vent cap.
Demolish and Install	Ea.	35	27.45		62.45	
Clean	Ea.	.43	5.75		6.18	
Paint	Ea.	.17	6.95		7.12	
Minimum Charge	Job		97		97	
Medium						
Demolish	Ea.		9.80		9.80	Includes material and labor to install a
Install	Ea.	30.50	16.90		47.40	PVC plumbing pipe vent cap.
Demolish and Install	Ea.	30.50	26.70		57.20	
Clean	Ea.	.43	5.75		6.18	
Paint	Ea.	.17	6.95		7.12	
Minimum Charge	Job		97		97	
Small						
Demolish	Ea.		9.80		9.80	Includes material and labor to install a
Install	Ea.	8.05	16.20		24.25	PVC plumbing pipe vent cap.
Demolish and Install	Ea.	8.05	26		34.05	
Clean	Ea.	.43	5.75		6.18	
Paint	Ea.	.17	6.95		7.12	
Minimum Charge	Job		97		97	

Plumbing	Unit	Material	Labor	Equip.	Total	Specification
Sump Pump						
1/4 HP						
Demolish	Ea.		39		39	Includes material and labor to install a
Install	Ea.	134	65		199	submersible sump pump.
Demolish and Install	Ea.	134	104		238	
Reinstall	Ea.		51.84		51.84	
Clean	Ea.	1.09	15.90		16.99	
Minimum Charge	Job		194		194	
1/3 HP						
Demolish	Ea.		39		39	Includes material and labor to install a
Install	Ea.	161	78		239	submersible sump pump.
Demolish and Install	Ea.	161	117		278	
Reinstall	Ea.		62.21		62.21	
Clean	Ea.	1.09	15.90		16.99	
Minimum Charge	Job		194		194	
1/2 HP						
Demolish	Ea.		39		39	Includes material and labor to install a
Install	Ea.	198	78		276	submersible sump pump.
Demolish and Install	Ea.	198	117		315	
Reinstall	Ea.		62.21		62.21	
Clean	Ea.	1.09	15.90		16.99	
Minimum Charge	Job		194		194	
Clean and Service						
Clean	Ea.		24.50		24.50	Includes labor to clean and service a
Minimum Charge	Job		194		194	sump pump.

Rough Mechanical

Water Heater	Unit	Material	Labor	Equip.	Total	Specification
30 Gallon						
Demolish	Ea.		122		122	Includes material and labor to install a
Install	Ea.	425	177		602	30 gallon electric water heater.
Demolish and Install	Ea.	425	299		724	
Reinstall	Ea.		141.38		141.38	
Clean	Ea.	1.09	15.90		16.99	
Minimum Charge	Job		194		194	
40 Gallon						
Demolish	Ea.		122		122	Includes material and labor to install a
Install	Ea.	450	194		644	40 gallon electric water heater.
Demolish and Install	Ea.	450	316		766	
Reinstall	Ea.		155.52		155.52	
Clean	Ea.	1.09	15.90		16.99	
Minimum Charge	Job		194		194	
52 Gallon						
Demolish	Ea.		122		122	Includes material and labor to install a
Install	Ea.	505	194		699	52 gallon electric water heater.
Demolish and Install	Ea.	505	316		821	
Reinstall	Ea.		155.52		155.52	
Clean	Ea.	1.09	15.90		16.99	
Minimum Charge	Job		194		194	
82 Gallon						
Demolish	Ea.		122		122	Includes material and labor to install
Install	Ea.	770	243		1013	an 82 gallon electric water heater.
Demolish and Install	Ea.	770	365		1135	
Reinstall	Ea.		194.40		194.40	
Clean	Ea.	1.09	15.90		16.99	
Minimum Charge	Job		194		194	
Gas-fired						
30 Gallon Gas-fired						
Demolish	Ea.		150		150	Includes material and labor to install a
Install	Ea.	690	194		884	30 gallon gas water heater.
Demolish and Install	Ea.	690	344		1034	
Reinstall	Ea.		155.52		155.52	
Clean	Ea.	1.09	15.90		16.99	
Minimum Charge	Job		194		194	
30 Gallon Gas-fired, Flue						
Install	Ea.	37	55.50		92.50	Includes labor and materials to install
Minimum Charge	Job		194		194	a flue for a water heater (up to 50 gallon size).
40 Gallon Gas-fired						
Demolish	Ea.		150		150	Includes material and labor to install a
Install	Ea.	700	205		905	40 gallon gas water heater.
Demolish and Install	Ea.	700	355		1055	
Reinstall	Ea.		163.71		163.71	
Clean	Ea.	1.09	15.90		16.99	
Minimum Charge	Job		194		194	
40 Gallon Gas-fired, Flue						
Install	Ea.	37	55.50		92.50	Includes labor and materials to install
Minimum Charge	Job		194		194	a flue for a water heater (up to 50 gallon size).
50 Gallon Gas-fired						
Demolish	Ea.		150		150	Includes material and labor to install a
Install	Ea.	730	216		946	50 gallon gas water heater.
Demolish and Install	Ea.	730	366		1096	
Reinstall	Ea.		172.80		172.80	
Clean	Ea.	1.09	15.90		16.99	
Minimum Charge	Job		194		194	

Rough Mechanical

Water Heater	Unit	Material	Labor	Equip.	Total	Specification
50 Gallon Gas-fired, Flue						
Install	Ea.	37	55.50		92.50	Includes labor and materials to install
Minimum Charge	Job		194		194	a flue for a water heater (up to 50
						gallon size).
Solar Hot Water System						
Water Heater						
Demolish	Ea.		97		97	Includes material and labor to install a
Install	Ea.	965	259		1224	solar hot water heat exchanger
Demolish and Install	Ea.	965	356		1321	storage tank, 100 gallon.
Reinstall	Ea.		207.36		207.36	
Minimum Charge	Job		196		196	
Solar Panel						
Demolish	S.F.		1.42		1.42	Includes material and labor to install
Install	S.F.	23	2.56		25.56	solar panel.
Demolish and Install	S.F.	23	3.98		26.98	
Reinstall	S.F.		2.05		2.05	
Minimum Charge	Job		196		196	
Repair Plumbing						
Install	Job		196		196	Includes 4 hour minimum labor charge
Minimum Charge	Job		196		196	for a steamfitter or pipefitter.
Insulating Wrap						
Demolish	Ea.		4.50		4.50	Includes material and labor to install
Install	Ea.	71.50	12.15		83.65	insulating wrap around a 100 gallon
Demolish and Install	Ea.	71.50	16.65		88.15	solar hot water heat exchanger
Reinstall	Ea.		9.72		9.72	storage tank.
Minimum Charge	Job		194		194	
Water Softener						
Demolish	Ea.		72		72	Includes material and labor to install a
Install	Ea.	1350	175		1525	14 GPM water softener.
Demolish and Install	Ea.	1350	247		1597	
Reinstall	Ea.		140		140	
Clean	Ea.	1.09	15.90		16.99	
Minimum Charge	Ea.		194		194	

Ductwork	Unit	Material	Labor	Equip.	Total	Specification
Galvanized, Square or Rectangular						
6"						
Demolish	L.F.		1.79		1.79	Includes material and labor to install
Install	L.F.	2.42	16.10		18.52	rectangular ductwork with a greatest
Demolish and Install	L.F.	2.42	17.89		20.31	dimension of 6".
Reinstall	L.F.		12.89		12.89	
Clean	L.F.		2.06		2.06	
Minimum Charge	Job		193		193	
8"						
Demolish	L.F.		1.79		1.79	Includes material and labor to install
Install	L.F.	3.21	16.10		19.31	rectangular ductwork with a greatest
Demolish and Install	L.F.	3.21	17.89		21.10	dimension of 8".
Reinstall	L.F.		12.89		12.89	
Clean	L.F.		2.75		2.75	
Minimum Charge	Job		193		193	

Rough Mechanical

Ductwork

	Unit	Material	Labor	Equip.	Total	Specification
10"						
Demolish	L.F.		1.79		1.79	Includes material and labor to install
Install	L.F.	4.03	16.10		20.13	rectangular ductwork with a greatest
Demolish and Install	L.F.	4.03	17.89		21.92	dimension of 10".
Reinstall	L.F.		12.89		12.89	
Clean	L.F.		3.44		3.44	
Minimum Charge	Job		193		193	
Galvanized, Spiral						
4"						
Demolish	L.F.		2.15		2.15	Cost includes material and labor to
Install	L.F.	1.33	1.93		3.26	install 4" diameter spiral pre-formed
Demolish and Install	L.F.	1.33	4.08		5.41	steel galvanized ductwork.
Reinstall	L.F.		1.55		1.55	
Clean	L.F.		1.08		1.08	
Minimum Charge	Job		193		193	
6"						
Demolish	L.F.		2.15		2.15	Cost includes material and labor to
Install	L.F.	1.94	2.49		4.43	install 6" diameter spiral pre-formed
Demolish and Install	L.F.	1.94	4.64		6.58	steel galvanized ductwork.
Reinstall	L.F.		1.99		1.99	
Clean	L.F.		1.62		1.62	
Minimum Charge	Job		193		193	
8"						
Demolish	L.F.		2.15		2.15	Cost includes material and labor to
Install	L.F.	2.54	3.48		6.02	install 8" diameter spiral pre-formed
Demolish and Install	L.F.	2.54	5.63		8.17	steel galvanized ductwork.
Reinstall	L.F.		2.79		2.79	
Clean	L.F.		2.16		2.16	
Minimum Charge	Job		193		193	
10"						
Demolish	L.F.		2.69		2.69	Cost includes material and labor to
Install	L.F.	3.15	4.35		7.50	install 10" diameter spiral pre-formed
Demolish and Install	L.F.	3.15	7.04		10.19	steel galvanized ductwork.
Reinstall	L.F.		3.48		3.48	
Clean	L.F.		2.70		2.70	
Minimum Charge	Job		193		193	
12"						
Demolish	L.F.		3.58		3.58	Cost includes material and labor to
Install	L.F.	3.87	5.80		9.67	install 12" diameter spiral pre-formed
Demolish and Install	L.F.	3.87	9.38		13.25	steel galvanized ductwork.
Reinstall	L.F.		4.64		4.64	
Clean	L.F.		3.24		3.24	
Minimum Charge	Job		193		193	
16"						
Demolish	L.F.		7.15		7.15	Cost includes material and labor to
Install	L.F.	6.65	11.60		18.25	install 16" diameter spiral pre-formed
Demolish and Install	L.F.	6.65	18.75		25.40	steel galvanized ductwork.
Reinstall	L.F.		9.29		9.29	
Clean	L.F.		4.32		4.32	
Minimum Charge	Job		193		193	
Flexible Fiberglass						
6" Non-insulated						
Demolish	L.F.		1.65		1.65	Cost includes material and labor to
Install	L.F.	1.76	2.49		4.25	install 6" diameter non-insulated
Demolish and Install	L.F.	1.76	4.14		5.90	flexible fiberglass fabric on corrosion
Reinstall	L.F.		1.99		1.99	resistant metal ductwork.
Clean	L.F.		1.62		1.62	
Minimum Charge	Job		193		193	

Rough Mechanical

Ductwork		Unit	Material	Labor	Equip.	Total	Specification
8″ Non-insulated							
	Demolish	L.F.		2.39		2.39	Cost includes material and labor to
	Install	L.F.	2.42	3.48		5.90	install 8″ diameter non-insulated
	Demolish and Install	L.F.	2.42	5.87		8.29	flexible fiberglass fabric on corrosion
	Reinstall	L.F.		2.79		2.79	resistant metal ductwork.
	Clean	L.F.		2.16		2.16	
	Minimum Charge	Job		193		193	
10″ Non-insulated							
	Demolish	L.F.		3.07		3.07	Cost includes material and labor to
	Install	L.F.	3.04	4.35		7.39	install 10″ diameter non-insulated
	Demolish and Install	L.F.	3.04	7.42		10.46	flexible fiberglass fabric on corrosion
	Reinstall	L.F.		3.48		3.48	resistant metal ductwork.
	Clean	L.F.		2.70		2.70	
	Minimum Charge	Job		193		193	
12″ Non-insulated							
	Demolish	L.F.		4.30		4.30	Cost includes material and labor to
	Install	L.F.	3.67	5.80		9.47	install 12″ diameter non-insulated
	Demolish and Install	L.F.	3.67	10.10		13.77	flexible fiberglass fabric on corrosion
	Reinstall	L.F.		4.64		4.64	resistant metal ductwork.
	Clean	L.F.		3.24		3.24	
	Minimum Charge	Job		193		193	
4″ Insulated							
	Demolish	L.F.		1.26		1.26	Cost includes material and labor to
	Install	L.F.	2.44	2.05		4.49	install 4″ diameter insulated flexible
	Demolish and Install	L.F.	2.44	3.31		5.75	fiberglass fabric on corrosion resistant
	Reinstall	L.F.		1.64		1.64	metal ductwork.
	Clean	L.F.		1.08		1.08	
	Minimum Charge	Job		193		193	
6″ Insulated							
	Demolish	L.F.		1.65		1.65	Cost includes material and labor to
	Install	L.F.	2.79	2.68		5.47	install 6″ diameter insulated flexible
	Demolish and Install	L.F.	2.79	4.33		7.12	fiberglass fabric on corrosion resistant
	Reinstall	L.F.		2.14		2.14	metal ductwork.
	Clean	L.F.		1.62		1.62	
	Minimum Charge	Job		193		193	
8″ Insulated							
	Demolish	L.F.		2.39		2.39	Cost includes material and labor to
	Install	L.F.	3.47	3.87		7.34	install 8″ diameter insulated flexible
	Demolish and Install	L.F.	3.47	6.26		9.73	fiberglass fabric on corrosion resistant
	Reinstall	L.F.		3.10		3.10	metal ductwork.
	Clean	L.F.		2.16		2.16	
	Minimum Charge	Job		193		193	
10″ Insulated							
	Demolish	L.F.		3.07		3.07	Cost includes material and labor to
	Install	L.F.	4.08	4.98		9.06	install 10″ diameter insulated flexible
	Demolish and Install	L.F.	4.08	8.05		12.13	fiberglass fabric on corrosion resistant
	Reinstall	L.F.		3.98		3.98	metal ductwork.
	Clean	L.F.		2.70		2.70	
	Minimum Charge	Job		193		193	
12″ Insulated							
	Demolish	L.F.		4.30		4.30	Cost includes material and labor to
	Install	L.F.	5.10	6.95		12.05	install 12″ diameter insulated flexible
	Demolish and Install	L.F.	5.10	11.25		16.35	fiberglass fabric on corrosion resistant
	Reinstall	L.F.		5.57		5.57	metal ductwork.
	Clean	L.F.		3.24		3.24	
	Minimum Charge	Job		193		193	

Rough Mechanical

Heating		Unit	Material	Labor	Equip.	Total	Specification
Hot Water Baseboard							
	Demolish	L.F.		6.70		6.70	Includes material and labor to install
	Install	L.F.	7.30	11.80		19.10	baseboard radiation.
	Demolish and Install	L.F.	7.30	18.50		25.80	
	Reinstall	L.F.		9.43		9.43	
	Clean	L.F.	1.09	3.97		5.06	
	Minimum Charge	Job		194		194	
3'							
	Demolish	Ea.		21.50		21.50	Includes material and labor to install
	Install	Ea.	23	39.50		62.50	fin tube baseboard radiation in 3'
	Demolish and Install	Ea.	23	61		84	sections.
	Reinstall	Ea.		31.42		31.42	
	Clean	Ea.	.31	12.10		12.41	
	Minimum Charge	Job		194		194	
4'							
	Demolish	Ea.		29		29	Includes material and labor to install
	Install	Ea.	31.50	44		75.50	fin tube baseboard radiation in 4'
	Demolish and Install	Ea.	31.50	73		104.50	sections.
	Reinstall	Ea.		35.34		35.34	
	Clean	Ea.	.63	15.90		16.53	
	Minimum Charge	Job		194		194	
5'							
	Demolish	Ea.		36		36	Includes material and labor to install
	Install	Ea.	39	50.50		89.50	fin tube baseboard radiation in 5'
	Demolish and Install	Ea.	39	86.50		125.50	sections.
	Reinstall	Ea.		40.39		40.39	
	Clean	Ea.	.63	19.55		20.18	
	Minimum Charge	Job		194		194	
6'							
	Demolish	Ea.		43		43	Includes material and labor to install
	Install	Ea.	47	59		106	fin tube baseboard radiation in 6'
	Demolish and Install	Ea.	47	102		149	sections.
	Reinstall	Ea.		47.13		47.13	
	Clean	Ea.	.63	23		23.63	
	Minimum Charge	Job		194		194	
7'							
	Demolish	Ea.		48		48	Includes material and labor to install
	Install	Ea.	54.50	78		132.50	fin tube baseboard radiation in 7'
	Demolish and Install	Ea.	54.50	126		180.50	sections.
	Reinstall	Ea.		62.21		62.21	
	Clean	Ea.	.78	28		28.78	
	Minimum Charge	Job		194		194	
9'							
	Demolish	Ea.		72		72	Includes material and labor to install
	Install	Ea.	70	111		181	fin tube baseboard radiation in 9'
	Demolish and Install	Ea.	70	183		253	sections.
	Reinstall	Ea.		88.87		88.87	
	Clean	Ea.	1.24	36.50		37.74	
	Minimum Charge	Job		194		194	

Heating		Unit	Material	Labor	Equip.	Total	Specification
50 MBH							
	Demolish	Ea.		35.50		35.50	Includes material and labor to install a
	Install	Ea.	665	101		766	cabinet space heater with an output of
	Demolish and Install	Ea.	665	136.50		801.50	50 MBH.
	Reinstall	Ea.		80.79		80.79	
	Clean	Ea.	.78	32		32.78	
	Minimum Charge	Job		196		196	

Rough Mechanical

Heating	Unit	Material	Labor	Equip.	Total	Specification
75 MBH						
Demolish	Ea.		35.50		35.50	Includes material and labor to install a
Install	Ea.	765	118		883	cabinet space heater with an output of
Demolish and Install	Ea.	765	153.50		918.50	75 MBH.
Reinstall	Ea.		94.25		94.25	
Clean	Ea.	.78	32		32.78	
Minimum Charge	Job		196		196	
125 MBH						
Demolish	Ea.		42.50		42.50	Includes material and labor to install a
Install	Ea.	990	141		1131	cabinet space heater with an output of
Demolish and Install	Ea.	990	183.50		1173.50	125 MBH.
Reinstall	Ea.		113.10		113.10	
Clean	Ea.	.78	32		32.78	
Minimum Charge	Job		196		196	
175 MBH						
Demolish	Ea.		53		53	Includes material and labor to install a
Install	Ea.	1200	202		1402	cabinet space heater with an output of
Demolish and Install	Ea.	1200	255		1455	175 MBH.
Reinstall	Ea.		161.57		161.57	
Clean	Ea.	.78	32		32.78	
Minimum Charge	Job		196		196	

Furnace Unit	Unit	Material	Labor	Equip.	Total	Specification
Wall (gas-fired)						
25 MBH w / Thermostat						
Demolish	Ea.		87		87	Includes material and labor to install a
Install	Ea.	445	139		584	gas fired hot air furnace rated at 25
Demolish and Install	Ea.	445	226		671	MBH.
Reinstall	Ea.		111.44		111.44	
Clean	Ea.	1.56	70.50		72.06	
Minimum Charge	Job		196		196	
35 MBH w / Thermostat						
Demolish	Ea.		87		87	Includes material and labor to install a
Install	Ea.	625	155		780	gas fired hot air furnace rated at 35
Demolish and Install	Ea.	625	242		867	MBH.
Reinstall	Ea.		123.82		123.82	
Clean	Ea.	1.56	70.50		72.06	
Minimum Charge	Job		196		196	
50 MBH w / Thermostat						
Demolish	Ea.		87		87	Includes material and labor to install a
Install	Ea.	875	174		1049	gas fired hot air furnace rated at 50
Demolish and Install	Ea.	875	261		1136	MBH.
Reinstall	Ea.		139.30		139.30	
Clean	Ea.	1.56	70.50		72.06	
Minimum Charge	Job		196		196	
60 MBH w / Thermostat						
Demolish	Ea.		87		87	Includes material and labor to install a
Install	Ea.	965	193		1158	gas fired hot air furnace rated at 60
Demolish and Install	Ea.	965	280		1245	MBH.
Reinstall	Ea.		154.77		154.77	
Clean	Ea.	1.56	70.50		72.06	
Minimum Charge	Job		196		196	

Rough Mechanical

Furnace Unit	Unit	Material	Labor	Equip.	Total	Specification
Floor (gas-fired)						
30 MBH w / Thermostat						
Demolish	Ea.		87		87	Includes material and labor to install a
Install	Ea.	700	139		839	gas fired hot air floor furnace rated at
Demolish and Install	Ea.	700	226		926	30 MBH.
Reinstall	Ea.		111.44		111.44	
Clean	Ea.	1.56	70.50		72.06	
Minimum Charge	Job		196		196	
50 MBH w / Thermostat						
Demolish	Ea.		87		87	Includes material and labor to install a
Install	Ea.	760	174		934	gas fired hot air floor furnace rated at
Demolish and Install	Ea.	760	261		1021	50 MBH.
Reinstall	Ea.		139.30		139.30	
Clean	Ea.	1.56	70.50		72.06	
Minimum Charge	Job		196		196	
Gas Forced Air						
50 MBH						
Demolish	Ea.		87		87	Includes material and labor to install
Install	Ea.	615	174		789	an up flow gas fired hot air furnace
Demolish and Install	Ea.	615	261		876	rated at 50 MBH.
Reinstall	Ea.		139.30		139.30	
Clean	Ea.	1.56	70.50		72.06	
Minimum Charge	Job		196		196	
80 MBH						
Demolish	Ea.		87		87	Includes material and labor to install
Install	Ea.	675	193		868	an up flow gas fired hot air furnace
Demolish and Install	Ea.	675	280		955	rated at 80 MBH.
Reinstall	Ea.		154.77		154.77	
Clean	Ea.	1.56	70.50		72.06	
Minimum Charge	Job		196		196	
100 MBH						
Demolish	Ea.		87		87	Includes material and labor to install
Install	Ea.	730	218		948	an up flow gas fired hot air furnace
Demolish and Install	Ea.	730	305		1035	rated at 100 MBH.
Reinstall	Ea.		174.12		174.12	
Clean	Ea.	1.56	70.50		72.06	
Minimum Charge	Job		196		196	
120 MBH						
Demolish	Ea.		87		87	Includes material and labor to install
Install	Ea.	815	232		1047	an up flow gas fired hot air furnace
Demolish and Install	Ea.	815	319		1134	rated at 120 MBH.
Reinstall	Ea.		185.73		185.73	
Clean	Ea.	1.56	70.50		72.06	
Minimum Charge	Job		196		196	
Electric Forced Air						
31.4 MBH						
Demolish	Ea.		87		87	Includes material and labor to install
Install	Ea.	1100	197		1297	an electric hot air furnace rated at
Demolish and Install	Ea.	1100	284		1384	31.4 MBH.
Reinstall	Ea.		157.87		157.87	
Clean	Ea.	1.56	70.50		72.06	
Minimum Charge	Job		196		196	

Furnace Unit		Unit	Material	Labor	Equip.	Total	Specification
47.1 MBH							
	Demolish	Ea.		87		87	Includes material and labor to install
	Install	Ea.	1300	207		1507	an electric hot air furnace rated at
	Demolish and Install	Ea.	1300	294		1594	47.1 MBH.
	Reinstall	Ea.		165.21		165.21	
	Clean	Ea.	1.56	70.50		72.06	
	Minimum Charge	Job		196		196	
62.9 MBH							
	Demolish	Ea.		87		87	Includes material and labor to install
	Install	Ea.	1550	222		1772	an electric hot air furnace rated at
	Demolish and Install	Ea.	1550	309		1859	62.9 MBH.
	Reinstall	Ea.		177.60		177.60	
	Clean	Ea.	1.56	70.50		72.06	
	Minimum Charge	Job		196		196	
78.5 MBH							
	Demolish	Ea.		87		87	Includes material and labor to install
	Install	Ea.	1725	228		1953	an electric hot air furnace rated at
	Demolish and Install	Ea.	1725	315		2040	78.5 MBH.
	Reinstall	Ea.		182.15		182.15	
	Clean	Ea.	1.56	70.50		72.06	
	Minimum Charge	Job		196		196	
110 MBH							
	Demolish	Ea.		87		87	Includes material and labor to install
	Install	Ea.	2375	240		2615	an electric hot air furnace rated at
	Demolish and Install	Ea.	2375	327		2702	110 MBH.
	Reinstall	Ea.		192		192	
	Clean	Ea.	1.56	70.50		72.06	
	Minimum Charge	Job		196		196	
A/C Junction Box							
	Install	Ea.	61	96		157	Includes labor and materials to install
	Minimum Charge	Job		196		196	a heating system emergency shut-off
							switch. Cost includes up to 40 feet of
							non-metallic cable.

Boiler		Unit	Material	Labor	Equip.	Total	Specification
Gas-Fired							
60 MBH							
	Demolish	Ea.		715		715	Includes material and labor to install a
	Install	Ea.	2125	1275		3400	gas fired cast iron boiler rated at 60
	Demolish and Install	Ea.	2125	1990		4115	MBH.
	Reinstall	Ea.		1028.19		1028.19	
	Clean	Ea.	3.11	42.50		45.61	
	Minimum Charge	Job		196		196	
Gas-Fired							
125 MBH							
	Demolish	Ea.		1975		1975	Includes material and labor to install a
	Install	Ea.	3050	1575		4625	gas fired cast iron boiler rated at 125
	Demolish and Install	Ea.	3050	3550		6600	MBH.
	Reinstall	Ea.		1256.68		1256.68	
	Clean	Ea.	3.11	42.50		45.61	
	Minimum Charge	Job		196		196	

Rough Mechanical

Boiler		Unit	Material	Labor	Equip.	Total	Specification
Gas-Fired							
400 MBH							
	Demolish	Ea.		1975		1975	Includes material and labor to install a
	Install	Ea.	4300	2200		6500	gas fired cast iron boiler rated at 400
	Demolish and Install	Ea.	4300	4175		8475	MBH.
	Reinstall	Ea.		2209		2209	
	Clean	Ea.	3.11	42.50		45.61	
	Minimum Charge	Job		196		196	

Heat Pump		Unit	Material	Labor	Equip.	Total	Specification
2 Ton							
	Demolish	Ea.		92.50		92.50	Includes material and labor to install a
	Install	Ea.	1750	590		2340	2 ton heat pump.
	Demolish and Install	Ea.	1750	682.50		2432.50	
	Reinstall	Ea.		471.25		471.25	
	Clean	Ea.	3.11	32		35.11	
	Minimum Charge	Job		196		196	
3 Ton							
	Demolish	Ea.		175		175	Includes material and labor to install a
	Install	Ea.	2200	885		3085	3 ton heat pump.
	Demolish and Install	Ea.	2200	1060		3260	
	Reinstall	Ea.		706.88		706.88	
	Clean	Ea.	3.11	32		35.11	
	Minimum Charge	Job		196		196	
4 Ton							
	Demolish	Ea.		320		320	Includes material and labor to install a
	Install	Ea.	2525	1175		3700	4 ton heat pump.
	Demolish and Install	Ea.	2525	1495		4020	
	Reinstall	Ea.		942.51		942.51	
	Clean	Ea.	3.11	32		35.11	
	Minimum Charge	Job		196		196	
5 Ton							
	Demolish	Ea.		465		465	Includes material and labor to install a
	Install	Ea.	2750	1425		4175	5 ton heat pump.
	Demolish and Install	Ea.	2750	1890		4640	
	Reinstall	Ea.		1131.01		1131.01	
	Clean	Ea.	3.11	32		35.11	
	Minimum Charge	Job		196		196	
Thermostat							
	Demolish	Ea.		10.90		10.90	Includes material and labor to install
	Install	Ea.	230	48.50		278.50	an electric thermostat, 2 set back.
	Demolish and Install	Ea.	230	59.40		289.40	
	Reinstall	Ea.		38.68		38.68	
	Clean	Ea.	.07	7.95		8.02	
	Minimum Charge	Job		196		196	
A/C Junction Box							
	Install	Ea.	670	295		965	Includes labor and materials to install
	Minimum Charge	Job		192		192	a heat pump circuit including up to 40
							feet of non-metallic cable.

Heating	Unit	Material	Labor	Equip.	Total	Specification
Electric Baseboard						
Demolish	L.F.		5.30		5.30	Includes material and labor to install
Install	L.F.	12.80	12.75		25.55	electric baseboard heating elements.
Demolish and Install	L.F.	12.80	18.05		30.85	
Reinstall	L.F.		10.22		10.22	
Clean	L.F.	1.09	3.97		5.06	
Minimum Charge	Job		192		192	
2' 6"						
Demolish	Ea.		17.75		17.75	Includes material and labor to install a
Install	Ea.	38.50	48		86.50	2' 6" long section of electric
Demolish and Install	Ea.	38.50	65.75		104.25	baseboard heating elements.
Reinstall	Ea.		38.32		38.32	
Clean	Ea.	.31	9.75		10.06	
Minimum Charge	Job		192		192	
3'						
Demolish	Ea.		19.35		19.35	Includes material and labor to install a
Install	Ea.	43	48		91	3' long section of electric baseboard
Demolish and Install	Ea.	43	67.35		110.35	heating elements.
Reinstall	Ea.		38.32		38.32	
Clean	Ea.	.31	12.10		12.41	
Minimum Charge	Job		192		192	
4'						
Demolish	Ea.		21.50		21.50	Includes material and labor to install a
Install	Ea.	54	57		111	4' long section of electric baseboard
Demolish and Install	Ea.	54	78.50		132.50	heating elements.
Reinstall	Ea.		45.76		45.76	
Clean	Ea.	.12	2.65		2.77	
Minimum Charge	Job		192		192	
5'						
Demolish	Ea.		23.50		23.50	Includes material and labor to install a
Install	Ea.	64	67		131	5' long section of electric baseboard
Demolish and Install	Ea.	64	90.50		154.50	heating elements.
Reinstall	Ea.		53.78		53.78	
Clean	Ea.	.63	19.55		20.18	
Minimum Charge	Job		192		192	
6'						
Demolish	Ea.		26.50		26.50	Includes material and labor to install a
Install	Ea.	71	76.50		147.50	6' long section of electric baseboard
Demolish and Install	Ea.	71	103		174	heating elements.
Reinstall	Ea.		61.31		61.31	
Clean	Ea.	.12	3.18		3.30	
Minimum Charge	Job		192		192	
8'						
Demolish	Ea.		35.50		35.50	Includes material and labor to install
Install	Ea.	89.50	96		185.50	an 8' long section of electric
Demolish and Install	Ea.	89.50	131.50		221	baseboard heating elements.
Reinstall	Ea.		76.64		76.64	
Clean	Ea.	.12	3.97		4.09	
Minimum Charge	Job		192		192	
10'						
Demolish	Ea.		42.50		42.50	Includes material and labor to install a
Install	Ea.	111	116		227	10' long section of electric baseboard
Demolish and Install	Ea.	111	158.50		269.50	heating elements.
Reinstall	Ea.		92.90		92.90	
Clean	Ea.	1.24	39.50		40.74	
Minimum Charge	Job		192		192	

Rough Mechanical

Heating	Unit	Material	Labor	Equip.	Total	Specification
Electric Radiant Wall Panel						
Demolish	Ea.		28.50		28.50	Includes material and labor to install
Install	Ea.	405	76.50		481.50	an 18″ x 23″ electric radiant heat wall
Demolish and Install	Ea.	405	105		510	panel.
Reinstall	Ea.		61.31		61.31	
Clean	Ea.	.15	7.95		8.10	
Minimum Charge	Job		192		192	

Heating Control	Unit	Material	Labor	Equip.	Total	Specification
Thermostat						
Demolish	Ea.		10.90		10.90	Includes material and labor to install
Install	Ea.	230	48.50		278.50	an electric thermostat, 2 set back.
Demolish and Install	Ea.	230	59.40		289.40	
Reinstall	Ea.		38.68		38.68	
Clean	Ea.	.07	7.95		8.02	
Minimum Charge	Job		196		196	

Air Conditioning	Unit	Material	Labor	Equip.	Total	Specification
2 Ton						
Demolish	Ea.		360		360	Includes material and labor to install a
Install	Ea.	1800	610		2410	rooftop air conditioning unit, 2 ton.
Demolish and Install	Ea.	1800	970		2770	No hoisting is included.
Reinstall	Ea.		609.91		609.91	
Clean	Ea.	3.11	32		35.11	
Minimum Charge	Job		196		196	
3 Ton						
Demolish	Ea.		360		360	Includes material and labor to install a
Install	Ea.	1875	805		2680	rooftop air conditioning unit, 3 ton.
Demolish and Install	Ea.	1875	1165		3040	No hoisting is included.
Reinstall	Ea.		804.19		804.19	
Clean	Ea.	3.11	32		35.11	
Minimum Charge	Job		196		196	
4 Ton						
Demolish	Ea.		360		360	Includes material and labor to install a
Install	Ea.	2225	935		3160	rooftop air conditioning unit, 4 ton.
Demolish and Install	Ea.	2225	1295		3520	No hoisting is included.
Reinstall	Ea.		932.56		932.56	
Clean	Ea.	3.11	32		35.11	
Minimum Charge	Job		196		196	
5 Ton						
Demolish	Ea.		360		360	Includes material and labor to install a
Install	Ea.	2375	1000		3375	rooftop air conditioning unit, 5 ton.
Demolish and Install	Ea.	2375	1360		3735	No hoisting is included.
Reinstall	Ea.		1006.95		1006.95	
Clean	Ea.	3.11	32		35.11	
Minimum Charge	Job		196		196	
Compressor						
2 Ton						
Demolish	Ea.		325		325	Includes material and labor to install a
Install	Ea.	268	196		464	refrigeration compressor.
Demolish and Install	Ea.	268	521		789	
Reinstall	Ea.		156.96		156.96	
Clean	Ea.	3.11	32		35.11	
Minimum Charge	Job		196		196	

Rough Mechanical

Air Conditioning	Unit	Material	Labor	Equip.	Total	Specification
2.5 Ton						
Demolish	Ea.		325		325	Includes material and labor to install a
Install	Ea.	325	213		538	refrigeration compressor.
Demolish and Install	Ea.	325	538		863	
Reinstall	Ea.		170.05		170.05	
Clean	Ea.	3.11	32		35.11	
Minimum Charge	Job		196		196	
3 Ton						
Demolish	Ea.		325		325	Includes material and labor to install a
Install	Ea.	360	222		582	refrigeration compressor.
Demolish and Install	Ea.	360	547		907	
Reinstall	Ea.		177.46		177.46	
Clean	Ea.	3.11	32		35.11	
Minimum Charge	Job		196		196	
4 Ton						
Demolish	Ea.		325		325	Includes material and labor to install a
Install	Ea.	455	178		633	refrigeration compressor.
Demolish and Install	Ea.	455	503		958	
Reinstall	Ea.		142.69		142.69	
Clean	Ea.	3.11	32		35.11	
Minimum Charge	Job		196		196	
5 Ton						
Demolish	Ea.		325		325	Includes material and labor to install a
Install	Ea.	500	243		743	refrigeration compressor.
Demolish and Install	Ea.	500	568		1068	
Reinstall	Ea.		194.38		194.38	
Clean	Ea.	3.11	32		35.11	
Minimum Charge	Job		196		196	

Condensing Unit

	Unit	Material	Labor	Equip.	Total	Specification
2 Ton Air-cooled						
Demolish	Ea.		93.50		93.50	Includes material and labor to install a
Install	Ea.	915	335		1250	2 ton condensing unit.
Demolish and Install	Ea.	915	428.50		1343.50	
Reinstall	Ea.		269.29		269.29	
Clean	Ea.	3.11	32		35.11	
Minimum Charge	Job		196		196	
2.5 Ton Air-cooled						
Demolish	Ea.		109		109	Includes material and labor to install a
Install	Ea.	950	415		1365	2.5 ton condensing unit.
Demolish and Install	Ea.	950	524		1474	
Reinstall	Ea.		332.65		332.65	
Clean	Ea.	3.11	32		35.11	
Minimum Charge	Job		196		196	
3 Ton Air-cooled						
Demolish	Ea.		131		131	Includes material and labor to install a
Install	Ea.	1100	545		1645	3 ton condensing unit.
Demolish and Install	Ea.	1100	676		1776	
Reinstall	Ea.		435		435	
Clean	Ea.	3.11	32		35.11	
Minimum Charge	Job		196		196	
4 Ton Air-cooled						
Demolish	Ea.		196		196	Includes material and labor to install a
Install	Ea.	1425	785		2210	4 ton condensing unit.
Demolish and Install	Ea.	1425	981		2406	
Reinstall	Ea.		628.34		628.34	
Clean	Ea.	3.11	32		35.11	
Minimum Charge	Job		196		196	

Rough Mechanical

Air Conditioning		Unit	Material	Labor	Equip.	Total	Specification
5 Ton Air-cooled							
	Demolish	Ea.		295		295	Includes material and labor to install a
	Install	Ea.	1725	1175		2900	5 ton condensing unit.
	Demolish and Install	Ea.	1725	1470		3195	
	Reinstall	Ea.		942.51		942.51	
	Clean	Ea.	3.11	32		35.11	
	Minimum Charge	Job		196		196	
Air Handler							
1,200 CFM							
	Demolish	Ea.		236		236	Includes material and labor to install a
	Install	Ea.	2300	545		2845	1200 CFM capacity air handling unit.
	Demolish and Install	Ea.	2300	781		3081	
	Reinstall	Ea.		435		435	
	Clean	Ea.	3.11	32		35.11	
	Minimum Charge	Job		196		196	
2,000 CFM							
	Demolish	Ea.		236		236	Includes material and labor to install a
	Install	Ea.	3850	645		4495	2000 CFM capacity air handling unit.
	Demolish and Install	Ea.	3850	881		4731	
	Reinstall	Ea.		514.09		514.09	
	Clean	Ea.	3.11	32		35.11	
	Minimum Charge	Job		196		196	
3,000 CFM							
	Demolish	Ea.		262		262	Includes material and labor to install a
	Install	Ea.	4825	705		5530	3000 CFM capacity air handling unit.
	Demolish and Install	Ea.	4825	967		5792	
	Reinstall	Ea.		565.50		565.50	
	Clean	Ea.	3.11	32		35.11	
	Minimum Charge	Job		196		196	
Evaporative Cooler							
4,300 CFM, 1/3 HP							
	Demolish	Ea.		96		96	Includes material and labor to install
	Install	Ea.	760	218		978	an evaporative cooler.
	Demolish and Install	Ea.	760	314		1074	
	Reinstall	Ea.		174.12		174.12	
	Clean	Ea.	1.24	32		33.24	
	Minimum Charge	Job		196		196	
4,700 CFM, 1/2 HP							
	Demolish	Ea.		96		96	Includes material and labor to install
	Install	Ea.	710	225		935	an evaporative cooler.
	Demolish and Install	Ea.	710	321		1031	
	Reinstall	Ea.		179.74		179.74	
	Clean	Ea.	1.24	32		33.24	
	Minimum Charge	Job		196		196	
5,600 CFM, 3/4 HP							
	Demolish	Ea.		96		96	Includes material and labor to install
	Install	Ea.	1250	232		1482	an evaporative cooler.
	Demolish and Install	Ea.	1250	328		1578	
	Reinstall	Ea.		185.73		185.73	
	Clean	Ea.	1.24	32		33.24	
	Minimum Charge	Job		196		196	
Master Cool Brand							
	Demolish	Ea.		96		96	Includes material and labor to install
	Install	Ea.	760	240		1000	an evaporative cooler.
	Demolish and Install	Ea.	760	336		1096	
	Reinstall	Ea.		192.13		192.13	
	Clean	Ea.	1.24	32		33.24	
	Minimum Charge	Job		196		196	

Rough Mechanical

Air Conditioning	Unit	Material	Labor	Equip.	Total	Specification
Motor						
Demolish	Ea.		54		54	Includes material and labor to install
Install	Ea.	257	169		426	an evaporative cooler motor.
Demolish and Install	Ea.	257	223		480	
Reinstall	Ea.		135.36		135.36	
Clean	Ea.	1.24	32		33.24	
Minimum Charge	Job		196		196	
Water Pump						
Demolish	Ea.		54		54	Includes material and labor to install
Install	Ea.	48.50	96.50		145	an evaporative cooler water pump.
Demolish and Install	Ea.	48.50	150.50		199	
Reinstall	Ea.		77.36		77.36	
Clean	Ea.	1.24	32		33.24	
Minimum Charge	Job		196		196	
Fan Belt						
Demolish	Ea.		10.80		10.80	Includes material and labor to install
Install	Ea.	32.50	19.35		51.85	an evaporative cooler fan belt.
Demolish and Install	Ea.	32.50	30.15		62.65	
Reinstall	Ea.		15.47		15.47	
Minimum Charge	Job		196		196	
Pads (4 ea.)						
Demolish	Ea.		10.80		10.80	Includes material and labor to install
Install	Ea.	38	48.50		86.50	pre-fabricated roof pads for an
Demolish and Install	Ea.	38	59.30		97.30	evaporative cooler.
Reinstall	Ea.		38.68		38.68	
Clean	Ea.	1.24	32		33.24	
Minimum Charge	Job		196		196	
Service call						
Install	Job		193		193	Includes 4 hour minimum labor charge for an sheetmetal worker.
Window Mounted						
Low Output BTU						
Install	Ea.	208	43.50		251.50	Includes material and labor to install a
Reinstall	Ea.		34.88		34.88	window air conditioning unit.
Clean	Ea.	.46	7.95		8.41	
Minimum Charge	Job		196		196	
Mid-range Output BTU						
Demolish	Ea.		30.50		30.50	Includes material and labor to install a
Install	Ea.	330	58		388	window air conditioning unit.
Demolish and Install	Ea.	330	88.50		418.50	
Reinstall	Ea.		46.51		46.51	
Clean	Ea.	.46	7.95		8.41	
Minimum Charge	Job		196		196	
High Output BTU						
Demolish	Ea.		39		39	Includes material and labor to install a
Install	Ea.	385	58		443	window air conditioning unit.
Demolish and Install	Ea.	385	97		482	
Reinstall	Ea.		46.51		46.51	
Clean	Ea.	.46	7.95		8.41	
Minimum Charge	Job		196		196	
Hotel-type Hot / Cold						
Demolish	Ea.		64.50		64.50	Includes material and labor to install
Install	Ea.	1525	236		1761	packaged air conditioning unit,
Demolish and Install	Ea.	1525	300.50		1825.50	15,000 BTU cooling, 13,900 BTU
Reinstall	Ea.		188.50		188.50	heating.
Clean	Ea.	.63	15.90		16.53	
Minimum Charge	Job		196		196	

Rough Mechanical

Air Conditioning

Air Conditioning	Unit	Material	Labor	Equip.	Total	Specification
Thermostat						
Demolish	Ea.		10.90		10.90	Includes material and labor to install
Install	Ea.	230	48.50		278.50	an electric thermostat, 2 set back.
Demolish and Install	Ea.	230	59.40		289.40	
Reinstall	Ea.		38.68		38.68	
Clean	Ea.	.07	7.95		8.02	
Minimum Charge	Job		196		196	
Concrete Pad						
Demolish	S.F.		1.76		1.76	Includes material and labor to install
Install	S.F.	2.57	4.01		6.58	air conditioner pad.
Demolish and Install	S.F.	2.57	5.77		8.34	
Reinstall	S.F.		3.20		3.20	
Clean	S.F.	.07	.51		.58	
Paint	S.F.	.14	.31		.45	
Minimum Charge	Job		196		196	
A/C Junction Box						
Install	Ea.	231	109		340	Includes labor and materials to install
Minimum Charge	Job		192		192	an air conditioning circuit including up to 40 feet of non-metallic cable.

Sprinkler System

Sprinkler System	Unit	Material	Labor	Equip.	Total	Specification
Exposed						
Wet Type						
Demolish	SF Flr.		.17		.17	Includes material and labor to install
Install	SF Flr.	1.27	1.69		2.96	an exposed wet pipe fire sprinkler
Demolish and Install	SF Flr.	1.27	1.86		3.13	system.
Reinstall	SF Flr.		1.69		1.69	
Minimum Charge	Job		194		194	
Dry Type						
Demolish	SF Flr.		.17		.17	Includes material and labor to install
Install	SF Flr.	1.45	1.69		3.14	an exposed dry pipe fire sprinkler
Demolish and Install	SF Flr.	1.45	1.86		3.31	system.
Reinstall	SF Flr.		1.69		1.69	
Minimum Charge	Job		194		194	
Concealed						
Wet Type						
Demolish	SF Flr.		.35		.35	Includes material and labor to install a
Install	SF Flr.	1.27	2.31		3.58	concealed wet pipe fire sprinkler
Demolish and Install	SF Flr.	1.27	2.66		3.93	system.
Reinstall	SF Flr.		2.31		2.31	
Minimum Charge	Job		194		194	
Dry Type						
Demolish	SF Flr.		.35		.35	Includes material and labor to install a
Install	SF Flr.	.01	2.31		2.32	concealed dry pipe fire sprinkler
Demolish and Install	SF Flr.	.01	2.66		2.67	system.
Reinstall	SF Flr.		2.31		2.31	
Minimum Charge	Job		194		194	

Junction Boxes	Cable	Switch Box

Electrical Per S.F.		Unit	Material	Labor	Equip.	Total	Specification
Residential							
Wiring, Lighting and Fixtures							
	Install	S.F.	1.64	1.60		3.24	Includes labor and material to replace house wiring, lighting fixtures, and built-in appliances. Apply S.F. cost to floor area including garages but not basements.
	Minimum Charge	Job		192		192	
Wiring Only							
	Install	S.F.	.32	.63		.95	Includes labor and material to replace house wiring only. Apply S.F. cost to floor area including garages but not basements.

Main Service		Unit	Material	Labor	Equip.	Total	Specification
100 Amp.							
	Demolish	Ea.		158		158	Cost includes material and labor to install 100 Amp. residential service panel including 12 breaker interior panel, 10 single pole breakers and hook-up. Wiring is not included.
	Install	Ea.	565	320		885	
	Demolish and Install	Ea.	565	478		1043	
	Reinstall	Ea.		257.61		257.61	
	Clean	Ea.	.31	63.50		63.81	
	Minimum Charge	Job		192		192	
150 Amp.							
	Demolish	Ea.		158		158	Cost includes material and labor to install 150 Amp. residential service, incl. 20 breaker exterior panel, main switch, 1 GFI and 14 single pole breakers, meter socket including hookup. Wiring is not included.
	Install	Ea.	880	370		1250	
	Demolish and Install	Ea.	880	528		1408	
	Reinstall	Ea.		297.63		297.63	
	Clean	Ea.	.31	63.50		63.81	
	Minimum Charge	Job		192		192	
200 Amp.							
	Demolish	Ea.		158		158	Cost includes material and labor to install 200 Amp. residential service, incl. 20 breaker exterior panel, main switch, 1 GFI and 18 single pole breakers, meter socket including hookup. Wiring is not included.
	Install	Ea.	1150	425		1575	
	Demolish and Install	Ea.	1150	583		1733	
	Reinstall	Ea.		340.62		340.62	
	Clean	Ea.	.31	63.50		63.81	
	Minimum Charge	Job		192		192	

Rough Electrical

Main Service	Unit	Material	Labor	Equip.	Total	Specification
Masthead						
Demolish	Ea.		26.50		26.50	Includes material and labor to install
Install	Ea.	117	160		277	galvanized steel conduit including
Demolish and Install	Ea.	117	186.50		303.50	masthead.
Reinstall	Ea.		127.73		127.73	
Minimum Charge	Job		192		192	
Allowance to Repair						
Install	Job		192		192	Minimum labor charge for residential
Minimum Charge	Job		192		192	wiring.

Insulation

Wall Insulation

Floor Insulation		Unit	Material	Labor	Equip.	Total	Specification
Kraft Faced Batt							
3-1/2″ (R-11)							
	Demolish	S.F.		.12		.12	Cost includes material and labor to
	Install	S.F.	.50	.50		1	install 3-1/2″ (R-11) kraft paper or foil
	Demolish and Install	S.F.	.50	.62		1.12	faced roll and batt fiberglass
	Minimum Charge	Job		174		174	insulation, joists 16″ O.C.
6″ (R-19)							
	Demolish	S.F.		.12		.12	Cost includes material and labor to
	Install	S.F.	.56	.58		1.14	install 6″ (R-19) kraft paper or foil
	Demolish and Install	S.F.	.56	.70		1.26	faced roll and batt fiberglass
	Minimum Charge	Job		174		174	insulation, joists 16″ O.C.
10″ (R-30)							
	Demolish	S.F.		.12		.12	Cost includes material and labor to
	Install	S.F.	.86	.70		1.56	install 10″ (R-30) kraft paper-faced roll
	Demolish and Install	S.F.	.86	.82		1.68	and batt fiberglass insulation, joists
	Minimum Charge	Job		174		174	16″ O.C.
12″ (R-38)							
	Demolish	S.F.		.12		.12	Cost includes material and labor to
	Install	S.F.	1.66	.73		2.39	install 12″ (R-38) kraft paper-faced roll
	Demolish and Install	S.F.	1.66	.85		2.51	and batt fiberglass insulation, joists
	Minimum Charge	Job		174		174	16″ O.C.
Foil Faced Batt							
3-1/2″ (R-11)							
	Demolish	S.F.		.12		.12	Cost includes material and labor to
	Install	S.F.	.50	.50		1	install 3-1/2″ (R-11) kraft paper or foil
	Demolish and Install	S.F.	.50	.62		1.12	faced roll and batt fiberglass
	Minimum Charge	Job		174		174	insulation, joists 16″ O.C.
6″ (R-19)							
	Demolish	S.F.		.12		.12	Cost includes material and labor to
	Install	S.F.	.56	.58		1.14	install 6″ (R-19) kraft paper or foil
	Demolish and Install	S.F.	.56	.70		1.26	faced roll and batt fiberglass
	Minimum Charge	Job		174		174	insulation, joists 16″ O.C.
Unfaced Batt							
3-1/2″ (R-11)							
	Demolish	S.F.		.12		.12	Cost includes material and labor to
	Install	S.F.	.40	.58		.98	install 3-1/2″ (R-11) unfaced roll and
	Demolish and Install	S.F.	.40	.70		1.10	batt fiberglass insulation, joists 16″
	Minimum Charge	Job		174		174	O.C.

Insulation

Floor Insulation

Floor Insulation		Unit	Material	Labor	Equip.	Total	Specification
6″ (R-19)							
	Demolish	S.F.		.12		.12	Cost includes material and labor to
	Install	S.F.	.63	.70		1.33	install 6″ (R-19) unfaced roll and batt
	Demolish and Install	S.F.	.63	.82		1.45	fiberglass insulation, joists 16″ O.C.
	Minimum Charge	Job		174		174	
10″ (R-30)							
	Demolish	S.F.		.12		.12	Cost includes material and labor to
	Install	S.F.	.86	.78		1.64	install 10″ (R-30) unfaced roll and batt
	Demolish and Install	S.F.	.86	.90		1.76	fiberglass insulation, joists 16″ O.C.
	Minimum Charge	Job		174		174	
12″ (R-38)							
	Demolish	S.F.		.12		.12	Cost includes material and labor to
	Install	S.F.	.98	.82		1.80	install 12″ (R-38) unfaced roll and batt
	Demolish and Install	S.F.	.98	.94		1.92	fiberglass insulation, joists 16″ O.C.
	Minimum Charge	Job		174		174	
Rigid Board							
1/2″							
	Demolish	S.F.		.12		.12	Includes material and labor to install
	Install	S.F.	.33	.44		.77	rigid foam insulation board foil faced
	Demolish and Install	S.F.	.33	.56		.89	two sides.
	Minimum Charge	Job		174		174	
3/4″							
	Demolish	S.F.		.12		.12	Includes material and labor to install
	Install	S.F.	.39	.44		.83	rigid foam insulation board foil faced
	Demolish and Install	S.F.	.39	.56		.95	two sides.
	Minimum Charge	Job		174		174	
1-1/2″							
	Demolish	S.F.		.12		.12	Includes material and labor to install
	Install	S.F.	.70	.47		1.17	rigid foam insulation board foil faced
	Demolish and Install	S.F.	.70	.59		1.29	two sides.
	Minimum Charge	Job		174		174	
Support Wire							
	Demolish	S.F.		.05		.05	Includes material and labor to install
	Install	S.F.	.02	.10		.12	friction fit support wire.
	Demolish and Install	S.F.	.02	.15		.17	
	Minimum Charge	Job		174		174	
Vapor Barrier							
	Demolish	S.F.		.06		.06	Cost includes material and labor to
	Install	S.F.	.03	.09		.12	install 4 mil polyethylene vapor barrier
	Demolish and Install	S.F.	.03	.15		.18	10′ wide sheets with 6″ overlaps.
	Minimum Charge	Job		174		174	

Wall Insulation

Wall Insulation		Unit	Material	Labor	Equip.	Total	Specification
Kraft Faced Batt							
3-1/2″ (R-11)							
	Demolish	S.F.		.11		.11	Cost includes material and labor to
	Install	S.F.	.41	.22		.63	install 3-1/2″ (R-11) kraft paper-faced
	Demolish and Install	S.F.	.41	.33		.74	roll and batt fiberglass insulation, joist
	Minimum Charge	Job		174		174	or studs 16″ O.C.
6″ (R-19)							
	Demolish	S.F.		.11		.11	Cost includes material and labor to
	Install	S.F.	.47	.26		.73	install 6″ (R-19) kraft paper-faced roll
	Demolish and Install	S.F.	.47	.37		.84	and batt fiberglass insulation, joist or
	Minimum Charge	Job		174		174	studs 16″ O.C.

Insulation

Wall Insulation		Unit	Material	Labor	Equip.	Total	Specification
10″ (R-30)							
	Demolish	S.F.		.11		.11	Cost includes material and labor to
	Install	S.F.	1.07	.30		1.37	install 10″ (R-30) kraft paper-faced roll
	Demolish and Install	S.F.	1.07	.41		1.48	and batt fiberglass insulation, joist or
	Minimum Charge	Job		174		174	studs 16″ O.C.
12″ (R-38)							
	Demolish	S.F.		.11		.11	Cost includes material and labor to
	Install	S.F.	.99	.35		1.34	install 12″ (R-38) kraft paper-faced roll
	Demolish and Install	S.F.	.99	.46		1.45	and batt fiberglass insulation, joist or
	Minimum Charge	Job		174		174	studs 16″ O.C.
Foil Faced Batt							
3-1/2″ (R-11)							
	Demolish	S.F.		.11		.11	Cost includes material and labor to
	Install	S.F.	.59	.22		.81	install 3-1/2″ (R-11) foil faced roll and
	Demolish and Install	S.F.	.59	.33		.92	batt fiberglass insulation, joist or studs
	Minimum Charge	Job		174		174	16″ O.C.
6″ (R-19)							
	Demolish	S.F.		.11		.11	Cost includes material and labor to
	Install	S.F.	.58	.26		.84	install 6″ (R-19) foil faced roll and batt
	Demolish and Install	S.F.	.58	.37		.95	fiberglass insulation, joist or studs 16″
	Minimum Charge	Job		174		174	O.C.
Unfaced Batt							
3-1/2″ (R-11)							
	Demolish	S.F.		.11		.11	Cost includes material and labor to
	Install	S.F.	.40	.26		.66	install 3-1/2″ (R-11) unfaced roll and
	Demolish and Install	S.F.	.40	.37		.77	batt fiberglass insulation, joist or studs
	Minimum Charge	Job		174		174	16″ O.C.
6″ (R-19)							
	Demolish	S.F.		.11		.11	Cost includes material and labor to
	Install	S.F.	.63	.30		.93	install 6″ (R-19) unfaced roll and batt
	Demolish and Install	S.F.	.63	.41		1.04	fiberglass insulation, joist or studs 16″
	Minimum Charge	Job		174		174	O.C.
10″ (R-30)							
	Demolish	S.F.		.11		.11	Cost includes material and labor to
	Install	S.F.	.91	.35		1.26	install 10″ (R-30) unfaced roll and batt
	Demolish and Install	S.F.	.91	.46		1.37	fiberglass insulation, joist or studs 16″
	Minimum Charge	Job		174		174	O.C.
12″ (R-38)							
	Demolish	S.F.		.11		.11	Cost includes material and labor to
	Install	S.F.	.98	.35		1.33	install 12″ (R-38) unfaced roll and batt
	Demolish and Install	S.F.	.98	.46		1.44	fiberglass insulation, joist or studs 16″
	Minimum Charge	Job		174		174	O.C.
Blown							
R-11							
	Demolish	S.F.		.11		.11	Includes material and labor to install
	Install	S.F.	.30	1.14	.42	1.86	fiberglass blown-in insulation in wall
	Demolish and Install	S.F.	.30	1.25	.42	1.97	with wood siding.
	Minimum Charge	Job		395	145	540	
R-19							
	Demolish	S.F.		.11		.11	Includes material and labor to install
	Install	S.F.	.46	1.32	.48	2.26	fiberglass blown-in insulation in wall
	Demolish and Install	S.F.	.46	1.43	.48	2.37	with wood siding.
	Minimum Charge	Job		395	145	540	

Insulation

Wall Insulation

Wall Insulation	Unit	Material	Labor	Equip.	Total	Specification
Rigid Board						
1/2″						
Demolish	S.F.		.14		.14	Includes material and labor to install
Install	S.F.	.33	.44		.77	rigid foam insulation board, foil faced,
Demolish and Install	S.F.	.33	.58		.91	two sides.
Minimum Charge	Job		174		174	
3/4″						
Demolish	S.F.		.14		.14	Includes material and labor to install
Install	S.F.	.39	.44		.83	rigid foam insulation board, foil faced,
Demolish and Install	S.F.	.39	.58		.97	two sides.
Minimum Charge	Job		174		174	
1-1/2″						
Demolish	S.F.		.14		.14	Includes material and labor to install
Install	S.F.	.70	.48		1.18	rigid foam insulation board, foil faced,
Demolish and Install	S.F.	.70	.62		1.32	two sides.
Minimum Charge	Job		174		174	
Vapor Barrier						
Demolish	S.F.		.04		.04	Cost includes material and labor to
Install	S.F.	.03	.09		.12	install 4 mil polyethylene vapor barrier
Demolish and Install	S.F.	.03	.13		.16	10′ wide sheets with 6″ overlaps.
Minimum Charge	Job		174		174	

Ceiling Insulation

Ceiling Insulation	Unit	Material	Labor	Equip.	Total	Specification
Kraft Faced Batt						
3-1/2″ (R-11)						
Demolish	S.F.		.12		.12	Cost includes material and labor to
Install	S.F.	.41	.22		.63	install 3-1/2″ (R-11) kraft paper-faced
Demolish and Install	S.F.	.41	.34		.75	roll and batt fiberglass insulation, joist
Minimum Charge	Job		174		174	or studs 16″ O.C.
6″ (R-19)						
Demolish	S.F.		.12		.12	Cost includes material and labor to
Install	S.F.	.47	.26		.73	install 6″ (R-19) kraft paper-faced roll
Demolish and Install	S.F.	.47	.38		.85	and batt fiberglass insulation, joist or
Minimum Charge	Job		174		174	studs 16″ O.C.
10″ (R-30)						
Demolish	S.F.		.12		.12	Cost includes material and labor to
Install	S.F.	1.07	.30		1.37	install 10″ (R-30) kraft paper-faced roll
Demolish and Install	S.F.	1.07	.42		1.49	and batt fiberglass insulation, joist or
Minimum Charge	Job		174		174	studs 16″ O.C.
12″ (R-38)						
Demolish	S.F.		.12		.12	Cost includes material and labor to
Install	S.F.	.99	.35		1.34	install 12″ (R-38) kraft paper-faced roll
Demolish and Install	S.F.	.99	.47		1.46	and batt fiberglass insulation, joist or
Minimum Charge	Job		174		174	studs 16″ O.C.
Foil Faced Batt						
3-1/2″ (R-11)						
Demolish	S.F.		.12		.12	Cost includes material and labor to
Install	S.F.	.59	.22		.81	install 3-1/2″ (R-11) foil faced roll and
Demolish and Install	S.F.	.59	.34		.93	batt fiberglass insulation, joist or studs
Minimum Charge	Job		174		174	16″ O.C.

Insulation

Ceiling Insulation

		Unit	Material	Labor	Equip.	Total	Specification
6″ (R-19)							
	Demolish	S.F.		.12		.12	Cost includes material and labor to
	Install	S.F.	.58	.26		.84	install 6″ (R-19) foil faced roll and batt
	Demolish and Install	S.F.	.58	.38		.96	fiberglass insulation, joist or studs 16″
	Minimum Charge	Job		174		174	O.C.

Unfaced Batt

		Unit	Material	Labor	Equip.	Total	Specification
3-1/2″ (R-11)							
	Demolish	S.F.		.12		.12	Cost includes material and labor to
	Install	S.F.	.40	.26		.66	install 3-1/2″ (R-11) unfaced roll and
	Demolish and Install	S.F.	.40	.38		.78	batt fiberglass insulation, joist or studs
	Minimum Charge	Job		174		174	16″ O.C.
6″ (R-19)							
	Demolish	S.F.		.12		.12	Cost includes material and labor to
	Install	S.F.	.63	.30		.93	install 6″ (R-19) unfaced roll and batt
	Demolish and Install	S.F.	.63	.42		1.05	fiberglass insulation, joist or studs 16″
	Minimum Charge	Job		174		174	O.C.
10″ (R-30)							
	Demolish	S.F.		.12		.12	Cost includes material and labor to
	Install	S.F.	.91	.35		1.26	install 10″ (R-30) unfaced roll and batt
	Demolish and Install	S.F.	.91	.47		1.38	fiberglass insulation, joist or studs 16″
	Minimum Charge	Job		174		174	O.C.
12″ (R-38)							
	Demolish	S.F.		.12		.12	Cost includes material and labor to
	Install	S.F.	.98	.35		1.33	install 12″ (R-38) unfaced roll and batt
	Demolish and Install	S.F.	.98	.47		1.45	fiberglass insulation, joist or studs 16″
	Minimum Charge	Job		174		174	O.C.

Blown

		Unit	Material	Labor	Equip.	Total	Specification
5″							
	Demolish	S.F.		.35		.35	Includes material and labor to install
	Install	S.F.	.23	.21	.08	.52	fiberglass blown-in insulation.
	Demolish and Install	S.F.	.23	.56	.08	.87	
	Minimum Charge	Job		395	145	540	
6″							
	Demolish	S.F.		.35		.35	Includes material and labor to install
	Install	S.F.	.28	.26	.10	.64	fiberglass blown-in insulation.
	Demolish and Install	S.F.	.28	.61	.10	.99	
	Minimum Charge	Job		395	145	540	
9″							
	Demolish	S.F.		.35		.35	Includes material and labor to install
	Install	S.F.	.40	.36	.13	.89	fiberglass blown-in insulation.
	Demolish and Install	S.F.	.40	.71	.13	1.24	
	Minimum Charge	Job		395	145	540	
12″							
	Demolish	S.F.		.35		.35	Includes material and labor to install
	Install	S.F.	.55	.53	.19	1.27	fiberglass blown-in insulation.
	Demolish and Install	S.F.	.55	.88	.19	1.62	
	Minimum Charge	Job		395	145	540	

Rigid Board

		Unit	Material	Labor	Equip.	Total	Specification
1/2″							
	Demolish	S.F.		.20		.20	Includes material and labor to install
	Install	S.F.	.33	.44		.77	rigid foam insulation board, foil faced,
	Demolish and Install	S.F.	.33	.64		.97	two sides.
	Minimum Charge	Job		174		174	

Insulation

Ceiling Insulation

	Unit	Material	Labor	Equip.	Total	Specification
3/4"						
Demolish	S.F.		.20		.20	Includes material and labor to install
Install	S.F.	.39	.44		.83	rigid foam insulation board, foil faced,
Demolish and Install	S.F.	.39	.64		1.03	two sides.
Minimum Charge	Job		174		174	
1-1/2"						
Demolish	S.F.		.20		.20	Includes material and labor to install
Install	S.F.	.70	.48		1.18	rigid foam insulation board, foil faced,
Demolish and Install	S.F.	.70	.68		1.38	two sides.
Minimum Charge	Job		174		174	

Exterior Insulation

	Unit	Material	Labor	Equip.	Total	Specification
Rigid Board						
1/2"						
Demolish	S.F.		.11		.11	Includes material and labor to install
Install	S.F.	.33	.44		.77	rigid foam insulation board foil faced
Demolish and Install	S.F.	.33	.55		.88	two sides.
Minimum Charge	Job		174		174	
3/4"						
Demolish	S.F.		.11		.11	Includes material and labor to install
Install	S.F.	.39	.44		.83	rigid foam insulation board foil faced
Demolish and Install	S.F.	.39	.55		.94	two sides.
Minimum Charge	Job		174		174	
1-1/2"						
Demolish	S.F.		.11		.11	Includes material and labor to install
Install	S.F.	.70	.47		1.17	rigid foam insulation board foil faced
Demolish and Install	S.F.	.70	.58		1.28	two sides.
Minimum Charge	Job		174		174	
Vapor Barrier						
Demolish	S.F.		.06		.06	Cost includes material and labor to
Install	S.F.	.03	.09		.12	install 4 mil polyethylene vapor barrier
Demolish and Install	S.F.	.03	.15		.18	10' wide sheets with 6" overlaps.
Minimum Charge	Job		174		174	

Roof Insulation Board

	Unit	Material	Labor	Equip.	Total	Specification
Fiberglass						
3/4" Fiberglass						
Demolish	Sq.		37		37	Includes material and labor to install
Install	Sq.	60.50	26		86.50	fiberglass board insulation 3/4" thick,
Demolish and Install	Sq.	60.50	63		123.50	R-2.80 and C-0.36 values.
Minimum Charge	Job		174		174	
1" Fiberglass						
Demolish	Sq.		37		37	Includes material and labor to install
Install	Sq.	86	29		115	fiberglass board insulation 1" thick,
Demolish and Install	Sq.	86	66		152	R-4.20 and C-0.24 values.
Minimum Charge	Job		174		174	
1-3/8" Fiberglass						
Demolish	Sq.		37		37	Includes material and labor to install
Install	Sq.	138	31.50		169.50	fiberglass board insulation 1-3/8"
Demolish and Install	Sq.	138	68.50		206.50	thick, R-5.30 and C-0.19 values.
Minimum Charge	Job		174		174	

Insulation

Roof Insulation Board

	Unit	Material	Labor	Equip.	Total	Specification
1-5/8" Fiberglass						
Demolish	Sq.		37		37	Includes material and labor to install
Install	Sq.	141	33		174	fiberglass board insulation 1-5/8"
Demolish and Install	Sq.	141	70		211	thick, R-6.70 and C-0.15 values.
Minimum Charge	Job		174		174	
2-1/4" Fiberglass						
Demolish	Sq.		37		37	Includes material and labor to install
Install	Sq.	168	36.50		204.50	fiberglass board insulation 2-1/4"
Demolish and Install	Sq.	168	73.50		241.50	thick, R-8.30 and C-0.12 values.
Minimum Charge	Job		174		174	
Perlite						
1" Perlite						
Demolish	Sq.		37		37	Includes material and labor to install
Install	Sq.	49.50	29		78.50	perlite board insulation 1" thick,
Demolish and Install	Sq.	49.50	66		115.50	R-2.80 and C-0.36 values.
Minimum Charge	Job		174		174	
4" Perlite						
Demolish	Sq.		37		37	Includes material and labor to install
Install	Sq.	198	56		254	perlite board insulation 4" thick,
Demolish and Install	Sq.	198	93		291	R-10.0 and C-0.10 values.
Minimum Charge	Job		174		174	
1-1/2" Perlite						
Demolish	Sq.		37		37	Includes material and labor to install
Install	Sq.	51.50	31.50		83	perlite board insulation 1-1/2" thick,
Demolish and Install	Sq.	51.50	68.50		120	R-4.20 and C-0.24 values.
Minimum Charge	Job		174		174	
2" Perlite						
Demolish	Sq.		37		37	Includes material and labor to install
Install	Sq.	86	39		125	perlite board insulation 2" thick,
Demolish and Install	Sq.	86	76		162	R-5.30 and C-0.19 values.
Minimum Charge	Job		174		174	
2-1/2" Perlite						
Demolish	Sq.		37		37	Includes material and labor to install
Install	Sq.	101	43.50		144.50	perlite board insulation 2-1/2" thick,
Demolish and Install	Sq.	101	80.50		181.50	R-6.70 and C-0.15 values.
Minimum Charge	Job		174		174	
3" Perlite						
Demolish	Sq.		37		37	Includes material and labor to install
Install	Sq.	149	46.50		195.50	perlite board insulation 3" thick,
Demolish and Install	Sq.	149	83.50		232.50	R-8.30 and C-0.12 values.
Minimum Charge	Job		174		174	
Polystyrene						
2" Polystyrene						
Demolish	Sq.		37		37	Includes material and labor to install
Install	Sq.	39.50	36.50		76	extruded polystyrene board, R5.0/ 2"
Demolish and Install	Sq.	39.50	73.50		113	in thickness @25PSI compressive
Minimum Charge	Job		174		174	strength.
1" Polystyrene						
Demolish	Sq.		37		37	Includes material and labor to install
Install	Sq.	38	25		63	extruded polystyrene board, R5.0/ 1"
Demolish and Install	Sq.	38	62		100	in thickness @25PSI compressive
Minimum Charge	Job		174		174	strength.
1-1/2" Polystyrene						
Demolish	Sq.		37		37	Includes material and labor to install
Install	Sq.	38.50	29		67.50	extruded polystyrene board, R5.0/
Demolish and Install	Sq.	38.50	66		104.50	1-1/2" in thickness @25PSI
Minimum Charge	Job		174		174	compressive strength.

Insulation

Roof Insulation Board

	Unit	Material	Labor	Equip.	Total	Specification
Urethane						
3/4″ Urethane						
Demolish	Sq.		37		37	Includes material and labor to install
Install	Sq.	55.50	22		77.50	urethane board insulation 3/4″ thick,
Demolish and Install	Sq.	55.50	59		114.50	R-5.30 and C-0.19 values.
Minimum Charge	Job		174		174	
1″ Urethane						
Demolish	Sq.		37		37	Includes material and labor to install
Install	Sq.	61	23.50		84.50	urethane board insulation 1″ thick,
Demolish and Install	Sq.	61	60.50		121.50	R-6.70 and C-0.15 values.
Minimum Charge	Job		174		174	
1-1/2″ Urethane						
Demolish	Sq.		37		37	Includes material and labor to install
Install	Sq.	65	27		92	urethane board insulation 1-1/2″
Demolish and Install	Sq.	65	64		129	thick, R-10.1 and C-0.09 values.
Minimum Charge	Job		174		174	
2″ Urethane						
Demolish	Sq.		37		37	Includes material and labor to install
Install	Sq.	81.50	31.50		113	urethane board insulation 2″ thick,
Demolish and Install	Sq.	81.50	68.50		150	R-13.4 and C-0.06 values.
Minimum Charge	Job		174		174	

Gypsum Wallboard   **Tile Ceiling**   **Gypsum Wallboard Ceiling**

Wall Plaster		Unit	Material	Labor	Equip.	Total	Specification
Interior							
Cement							
	Demolish	S.Y.		4.70		4.70	Includes material and labor to install
	Install	S.Y.	14.85	19.80	1.69	36.34	cement plaster on interior wood
	Demolish and Install	S.Y.	14.85	24.50	1.69	41.04	framing with metal lath.
	Clean	S.Y.		2.31		2.31	
	Paint	S.Y.	1.72	4.07		5.79	
	Minimum Charge	Job		155		155	
Acoustic							
	Demolish	S.Y.		4.70		4.70	Includes material and labor to install
	Install	S.Y.	9.95	24	1.69	35.64	acoustical cement plaster on interior
	Demolish and Install	S.Y.	9.95	28.70	1.69	40.34	wood framing with metal lath.
	Clean	S.Y.		2.31		2.31	
	Paint	S.Y.	1.72	4.07		5.79	
	Minimum Charge	Job		155		155	
Gypsum							
	Demolish	S.Y.		4.70		4.70	Includes material and labor to install
	Install	S.Y.	8.40	20.50	1.43	30.33	gypsum wall or ceiling plaster on
	Demolish and Install	S.Y.	8.40	25.20	1.43	35.03	interior wood framing with metal lath.
	Clean	S.Y.		2.31		2.31	
	Paint	S.Y.	1.72	4.07		5.79	
	Minimum Charge	Job		155		155	
Stucco							
	Demolish	S.Y.		11.30		11.30	Includes material and labor to install
	Install	S.Y.	6.20	28	1.98	36.18	stucco on wire mesh over wood
	Demolish and Install	S.Y.	6.20	39.30	1.98	47.48	framing.
	Clean	S.Y.		2.31		2.31	
	Paint	S.Y.	1.72	4.07		5.79	
	Minimum Charge	Job		155		155	
Thin Coat							
Veneer							
	Demolish	S.Y.		4.70		4.70	Includes materials and labor to install
	Install	S.Y.	.69	3.66	.31	4.66	thin coat veneer plaster.
	Demolish and Install	S.Y.	.69	8.36	.31	9.36	
	Clean	S.Y.		2.31		2.31	
	Paint	S.Y.	1.72	4.07		5.79	
	Minimum Charge	Job		155		155	

Walls / Ceilings

Wall Plaster

	Unit	Material	Labor	Equip.	Total	Specification
Lath						
Metal						
Demolish	S.Y.		1.11		1.11	Includes material and labor to install
Install	S.Y.	5.25	4.10		9.35	wire lath on steel framed walls.
Demolish and Install	S.Y.	5.25	5.21		10.46	
Clean	S.Y.		2.31		2.31	
Minimum Charge	Job		72.50		72.50	
Gypsum						
Demolish	S.Y.		1.11		1.11	Includes material and labor to install
Install	S.Y.	4.65	3.85		8.50	gypsum lath on walls.
Demolish and Install	S.Y.	4.65	4.96		9.61	
Clean	S.Y.		2.31		2.31	
Minimum Charge	Job		155		155	
Repairs						
Fill in Small Cracks						
Install	L.F.	.06	2.49		2.55	Includes labor and material to fill-in
Minimum Charge	Job		155		155	hairline cracks up to 1/8" with filler, sand and prep for paint.
Repair up to 1" Crack						
Install	L.F.	.23	3.11		3.34	Includes labor and material to fill-in
Minimum Charge	Job		155		155	hairline cracks up to 1" with filler, sand and prep for paint.
Patch Small Hole (2")						
Install	Ea.	.06	3.27		3.33	Includes labor and material to patch
Minimum Charge	Job		155		155	up to a 2" diameter hole in plaster.
Patch Large Hole						
Install	Ea.	7	52		59	Includes labor and material to patch
Minimum Charge	Job		155		155	up to a 20" diameter hole in plaster.
Patch Plaster Section						
Install	Ea.	2.38	3.11		5.49	Includes labor and material to repair
Minimum Charge	Job		155		155	section of plaster wall, area less than 20 S.F.
Repair Inside Corner						
Install	L.F.	2.78	7.75		10.53	Includes material and labor to patch
Minimum Charge	Job		155		155	plaster corners.

Ceiling Plaster

	Unit	Material	Labor	Equip.	Total	Specification
Interior						
Cement						
Demolish	S.Y.		4.70		4.70	Includes material and labor to install
Install	S.Y.	14.85	19.80	1.69	36.34	cement plaster on interior wood
Demolish and Install	S.Y.	14.85	24.50	1.69	41.04	framing with metal lath.
Clean	S.Y.		2.42		2.42	
Paint	S.Y.	1.72	4.07		5.79	
Minimum Charge	Job		155		155	
Acoustic						
Demolish	S.Y.		4.70		4.70	Includes material and labor to install
Install	S.Y.	9.95	24	1.69	35.64	acoustical cement plaster on interior
Demolish and Install	S.Y.	9.95	28.70	1.69	40.34	wood framing with metal lath.
Clean	S.Y.		2.42		2.42	
Paint	S.Y.	1.72	4.07		5.79	
Minimum Charge	Job		155		155	

Walls / Ceilings

Ceiling Plaster		Unit	Material	Labor	Equip.	Total	Specification
Gypsum							
	Demolish	S.Y.		4.70		4.70	Includes material and labor to install
	Install	S.Y.	8.40	20.50	1.43	30.33	gypsum wall or ceiling plaster on
	Demolish and Install	S.Y.	8.40	25.20	1.43	35.03	interior wood framing with metal lath.
	Clean	S.Y.		2.42		2.42	
	Paint	S.Y.	1.72	4.07		5.79	
	Minimum Charge	Job		155		155	
Stucco							
	Demolish	S.Y.		11.30		11.30	Includes material and labor to install
	Install	S.Y.	6.20	28	1.98	36.18	stucco on wire mesh over wood
	Demolish and Install	S.Y.	6.20	39.30	1.98	47.48	framing.
	Clean	S.Y.		2.42		2.42	
	Paint	S.Y.	1.72	4.07		5.79	
	Minimum Charge	Job		155		155	
Thin Coat							
Veneer							
	Demolish	S.Y.		4.70		4.70	Includes materials and labor to install
	Install	S.Y.	.69	3.66	.31	4.66	thin coat veneer plaster.
	Demolish and Install	S.Y.	.69	8.36	.31	9.36	
	Clean	S.Y.		2.42		2.42	
	Paint	S.Y.	1.72	4.07		5.79	
	Minimum Charge	Job		155		155	
Lath							
Metal							
	Demolish	S.Y.		1.11		1.11	Includes material and labor to install
	Install	S.Y.	5.25	5.15		10.40	wire lath on steel framed ceilings.
	Demolish and Install	S.Y.	5.25	6.26		11.51	
	Minimum Charge	Job		155		155	
Gypsum							
	Demolish	S.Y.		1.11		1.11	Includes material and labor to install
	Install	S.Y.	4.65	3.20		7.85	gypsum lath on ceilings.
	Demolish and Install	S.Y.	4.65	4.31		8.96	
	Minimum Charge	Job		155		155	
Repairs							
Fill in Small Cracks							
	Install	S.F.	.06	.57		.63	Includes labor and material to fill in
	Minimum Charge	Job		310		310	hairline cracks up to 1/8" with filler,
							sand and prep for paint.

Gypsum Wallboard		Unit	Material	Labor	Equip.	Total	Specification
Finished							
1/4"							
	Demolish	S.F.		.28		.28	Cost includes material and labor to
	Install	S.F.	.31	1.05		1.36	install 1/4" drywall, including joint
	Demolish and Install	S.F.	.31	1.33		1.64	taping, premixed compound and
	Clean	S.F.		.23		.23	screws.
	Paint	S.F.	.22	.27		.49	
	Minimum Charge	Job		174		174	
3/8"							
	Demolish	S.F.		.28		.28	Cost includes material and labor to
	Install	S.F.	.31	1.06		1.37	install 3/8" drywall including joint
	Demolish and Install	S.F.	.31	1.34		1.65	tape, premixed compound and
	Clean	S.F.		.23		.23	screws.
	Paint	S.F.	.22	.27		.49	
	Minimum Charge	Job		174		174	

Gypsum Wallboard

Gypsum Wallboard	Unit	Material	Labor	Equip.	Total	Specification
1/2″						
Demolish	S.F.		.28		.28	Cost includes material and labor to
Install	S.F.	.30	1.07		1.37	install 1/2″ drywall including joint
Demolish and Install	S.F.	.30	1.35		1.65	tape, premixed compound and
Clean	S.F.		.23		.23	screws.
Paint	S.F.	.22	.27		.49	
Minimum Charge	Job		174		174	
1/2″ Water Resistant						
Demolish	S.F.		.28		.28	Cost includes material and labor to
Install	S.F.	.26	1.07		1.33	install 1/2″ WR drywall including joint
Demolish and Install	S.F.	.26	1.35		1.61	tape, premixed compound and
Clean	S.F.		.23		.23	screws.
Paint	S.F.	.22	.27		.49	
Minimum Charge	Job		174		174	
1/2″ Fire-rated						
Demolish	S.F.		.28		.28	Cost includes material and labor to
Install	S.F.	.35	1.07		1.42	install 1/2″ FR drywall including
Demolish and Install	S.F.	.35	1.35		1.70	screws. Taping and finishing not
Clean	S.F.		.23		.23	included.
Paint	S.F.	.22	.27		.49	
Minimum Charge	Job		174		174	
5/8″						
Demolish	S.F.		.28		.28	Cost includes material and labor to
Install	S.F.	.39	1.21		1.60	install 5/8″ drywall including joint
Demolish and Install	S.F.	.39	1.49		1.88	tape, premixed compound and
Clean	S.F.		.23		.23	screws.
Paint	S.F.	.22	.27		.49	
Minimum Charge	Job		174		174	
5/8″ Fire-rated						
Demolish	S.F.		.28		.28	Cost includes material and labor to
Install	S.F.	.38	1.21		1.59	install 5/8″ FR drywall, including joint
Demolish and Install	S.F.	.38	1.49		1.87	taping, premixed compound and
Clean	S.F.		.23		.23	screws.
Paint	S.F.	.22	.27		.49	
Minimum Charge	Job		174		174	
2-hour fire-rated						
Demolish	S.F.		.71		.71	Cost includes material and labor to
Install	S.F.	.90	2.41		3.31	install 2 layers of 5/8″ type X drywall
Demolish and Install	S.F.	.90	3.12		4.02	including joint taping, premixed
Clean	S.F.		.23		.23	compound and screws.
Paint	S.F.	.22	.27		.49	
Minimum Charge	Job		174		174	
Unfinished						
1/4″						
Demolish	S.F.		.28		.28	Cost includes material and labor to
Install	S.F.	.27	.36		.63	install 1/4″ drywall including screws.
Demolish and Install	S.F.	.27	.64		.91	Taping and finishing not included.
Clean	S.F.		.23		.23	
Minimum Charge	Job		174		174	
3/8″						
Demolish	S.F.		.28		.28	Cost includes material and labor to
Install	S.F.	.27	.37		.64	install 3/8″ drywall including screws.
Demolish and Install	S.F.	.27	.65		.92	Tape, finish or texture is not included.
Clean	S.F.		.23		.23	
Minimum Charge	Job		174		174	

Gypsum Wallboard		Unit	Material	Labor	Equip.	Total	Specification
1/2″							
	Demolish	S.F.		.28		.28	Cost includes material and labor to
	Install	S.F.	.26	.37		.63	install 1/2″ drywall including screws.
	Demolish and Install	S.F.	.26	.65		.91	Taping and finishing not included.
	Clean	S.F.		.23		.23	
	Minimum Charge	Job		174		174	
1/2″ Water Resistant							
	Demolish	S.F.		.28		.28	Cost includes material and labor to
	Install	S.F.	.23	.37		.60	install 1/2″ WR drywall including
	Demolish and Install	S.F.	.23	.65		.88	screws. Taping and finishing is not
	Clean	S.F.		.23		.23	included.
	Minimum Charge	Job		174		174	
1/2″ Fire-rated							
	Demolish	S.F.		.28		.28	Cost includes labor and material to
	Install	S.F.	.26	.25		.51	install 1/2″ FR drywall including
	Demolish and Install	S.F.	.26	.53		.79	screws. Tape, finish or texture is not
	Clean	S.F.		.23		.23	included.
	Minimum Charge	Job		174		174	
5/8″							
	Demolish	S.F.		.28		.28	Cost includes material and labor to
	Install	S.F.	.36	.41		.77	install 5/8″ drywall including screws.
	Demolish and Install	S.F.	.36	.69		1.05	Taping and finishing not included.
	Clean	S.F.		.23		.23	
	Minimum Charge	Job		174		174	
5/8″ Fire-rated							
	Demolish	S.F.		.28		.28	Cost includes material and labor to
	Install	S.F.	.35	.41		.76	install 5/8″ FR drywall including
	Demolish and Install	S.F.	.35	.69		1.04	screws. Taping and finishing not
	Clean	S.F.		.23		.23	included.
	Minimum Charge	Job		174		174	
2-hour Fire-rated							
	Demolish	S.F.		.71		.71	Cost includes material and labor to
	Install	S.F.	.90	2.41		3.31	install 2 layers of 5/8″ type X drywall
	Demolish and Install	S.F.	.90	3.12		4.02	including joint taping, premixed
	Clean	S.F.		.23		.23	compound and screws.
	Minimum Charge	Job		174		174	
Accessories							
Tape & Finish							
	Install	S.F.	.04	.35		.39	Includes labor and material for taping,
	Minimum Charge	Job		174		174	sanding and finish.
Texture							
Sprayed							
	Install	S.F.	.06	.38		.44	Includes labor and material to install
	Minimum Charge	Job		174		174	by spray, texture finish.
Trowel							
	Install	S.F.	.01	.20		.21	Includes labor and material to apply
	Minimum Charge	Job		174		174	hand troweled texture.
Patch Hole in Gypsum Wallboard							
	Install	S.F.	.51	46.50		47.01	Includes labor and material to patch a
	Minimum Charge	Job		174		174	one square foot area of damaged
							drywall.
Furring Strips							
	Demolish	S.F.		.07		.07	Cost includes material and labor to
	Install	S.F.	.10	.78		.88	install 1″ x 2″ furring strips, 16″ O.C.
	Demolish and Install	S.F.	.10	.85		.95	per S.F. of surface area to be covered.
	Minimum Charge	Job		87		87	

Walls / Ceilings

Ceramic Wall Tile		Unit	Material	Labor	Equip.	Total	Specification
Thin Set							
2″ x 2″							
	Demolish	S.F.		.94		.94	Cost includes material and labor to
	Install	S.F.	6.60	5.35		11.95	install 2″ x 2″ ceramic tile in organic
	Demolish and Install	S.F.	6.60	6.29		12.89	adhesive including cutting and grout.
	Clean	S.F.		.48		.48	
	Minimum Charge	Job		160		160	
4-1/4″ x 4-1/4″							
	Demolish	S.F.		.94		.94	Cost includes material and labor to
	Install	S.F.	2.38	2.98		5.36	install 4-1/4″ x 4-1/4″ ceramic tile in
	Demolish and Install	S.F.	2.38	3.92		6.30	organic adhesive including grout.
	Clean	S.F.		.48		.48	
	Minimum Charge	Job		160		160	
6″ x 6″							
	Demolish	S.F.		.94		.94	Cost includes material and labor to
	Install	S.F.	3.28	2.83		6.11	install 6″ x 6″ ceramic tile in organic
	Demolish and Install	S.F.	3.28	3.77		7.05	adhesive including and grout.
	Clean	S.F.		.48		.48	
	Minimum Charge	Job		160		160	
8″ x 8″							
	Demolish	S.F.		.94		.94	Cost includes material and labor to
	Install	S.F.	4.37	2.51		6.88	install 8″ x 8″ ceramic tile in organic
	Demolish and Install	S.F.	4.37	3.45		7.82	adhesive including cutting and grout.
	Clean	S.F.		.48		.48	
	Minimum Charge	Job		160		160	
Thick Set							
2″ x 2″							
	Demolish	S.F.		1.13		1.13	Cost includes material and labor to
	Install	S.F.	5.15	6.40		11.55	install 2″ x 2″ ceramic tile in mortar
	Demolish and Install	S.F.	5.15	7.53		12.68	bed including cutting and grout.
	Clean	S.F.		.48		.48	
	Minimum Charge	Job		160		160	
4-1/4″ x 4-1/4″							
	Demolish	S.F.		1.13		1.13	Cost includes material and labor to
	Install	S.F.	7.25	6.40		13.65	install 4-1/4″ x 4-1/4″ ceramic tile in
	Demolish and Install	S.F.	7.25	7.53		14.78	mortar bed including cutting and
	Clean	S.F.		.48		.48	grout.
	Minimum Charge	Job		160		160	
6″ x 6″							
	Demolish	S.F.		1.13		1.13	Cost includes material and labor to
	Install	S.F.	7.75	5.80		13.55	install 6″ x 6″ ceramic tile in mortar
	Demolish and Install	S.F.	7.75	6.93		14.68	bed including cutting and grout.
	Clean	S.F.		.48		.48	
	Minimum Charge	Job		160		160	
8″ x 8″							
	Demolish	S.F.		1.13		1.13	Cost includes material and labor to
	Install	S.F.	10.10	5.35		15.45	install 8″ x 8″ ceramic tile in mortar
	Demolish and Install	S.F.	10.10	6.48		16.58	bed including cutting and grout.
	Clean	S.F.		.48		.48	
	Minimum Charge	Job		160		160	
Re-grout							
	Install	S.F.	.15	3.20		3.35	Includes material and labor to regrout
	Minimum Charge	Job		160		160	ceramic tile walls.

Walls / Ceilings

Wall Tile		Unit	Material	Labor	Equip.	Total	Specification
Mirror							
Custom Cut Wall							
	Demolish	S.F.		.54		.54	Includes material and labor to install
	Install	S.F.	1.83	3.47		5.30	distortion-free float glass mirror, 1/8"
	Demolish and Install	S.F.	1.83	4.01		5.84	to 3/16" thick with cut and polished
	Clean	S.F.	.02	.15		.17	edges.
	Minimum Charge	Job		169		169	
Beveled Wall							
	Demolish	S.F.		.54		.54	Includes material and labor to install
	Install	S.F.	4.54	6.15		10.69	distortion-free float glass mirror, 1/8"
	Demolish and Install	S.F.	4.54	6.69		11.23	to 3/16" thick with beveled edges.
	Clean	S.F.	.02	.15		.17	
	Minimum Charge	Job		169		169	
Smoked Wall							
	Demolish	S.F.		.54		.54	Includes material and labor to install
	Install	S.F.	7.15	6.15		13.30	distortion-free smoked float glass
	Demolish and Install	S.F.	7.15	6.69		13.84	mirror, 1/8" to 3/16" thick with cut
	Clean	S.F.	.02	.15		.17	and polished edges.
	Minimum Charge	Job		169		169	
Marble							
Good Grade							
	Demolish	S.F.		1.76		1.76	Cost includes material and labor to
	Install	S.F.	7.90	13.90		21.80	install 3/8" x 12" x 12" marble tile in
	Demolish and Install	S.F.	7.90	15.66		23.56	mortar bed with grout.
	Clean	S.F.		.48		.48	
	Minimum Charge	Job		160		160	
Better Grade							
	Demolish	S.F.		1.76		1.76	Cost includes material and labor to
	Install	S.F.	10.45	13.90		24.35	install 3/8" x 12" x 12" marble, in
	Demolish and Install	S.F.	10.45	15.66		26.11	mortar bed with grout.
	Clean	S.F.		.48		.48	
	Minimum Charge	Job		160		160	
Premium Grade							
	Demolish	S.F.		1.76		1.76	Cost includes material and labor to
	Install	S.F.	20	13.90		33.90	install 3/8" x 12" x 12" marble, in
	Demolish and Install	S.F.	20	15.66		35.66	mortar bed with grout.
	Clean	S.F.		.48		.48	
	Minimum Charge	Job		160		160	
Glass Block							
	Demolish	S.F.		1.41		1.41	Cost includes material and labor to
	Install	S.F.	20.50	13.90		34.40	install 3-7/8" thick, 6" x 6" smooth
	Demolish and Install	S.F.	20.50	15.31		35.81	face glass block including mortar and
	Clean	S.F.		.48		.48	cleaning.
	Minimum Charge	Job		310		310	
Plastic							
	Demolish	S.F.		.43		.43	Cost includes material and labor to
	Install	S.F.	.99	1.83		2.82	install 4" plastic tile.
	Demolish and Install	S.F.	.99	2.26		3.25	
	Clean	S.F.		.48		.48	
	Minimum Charge	Job		160		160	
Cork							
	Demolish	S.F.		.22		.22	Cost includes material and labor to
	Install	S.F.	4.32	1.28		5.60	install 3/16" thick cork tile and
	Demolish and Install	S.F.	4.32	1.50		5.82	installation.
	Clean	S.F.		.48		.48	
	Minimum Charge	Job		154		154	

Walls / Ceilings

Wall Tile

Wall Tile		Unit	Material	Labor	Equip.	Total	Specification
Re-grout							
	Install	S.F.	.15	3.20		3.35	Includes material and labor to regrout
	Minimum Charge	Job		160		160	ceramic tile walls.

Wall Paneling

Wall Paneling		Unit	Material	Labor	Equip.	Total	Specification
Plywood							
Good Grade							
	Demolish	S.F.		.28		.28	Cost includes material and labor to
	Install	S.F.	.94	1.40		2.34	install 4' x 8' x 1/4" paneling,
	Demolish and Install	S.F.	.94	1.68		2.62	including nails and adhesive.
	Clean	S.F.		.23		.23	
	Paint	S.F.	.18	.40		.58	
	Minimum Charge	Job		174		174	
Better Grade							
	Demolish	S.F.		.28		.28	Cost includes material and labor to
	Install	S.F.	1.42	1.66		3.08	install 4' x 8' x 1/4" plywood wall
	Demolish and Install	S.F.	1.42	1.94		3.36	panel, including nails and adhesive.
	Clean	S.F.		.23		.23	
	Paint	S.F.	.18	.40		.58	
	Minimum Charge	Job		174		174	
Premium Grade							
	Demolish	S.F.		.28		.28	Cost includes material and labor to
	Install	S.F.	2.08	1.99		4.07	install 4' x 8' x 1/4" plywood wall
	Demolish and Install	S.F.	2.08	2.27		4.35	panel, including nails and adhesive.
	Clean	S.F.		.23		.23	
	Paint	S.F.	.18	.40		.58	
	Minimum Charge	Job		174		174	
Hardboard							
Embossed Paneling							
	Demolish	S.F.		.28		.28	Includes material and labor to install
	Install	S.F.	.65	1.40		2.05	hardboard embossed paneling
	Demolish and Install	S.F.	.65	1.68		2.33	including adhesives.
	Clean	S.F.		.23		.23	
	Minimum Charge	Job		174		174	
Plastic Molding							
	Demolish	L.F.		.28		.28	Includes material and labor to install
	Install	L.F.	.58	.87		1.45	plastic trim for paneling.
	Demolish and Install	L.F.	.58	1.15		1.73	
	Clean	L.F.	.02	.15		.17	
	Minimum Charge	Job		174		174	
Hardboard							
Embossed Brick							
	Demolish	S.F.		.28		.28	Cost includes material and labor to
	Install	S.F.	2.12	1.55		3.67	install 4' x 8' non-ceramic mineral
	Demolish and Install	S.F.	2.12	1.83		3.95	brick-like veneer wall panels including
	Clean	S.F.		.23		.23	adhesive.
	Paint	S.F.	.19	.46		.65	
	Minimum Charge	Job		174		174	
Embossed Ceramic Tile							
	Demolish	S.F.		.28		.28	Includes material and labor to install
	Install	S.F.	.65	1.40		2.05	hardboard embossed paneling
	Demolish and Install	S.F.	.65	1.68		2.33	including adhesives.
	Clean	S.F.		.23		.23	
	Paint	S.F.	.19	.46		.65	
	Minimum Charge	Job		174		174	

Walls / Ceilings

Wall Paneling

	Unit	Material	Labor	Equip.	Total	Specification
Pegboard						
Demolish	S.F.		.28		.28	Cost includes material and labor to
Install	S.F.	.43	1.40		1.83	install 4' x 8' x 1/4" pegboard panel,
Demolish and Install	S.F.	.43	1.68		2.11	including adhesive.
Clean	S.F.		.23		.23	
Paint	S.F.	.19	.46		.65	
Minimum Charge	Job		174		174	
Furring Strips						
Demolish	S.F.		.07		.07	Cost includes material and labor to
Install	S.F.	.10	.78		.88	install 1" x 2" furring strips, 16" O.C.
Demolish and Install	S.F.	.10	.85		.95	per S.F. of surface area to be covered.
Minimum Charge	Job		174		174	
Molding						
Unfinished Base						
Demolish	L.F.		.47		.47	Cost includes material and labor to
Install	L.F.	.59	1.07		1.66	install 1/2" x 2-1/2" pine or birch
Demolish and Install	L.F.	.59	1.54		2.13	base molding.
Reinstall	L.F.		.86		.86	
Clean	L.F.	.02	.15		.17	
Paint	L.F.	.03	.33		.36	
Minimum Charge	Job		174		174	
Unfinished Corner						
Demolish	L.F.		.47		.47	Includes material and labor to install
Install	L.F.	1.82	1.45		3.27	unfinished pine corner molding.
Demolish and Install	L.F.	1.82	1.92		3.74	
Reinstall	L.F.		1.16		1.16	
Clean	Ea.	.78	23		23.78	
Paint	L.F.	.13	.49		.62	
Minimum Charge	Job		174		174	
Unfinished Cove						
Demolish	L.F.		.47		.47	Cost includes material and labor to
Install	L.F.	1.56	1.29		2.85	install 3/4" x 3-1/2" unfinished pine
Demolish and Install	L.F.	1.56	1.76		3.32	cove molding.
Reinstall	L.F.		1.03		1.03	
Clean	Ea.	.78	23		23.78	
Paint	L.F.	.13	.49		.62	
Minimum Charge	Job		174		174	
Unfinished Crown						
Demolish	L.F.		.56		.56	Includes material and labor to install
Install	L.F.	2.05	1.40		3.45	unfinished pine crown molding.
Demolish and Install	L.F.	2.05	1.96		4.01	
Reinstall	L.F.		1.12		1.12	
Clean	L.F.	.03	.20		.23	
Paint	L.F.	.06	.33		.39	
Minimum Charge	Job		174		174	
Prefinished Base						
Demolish	L.F.		.47		.47	Cost includes material and labor to
Install	L.F.	1.69	1.12		2.81	install 9/16" x 3-5/16" prefinished
Demolish and Install	L.F.	1.69	1.59		3.28	base molding.
Reinstall	L.F.		.89		.89	
Clean	L.F.	.02	.15		.17	
Paint	L.F.	.03	.33		.36	
Minimum Charge	Job		174		174	

Walls / Ceilings

Wall Paneling

	Unit	Material	Labor	Equip.	Total	Specification
Prefinished Corner						
Demolish	L.F.		.47		.47	Includes material and labor to install
Install	L.F.	2.09	1.45		3.54	inside or outside prefinished corner
Demolish and Install	L.F.	2.09	1.92		4.01	molding.
Reinstall	L.F.		1.16		1.16	
Clean	L.F.		.10		.10	
Paint	L.F.	.13	.49		.62	
Minimum Charge	Job		174		174	
Prefinished Crown						
Demolish	L.F.		.56		.56	Cost includes material and labor to
Install	L.F.	2.24	1.40		3.64	install 3/4" x 3-13/16" prefinished
Demolish and Install	L.F.	2.24	1.96		4.20	crown molding.
Reinstall	L.F.		1.12		1.12	
Clean	L.F.	.03	.20		.23	
Paint	L.F.	.06	.33		.39	
Minimum Charge	Job		174		174	

Wallcovering

	Unit	Material	Labor	Equip.	Total	Specification
Surface Prep						
Patch Walls						
Install	S.F.	.12	3.11		3.23	Includes labor and material to repair
Minimum Charge	Job		154		154	section of plaster wall, area less than
						20 S.F.
Sizing						
Install	S.F.	.08	.15		.23	Includes labor and materials to apply
Minimum Charge	Job		154		154	wall sizing.
Good Grade						
Walls						
Demolish	S.F.		.57		.57	Includes material and labor to install
Install	S.F.	.87	.64		1.51	pre-pasted, vinyl-coated wallcovering.
Demolish and Install	S.F.	.87	1.21		2.08	
Clean	S.F.	.01	.06		.07	
Minimum Charge	Job		154		154	
Ceiling						
Demolish	S.F.		.57		.57	Includes material and labor to install
Install	S.F.	.87	.64		1.51	pre-pasted, vinyl-coated wallcovering.
Demolish and Install	S.F.	.87	1.21		2.08	
Clean	S.F.	.02	.15		.17	
Minimum Charge	Job		154		154	
Better Grade						
Walls						
Demolish	S.F.		.57		.57	Includes material and labor to install
Install	S.F.	.98	.64		1.62	pre-pasted, vinyl-coated wallcovering.
Demolish and Install	S.F.	.98	1.21		2.19	
Clean	S.F.	.01	.06		.07	
Minimum Charge	Job		154		154	
Ceiling						
Demolish	S.F.		.57		.57	Includes material and labor to install
Install	S.F.	.98	.64		1.62	pre-pasted, vinyl-coated wallcovering.
Demolish and Install	S.F.	.98	1.21		2.19	
Clean	S.F.	.02	.15		.17	
Minimum Charge	Job		154		154	

Walls / Ceilings

Wallcovering	Unit	Material	Labor	Equip.	Total	Specification
Premium Grade						
Walls						
Demolish	S.F.		.71		.71	Includes material and labor to install
Install	S.F.	1.23	.64		1.87	pre-pasted, vinyl-coated wallcovering.
Demolish and Install	S.F.	1.23	1.35		2.58	
Clean	S.F.	.01	.06		.07	
Minimum Charge	Job		154		154	
Ceiling						
Demolish	S.F.		.71		.71	Includes material and labor to install
Install	S.F.	1.23	.64		1.87	pre-pasted, vinyl-coated wallcovering.
Demolish and Install	S.F.	1.23	1.35		2.58	
Clean	S.F.	.02	.15		.17	
Minimum Charge	Job		154		154	
Mural						
Pre-printed						
Demolish	Ea.		88		88	Includes material and labor to install
Install	Ea.	231	231		462	wallcovering with a pre-printed mural.
Demolish and Install	Ea.	231	319		550	
Clean	Ea.		6.35		6.35	
Minimum Charge	Job		154		154	
Hand-painted						
Demolish	Ea.		88		88	Includes material and labor to install
Install	Ea.	460	465		925	wallcovering with a hand-painted
Demolish and Install	Ea.	460	553		1013	mural.
Clean	Ea.		6.35		6.35	
Minimum Charge	Job		154		154	
Vinyl-coated						
Good Grade						
Demolish	Ea.		15.70		15.70	Includes material and labor to install
Install	Ea.	19.20	21.50		40.70	pre-pasted, vinyl-coated wallcovering.
Demolish and Install	Ea.	19.20	37.20		56.40	Based upon 30 S.F. rolls.
Clean	Ea.		6.35		6.35	
Minimum Charge	Job		154		154	
Better Grade						
Demolish	Ea.		15.70		15.70	Includes material and labor to install
Install	Ea.	26	21.50		47.50	pre-pasted, vinyl-coated wallcovering.
Demolish and Install	Ea.	26	37.20		63.20	Based upon 30 S.F. rolls.
Clean	Ea.		6.35		6.35	
Minimum Charge	Job		154		154	
Premium Grade						
Demolish	Ea.		15.70		15.70	Includes material and labor to install
Install	Ea.	45	21.50		66.50	pre-pasted, vinyl-coated wallcovering.
Demolish and Install	Ea.	45	37.20		82.20	Based upon 30 S.F. rolls.
Clean	Ea.		6.35		6.35	
Minimum Charge	Job		154		154	
Blank Underliner						
Demolish	Ea.		15.70		15.70	Includes material and labor to install
Install	Ea.	11.35	22		33.35	blank stock liner paper or bridging
Demolish and Install	Ea.	11.35	37.70		49.05	paper. Based upon 30 S.F. rolls.
Clean	Ea.		6.35		6.35	
Minimum Charge	Job		154		154	
Solid Vinyl						
Good Grade						
Demolish	Roll		18.95		18.95	Includes material and labor to install
Install	Ea.	26	21.50		47.50	pre-pasted, vinyl wallcovering. Based
Demolish and Install	Ea.	26	40.45		66.45	upon 30 S.F. rolls.
Clean	Ea.		6.35		6.35	
Minimum Charge	Job		154		154	

Walls / Ceilings

Wallcovering		Unit	Material	Labor	Equip.	Total	Specification
Better Grade							
	Demolish	Roll		18.95		18.95	Includes material and labor to install
	Install	Ea.	36	21.50		57.50	pre-pasted, vinyl wallcovering. Based
	Demolish and Install	Ea.	36	40.45		76.45	upon 30 S.F. rolls.
	Clean	Ea.		6.35		6.35	
	Minimum Charge	Job		154		154	
Premium Grade							
	Demolish	Roll		18.95		18.95	Includes material and labor to install
	Install	Ea.	45.50	21.50		67	pre-pasted, vinyl wallcovering. Based
	Demolish and Install	Ea.	45.50	40.45		85.95	upon 30 S.F. rolls.
	Clean	Ea.		6.35		6.35	
	Minimum Charge	Job		154		154	
Blank Underliner							
	Demolish	Ea.		15.70		15.70	Includes material and labor to install
	Install	Ea.	11.35	22		33.35	blank stock liner paper or bridging
	Demolish and Install	Ea.	11.35	37.70		49.05	paper. Based upon 30 S.F. rolls.
	Clean	Ea.		6.35		6.35	
	Minimum Charge	Job		154		154	

Grasscloth / String
Good Grade

		Unit	Material	Labor	Equip.	Total	Specification
	Demolish	Ea.		15.70		15.70	Includes material and labor to install
	Install	Ea.	36.50	28		64.50	grasscloth or string wallcovering.
	Demolish and Install	Ea.	36.50	43.70		80.20	Based upon 30 S.F. rolls.
	Clean	Ea.		8.45		8.45	
	Minimum Charge	Job		154		154	
Better Grade							
	Demolish	Ea.		15.70		15.70	Includes material and labor to install
	Install	Ea.	47	28		75	grasscloth or string wallcovering.
	Demolish and Install	Ea.	47	43.70		90.70	Based upon 30 S.F. rolls.
	Clean	Ea.		8.45		8.45	
	Minimum Charge	Job		154		154	
Premium Grade							
	Demolish	Ea.		15.70		15.70	Includes material and labor to install
	Install	Ea.	60	28		88	grasscloth or string wallcovering.
	Demolish and Install	Ea.	60	43.70		103.70	Based upon 30 S.F. rolls.
	Clean	Ea.		8.45		8.45	
	Minimum Charge	Job		154		154	
Blank Underliner							
	Demolish	Ea.		15.70		15.70	Includes material and labor to install
	Install	Ea.	11.35	22		33.35	blank stock liner paper or bridging
	Demolish and Install	Ea.	11.35	37.70		49.05	paper. Based upon 30 S.F. rolls.
	Clean	Ea.		6.35		6.35	
	Minimum Charge	Job		154		154	

Designer
Premium Grade

		Unit	Material	Labor	Equip.	Total	Specification
	Demolish	S.F.		.71		.71	Includes material and labor to install
	Install	Ea.	112	32.50		144.50	paperback vinyl, untrimmed and
	Demolish and Install	Ea.	112	33.21		145.21	hand-painted. Based upon 30 S.F.
	Clean	Ea.		6.35		6.35	rolls.
	Minimum Charge	Job		154		154	
Blank Underliner							
	Demolish	Ea.		15.70		15.70	Includes material and labor to install
	Install	Ea.	11.35	22		33.35	blank stock liner paper or bridging
	Demolish and Install	Ea.	11.35	37.70		49.05	paper. Based upon 30 S.F. rolls.
	Clean	Ea.		6.35		6.35	
	Minimum Charge	Job		154		154	

Wallcovering

Wallcovering	Unit	Material	Labor	Equip.	Total	Specification
Foil						
Demolish	Ea.		15.70		15.70	Includes material and labor to install
Install	Ea.	38.50	28		66.50	foil coated wallcovering. Based upon
Demolish and Install	Ea.	38.50	43.70		82.20	30 S.F. rolls.
Clean	Ea.		6.35		6.35	
Minimum Charge	Job		154		154	
Wallcovering Border						
Good Grade						
Demolish	L.F.		1.03		1.03	Includes material and labor install
Install	L.F.	1.91	1.18		3.09	vinyl-coated border 4" to 6" wide.
Demolish and Install	L.F.	1.91	2.21		4.12	
Clean	L.F.		.10		.10	
Minimum Charge	Job		154		154	
Better Grade						
Demolish	L.F.		1.03		1.03	Includes material and labor install
Install	L.F.	2.55	1.18		3.73	vinyl-coated border 4" to 6" wide.
Demolish and Install	L.F.	2.55	2.21		4.76	
Clean	L.F.		.10		.10	
Minimum Charge	Job		154		154	
Premium Grade						
Demolish	L.F.		1.03		1.03	Includes material and labor install
Install	L.F.	3.20	1.18		4.38	vinyl-coated border 4" to 6" wide.
Demolish and Install	L.F.	3.20	2.21		5.41	
Clean	L.F.		.10		.10	
Minimum Charge	Job		154		154	

Acoustical Ceiling

Acoustical Ceiling	Unit	Material	Labor	Equip.	Total	Specification
Complete System						
2' x 2' Square Edge						
Demolish	S.F.		.47		.47	Includes material and labor to install
Install	S.F.	1.33	1.07		2.40	T-bar grid acoustical ceiling system,
Demolish and Install	S.F.	1.33	1.54		2.87	random pinhole 5/8" thick square
Reinstall	S.F.		.86		.86	edge tile.
Clean	S.F.	.03	.19		.22	
Minimum Charge	Job		174		174	
2' x 2' Reveal Edge						
Demolish	S.F.		.47		.47	Includes material and labor to install
Install	S.F.	2.29	1.40		3.69	T-bar grid acoustical ceiling system,
Demolish and Install	S.F.	2.29	1.87		4.16	random pinhole 5/8" thick reveal
Reinstall	S.F.		1.12		1.12	edge tile.
Clean	S.F.	.03	.19		.22	
Minimum Charge	Job		174		174	
2' x 2' Fire-rated						
Demolish	S.F.		.47		.47	Includes material and labor to install
Install	S.F.	2.62	1.07		3.69	T-bar grid FR acoustical ceiling system,
Demolish and Install	S.F.	2.62	1.54		4.16	random pinhole 5/8" thick square
Reinstall	S.F.		.86		.86	edge tile.
Clean	S.F.	.03	.19		.22	
Minimum Charge	Job		174		174	
2' x 4' Square Edge						
Demolish	S.F.		.47		.47	Cost includes material and labor to
Install	S.F.	1.12	.92		2.04	install 2' x 4' T-bar grid acoustical
Demolish and Install	S.F.	1.12	1.39		2.51	ceiling system, random pinhole 5/8"
Reinstall	S.F.		.73		.73	thick square edge tile.
Clean	S.F.	.03	.19		.22	
Minimum Charge	Job		174		174	

Walls / Ceilings

Acoustical Ceiling

	Unit	Material	Labor	Equip.	Total	Specification
2' x 4' Reveal Edge						
Demolish	S.F.		.47		.47	Includes material and labor to install
Install	S.F.	2.29	1.27		3.56	T-bar grid acoustical ceiling system,
Demolish and Install	S.F.	2.29	1.74		4.03	random pinhole 5/8″ thick reveal
Reinstall	S.F.		1.01		1.01	edge tile.
Clean	S.F.	.03	.19		.22	
Minimum Charge	Job		174		174	
2' x 4' Fire-rated						
Demolish	S.F.		.47		.47	Includes material and labor to install
Install	S.F.	2.62	1.07		3.69	T-bar grid FR acoustical ceiling system,
Demolish and Install	S.F.	2.62	1.54		4.16	random pinhole 5/8″ thick square
Reinstall	S.F.		.86		.86	edge tile.
Clean	S.F.	.03	.19		.22	
Minimum Charge	Job		174		174	

T-bar Grid

	Unit	Material	Labor	Equip.	Total	Specification
2' x 2'						
Demolish	S.F.		.55		.55	Includes material and labor to install
Install	S.F.	.75	.54		1.29	T-bar grid acoustical ceiling
Demolish and Install	S.F.	.75	1.09		1.84	suspension system.
Reinstall	S.F.		.43		.43	
Clean	S.F.	.03	.19		.22	
Minimum Charge	Job		155		155	
2' x 4'						
Demolish	S.F.		.55		.55	Includes material and labor to install
Install	S.F.	.61	.44		1.05	T-bar grid acoustical ceiling
Demolish and Install	S.F.	.61	.99		1.60	suspension system.
Reinstall	S.F.		.35		.35	
Clean	S.F.	.03	.19		.22	
Minimum Charge	Job		155		155	

Tile Panels

	Unit	Material	Labor	Equip.	Total	Specification
2' x 2' Square Edge						
Demolish	S.F.		.16		.16	Includes material and labor to install
Install	S.F.	.52	.56		1.08	square edge panels 5/8″ thick with
Demolish and Install	S.F.	.52	.72		1.24	pinhole pattern. T-bar grid not
Reinstall	S.F.		.45		.45	included.
Clean	S.F.	.03	.19		.22	
Minimum Charge	Job		174		174	
2' x 2' Reveal Edge						
Demolish	S.F.		.16		.16	Includes material and labor to install
Install	S.F.	1.36	.74		2.10	reveal edge panels 5/8″ thick with
Demolish and Install	S.F.	1.36	.90		2.26	pinhole pattern. T-bar grid not
Reinstall	S.F.		.59		.59	included.
Clean	S.F.	.03	.19		.22	
Minimum Charge	Job		174		174	
2' x 2' Fire-rated						
Demolish	S.F.		.16		.16	Includes material and labor to install
Install	S.F.	1.10	.52		1.62	FR square edge panels 5/8″ thick with
Demolish and Install	S.F.	1.10	.68		1.78	pinhole pattern. T-bar grid not
Reinstall	S.F.		.41		.41	included.
Clean	S.F.	.03	.19		.22	
Minimum Charge	Job		174		174	
2' x 4' Square Edge						
Demolish	S.F.		.16		.16	Includes material and labor to install
Install	S.F.	.52	.56		1.08	square edge panels 5/8″ thick with
Demolish and Install	S.F.	.52	.72		1.24	pinhole pattern. T-bar grid not
Reinstall	S.F.		.45		.45	included.
Clean	S.F.	.03	.19		.22	
Minimum Charge	Job		174		174	

Walls / Ceilings

Acoustical Ceiling		Unit	Material	Labor	Equip.	Total	Specification
2' x 4' Reveal Edge							
	Demolish	S.F.		.16		.16	Includes material and labor to install
	Install	S.F.	1.36	.74		2.10	reveal edge panels 5/8" thick with
	Demolish and Install	S.F.	1.36	.90		2.26	pinhole pattern. T-bar grid not
	Reinstall	S.F.		.59		.59	included.
	Clean	S.F.	.03	.19		.22	
	Minimum Charge	Job		174		174	
2' x 4' Fire-rated							
	Demolish	S.F.		.16		.16	Includes material and labor to install
	Install	S.F.	1.10	.52		1.62	FR square edge panels 5/8" thick with
	Demolish and Install	S.F.	1.10	.68		1.78	pinhole pattern. T-bar grid not
	Reinstall	S.F.		.41		.41	included.
	Clean	S.F.	.03	.19		.22	
	Minimum Charge	Job		174		174	
Bleach Out Stains							
	Install	S.F.	.08	.25		.33	Includes labor and material to apply
	Minimum Charge	Job		174		174	laundry type chlorine bleach to
							remove light staining of acoustic tile
							material.
Adhesive Tile							
Good Grade							
	Demolish	S.F.		.63		.63	Cost includes material and labor to
	Install	S.F.	.90	1.16		2.06	install 12" x 12" tile with embossed
	Demolish and Install	S.F.	.90	1.79		2.69	texture pattern. Furring strips not
	Clean	S.F.	.02	.15		.17	included.
	Minimum Charge	Job		174		174	
Better Grade							
	Demolish	S.F.		.63		.63	Cost includes material and labor to
	Install	S.F.	1.21	1.16		2.37	install 12" x 12" tile with embossed
	Demolish and Install	S.F.	1.21	1.79		3	texture pattern. Furring strips not
	Clean	S.F.	.02	.15		.17	included.
	Minimum Charge	Job		174		174	
Premium Grade							
	Demolish	S.F.		.63		.63	Cost includes material and labor to
	Install	S.F.	3.44	1.16		4.60	install 12" x 12" tile with embossed
	Demolish and Install	S.F.	3.44	1.79		5.23	texture pattern. Furring strips not
	Clean	S.F.	.02	.15		.17	included.
	Minimum Charge	Job		174		174	
Bleach Out Stains							
	Install	S.F.	.08	.25		.33	Includes labor and material to apply
	Minimum Charge	Job		174		174	laundry type chlorine bleach to
							remove light staining of acoustic tile
							material.
Stapled Tile							
Good Grade							
	Demolish	S.F.		.38		.38	Cost includes material and labor to
	Install	S.F.	.90	1.16		2.06	install 12" x 12" tile with embossed
	Demolish and Install	S.F.	.90	1.54		2.44	texture pattern. Furring strips not
	Clean	S.F.	.02	.15		.17	included.
	Minimum Charge	Job		174		174	
Better Grade							
	Demolish	S.F.		.38		.38	Cost includes material and labor to
	Install	S.F.	1.21	1.16		2.37	install 12" x 12" tile with embossed
	Demolish and Install	S.F.	1.21	1.54		2.75	texture pattern. Furring strips not
	Clean	S.F.	.02	.15		.17	included.
	Minimum Charge	Job		174		174	

Walls / Ceilings

Acoustical Ceiling		Unit	Material	Labor	Equip.	Total	Specification
Premium Grade							Cost includes material and labor to
	Demolish	S.F.		.38		.38	install 12" x 12" tile with embossed
	Install	S.F.	3.44	1.16		4.60	texture pattern. Furring strips not
	Demolish and Install	S.F.	3.44	1.54		4.98	included.
	Clean	S.F.	.02	.15		.17	
	Minimum Charge	Job		174		174	
Bleach Out Stains							Includes labor and material to apply
	Install	S.F.	.08	.25		.33	laundry type chlorine bleach to
	Minimum Charge	Job		174		174	remove light staining of acoustic tile
							material.
Blown							
Scrape Off Existing							Includes labor and material to remove
	Install	S.F.		.52		.52	existing blown acoustical ceiling
	Minimum Charge	Job		174		174	material.
Seal Ceiling Gypsum Wallboard							Includes labor and material to seal
	Install	S.F.	.06	.27		.33	existing drywall with shellac-based
	Minimum Charge	Job		153		153	material.
Reblow Existing							Includes labor and material to install
	Install	S.F.	.06	.38		.44	by spray, texture finish.
	Minimum Charge	Job		415	67	482	
Bleach Out Stains							Includes labor and material to apply
	Install	S.F.	.08	.25		.33	laundry type chlorine bleach to
	Minimum Charge	Job		174		174	remove light staining of acoustic tile
							material.
New Blown							Includes labor and material to install
	Install	S.F.	.06	.38		.44	by spray, texture finish.
	Minimum Charge	Job		153		153	
Apply Glitter To							Includes labor and material to apply
	Install	S.F.	.09	.13		.22	glitter material to existing or new
	Minimum Charge	Job		153		153	drywall.
Furring Strips							Cost includes material and labor to
	Demolish	S.F.		.07		.07	install 1" x 2" furring strips, 16" O.C.
	Install	S.F.	.10	.78		.88	per S.F. of surface area to be covered.
	Demolish and Install	S.F.	.10	.85		.95	
	Minimum Charge	Job		87		87	

Finished Stair

Bi-Fold Door

Raised Panel Door

Hollow Core Door	Unit	Material	Labor	Equip.	Total	Specification
Pre-hung Wood						
1' 6" x 6' 8"						
Demolish	Ea.		32.50		32.50	Cost includes material and labor to
Install	Ea.	146	16.60		162.60	install 1' 6" x 6' 8" pre-hung 1-3/8"
Demolish and Install	Ea.	146	49.10		195.10	lauan door with split pine jamb
Reinstall	Ea.		16.61		16.61	including casing, stops, and hinges.
Clean	Ea.	3.11	10.60		13.71	
Paint	Ea.	9.25	43.50		52.75	
Minimum Charge	Job		174		174	
2' x 6' 8"						
Demolish	Ea.		32.50		32.50	Cost includes material and labor to
Install	Ea.	152	17.45		169.45	install 2' x 6' 8" pre-hung 1-3/8"
Demolish and Install	Ea.	152	49.95		201.95	lauan door with split pine jamb
Reinstall	Ea.		17.44		17.44	including casing, stops, and hinges.
Clean	Ea.	3.11	10.60		13.71	
Paint	Ea.	9.25	43.50		52.75	
Minimum Charge	Job		174		174	
2' 4" x 6' 8"						
Demolish	Ea.		32.50		32.50	Cost includes material and labor to
Install	Ea.	157	17.45		174.45	install 2' 4" x 6' 8" pre-hung 1-3/8"
Demolish and Install	Ea.	157	49.95		206.95	lauan door with split pine jamb
Reinstall	Ea.		17.44		17.44	including casing, stops, and hinges.
Clean	Ea.	3.11	10.60		13.71	
Paint	Ea.	9.25	43.50		52.75	
Minimum Charge	Job		174		174	
2' 6" x 6' 8"						
Demolish	Ea.		32.50		32.50	Cost includes material and labor to
Install	Ea.	157	17.45		174.45	install 2' 6" x 6' 8" pre-hung 1-3/8"
Demolish and Install	Ea.	157	49.95		206.95	lauan door with split pine jamb
Reinstall	Ea.		17.44		17.44	including casing, stops, and hinges.
Clean	Ea.	3.11	10.60		13.71	
Paint	Ea.	9.25	43.50		52.75	
Minimum Charge	Job		174		174	
2' 8" x 6' 8"						
Demolish	Ea.		32.50		32.50	Cost includes material and labor to
Install	Ea.	168	17.45		185.45	install 2' 8" x 6' 8" pre-hung 1-3/8"
Demolish and Install	Ea.	168	49.95		217.95	lauan door with split pine jamb,
Reinstall	Ea.		17.44		17.44	including casing, stop and hinges.
Clean	Ea.	3.11	10.60		13.71	
Paint	Ea.	9.25	43.50		52.75	
Minimum Charge	Job		174		174	

Finish Carpentry

Hollow Core Door	Unit	Material	Labor	Equip.	Total	Specification
3' x 6' 8"						
Demolish	Ea.		32.50		32.50	Cost includes material and labor to
Install	Ea.	175	18.35		193.35	install 3' x 6' 8" pre-hung 1-3/8"
Demolish and Install	Ea.	175	50.85		225.85	lauan door with split pine jamb
Reinstall	Ea.		18.36		18.36	including casing, stops, and hinges.
Clean	Ea.	3.11	10.60		13.71	
Paint	Ea.	9.25	43.50		52.75	
Minimum Charge	Job		174		174	
Pocket Type						
Demolish	Ea.		48.50		48.50	Cost includes material and labor to
Install	Ea.	36	39		75	install 2' 8" x 6' 8" lauan door.
Demolish and Install	Ea.	36	87.50		123.50	
Reinstall	Ea.		38.76		38.76	
Clean	Ea.	3.11	10.60		13.71	
Paint	Ea.	9.25	43.50		52.75	
Minimum Charge	Job		174		174	
Stain						
Install	Ea.	4.44	25.50		29.94	Includes labor and material to stain
Minimum Charge	Job		153		153	single door and trim on both sides.
Pre-hung Masonite						
1' 6" x 6' 8"						
Demolish	Ea.		32.50		32.50	Cost includes material and labor to
Install	Ea.	202	16.60		218.60	install 1' 6" x 6' 8" pre-hung 1-3/8"
Demolish and Install	Ea.	202	49.10		251.10	masonite door with split pine jamb
Reinstall	Ea.		16.61		16.61	including casing, stops, and hinges.
Clean	Ea.	3.11	10.60		13.71	
Paint	Ea.	9.25	43.50		52.75	
Minimum Charge	Job		174		174	
2' x 6' 8"						
Demolish	Ea.		32.50		32.50	Cost includes material and labor to
Install	Ea.	212	17.45		229.45	install 2' x 6' 8" pre-hung 1-3/8"
Demolish and Install	Ea.	212	49.95		261.95	masonite door with split pine jamb
Reinstall	Ea.		17.44		17.44	including casing, stops, and hinges.
Clean	Ea.	3.11	10.60		13.71	
Paint	Ea.	9.25	43.50		52.75	
Minimum Charge	Job		174		174	
2' 4" x 6' 8"						
Demolish	Ea.		32.50		32.50	Cost includes material and labor to
Install	Ea.	215	17.45		232.45	install 2' 4" x 6' 8" pre-hung 1-3/8"
Demolish and Install	Ea.	215	49.95		264.95	masonite door with split pine jamb
Reinstall	Ea.		17.44		17.44	including casing, stops, and hinges.
Clean	Ea.	3.11	10.60		13.71	
Paint	Ea.	9.25	43.50		52.75	
Minimum Charge	Job		174		174	
2' 8" x 6' 8"						
Demolish	Ea.		32.50		32.50	Cost includes material and labor to
Install	Ea.	217	17.45		234.45	install 2' 8" x 6' 8" pre-hung 1-3/8"
Demolish and Install	Ea.	217	49.95		266.95	masonite door with split pine jamb
Reinstall	Ea.		17.44		17.44	including casing, stops, and hinges.
Clean	Ea.	3.11	10.60		13.71	
Paint	Ea.	9.25	43.50		52.75	
Minimum Charge	Job		174		174	
3' x 6' 8"						
Demolish	Ea.		32.50		32.50	Cost includes material and labor to
Install	Ea.	222	18.35		240.35	install 3' x 6' 8" pre-hung 1-3/8"
Demolish and Install	Ea.	222	50.85		272.85	masonite door with split pine jamb
Reinstall	Ea.		18.36		18.36	including casing, stops, and hinges.
Clean	Ea.	3.11	10.60		13.71	
Paint	Ea.	9.25	43.50		52.75	
Minimum Charge	Job		174		174	

Finish Carpentry

Hollow Core Door	Unit	Material	Labor	Equip.	Total	Specification
Pocket Type						
Demolish	Ea.		48.50		48.50	Cost includes material and labor to
Install	Ea.	36	39		75	install 2' 8" x 6' 8" lauan door.
Demolish and Install	Ea.	36	87.50		123.50	
Reinstall	Ea.		38.76		38.76	
Clean	Ea.	3.11	10.60		13.71	
Paint	Ea.	9.25	43.50		52.75	
Minimum Charge	Job		174		174	
Stain						
Install	Ea.	4.44	25.50		29.94	Includes labor and material to stain
Minimum Charge	Job		153		153	single door and trim on both sides.
Hollow Core Door Only						
1' 6" x 6' 8'						
Demolish	Ea.		6.90		6.90	Cost includes material and labor to
Install	Ea.	39	18.35		57.35	install 1' 6" x 6' 8" lauan hollow core
Demolish and Install	Ea.	39	25.25		64.25	door slab.
Reinstall	Ea.		18.36		18.36	
Clean	Ea.	1.30	7.95		9.25	
Paint	Ea.	2.82	38		40.82	
Minimum Charge	Job		174		174	
2' x 6' 8'						
Demolish	Ea.		6.90		6.90	Cost includes material and labor to
Install	Ea.	38	19.40		57.40	install 2' x 6' 8" lauan hollow core
Demolish and Install	Ea.	38	26.30		64.30	door slab.
Reinstall	Ea.		19.38		19.38	
Clean	Ea.	1.30	7.95		9.25	
Paint	Ea.	2.82	38		40.82	
Minimum Charge	Job		174		174	
2' 4" x 6' 8'						
Demolish	Ea.		6.90		6.90	Cost includes material and labor to
Install	Ea.	41.50	19.40		60.90	install 2' 4" x 6' 8" lauan hollow core
Demolish and Install	Ea.	41.50	26.30		67.80	door slab.
Reinstall	Ea.		19.38		19.38	
Clean	Ea.	1.30	7.95		9.25	
Paint	Ea.	2.82	38		40.82	
Minimum Charge	Job		174		174	
2' 8" x 6' 8'						
Demolish	Ea.		6.90		6.90	Cost includes material and labor to
Install	Ea.	43.50	19.40		62.90	install 2' 8" x 6' 8" lauan hollow core
Demolish and Install	Ea.	43.50	26.30		69.80	door slab.
Reinstall	Ea.		19.38		19.38	
Clean	Ea.	1.30	7.95		9.25	
Paint	Ea.	2.82	38		40.82	
Minimum Charge	Job		174		174	
3' x 6' 8'						
Demolish	Ea.		6.90		6.90	Cost includes material and labor to
Install	Ea.	45	20.50		65.50	install 3' x 6' 8" lauan hollow core
Demolish and Install	Ea.	45	27.40		72.40	door slab.
Reinstall	Ea.		20.52		20.52	
Clean	Ea.	1.30	7.95		9.25	
Paint	Ea.	2.82	38		40.82	
Minimum Charge	Job		174		174	
Pocket Type						
Demolish	Ea.		6.90		6.90	Cost includes material and labor to
Install	Ea.	36	39		75	install 2' 8" x 6' 8" lauan door.
Demolish and Install	Ea.	36	45.90		81.90	
Reinstall	Ea.		38.76		38.76	
Clean	Ea.	1.30	7.95		9.25	
Paint	Ea.	2.82	38		40.82	
Minimum Charge	Job		174		174	

Finish Carpentry

Hollow Core Door

Hollow Core Door	Unit	Material	Labor	Equip.	Total	Specification
Stain						
Install	Ea.	2.45	34		36.45	Includes labor and materials to stain a
Minimum Charge	Job		153		153	flush style door on both sides by
						brush.

Casing Trim
Single Width

	Unit	Material	Labor	Equip.	Total	Specification
Demolish	Ea.		6.15		6.15	Cost includes material and labor to
Install	Opng.	25.50	59		84.50	install 11/16" x 2-1/2" pine ranch
Demolish and Install	Opng.	25.50	65.15		90.65	style casing for one side of a standard
Reinstall	Opng.		47.29		47.29	door opening.
Clean	Opng.	.44	2.54		2.98	
Paint	Ea.	3.22	6.10		9.32	
Minimum Charge	Job		174		174	

Double Width

	Unit	Material	Labor	Equip.	Total	Specification
Demolish	Ea.		7.30		7.30	Cost includes material and labor to
Install	Opng.	29.50	70		99.50	install 11/16" x 2-1/2" pine ranch
Demolish and Install	Opng.	29.50	77.30		106.80	style door casing for one side of a
Reinstall	Opng.		55.81		55.81	double door opening.
Clean	Opng.	.52	5.10		5.62	
Paint	Ea.	3.22	6.95		10.17	
Minimum Charge	Job		174		174	

Hardware
Doorknob w / Lock

	Unit	Material	Labor	Equip.	Total	Specification
Demolish	Ea.		12.90		12.90	Includes material and labor to install
Install	Ea.	21.50	22		43.50	privacy lockset.
Demolish and Install	Ea.	21.50	34.90		56.40	
Reinstall	Ea.		17.44		17.44	
Clean	Ea.	.31	7.95		8.26	
Minimum Charge	Job		174		174	

Doorknob

	Unit	Material	Labor	Equip.	Total	Specification
Demolish	Ea.		12.90		12.90	Includes material and labor to install
Install	Ea.	18.75	22		40.75	residential passage lockset, keyless
Demolish and Install	Ea.	18.75	34.90		53.65	bored type.
Reinstall	Ea.		17.44		17.44	
Clean	Ea.	.31	7.95		8.26	
Minimum Charge	Job		174		174	

Deadbolt

	Unit	Material	Labor	Equip.	Total	Specification
Demolish	Ea.		12.10		12.10	Includes material and labor to install a
Install	Ea.	55	25		80	deadbolt.
Demolish and Install	Ea.	55	37.10		92.10	
Reinstall	Ea.		19.93		19.93	
Clean	Ea.	.31	7.95		8.26	
Minimum Charge	Job		174		174	

Lever Handle

	Unit	Material	Labor	Equip.	Total	Specification
Demolish	Ea.		12.90		12.90	Includes material and labor to install
Install	Ea.	28.50	35		63.50	residential passage lockset, keyless
Demolish and Install	Ea.	28.50	47.90		76.40	bored type with lever handle.
Reinstall	Ea.		27.90		27.90	
Clean	Ea.	.31	7.95		8.26	
Minimum Charge	Job		174		174	

Closer

	Unit	Material	Labor	Equip.	Total	Specification
Demolish	Ea.		7.25		7.25	Includes material and labor to install
Install	Ea.	98.50	53.50		152	pneumatic light duty closer for interior
Demolish and Install	Ea.	98.50	60.75		159.25	type doors.
Reinstall	Ea.		42.93		42.93	
Clean	Ea.	.15	10.60		10.75	
Minimum Charge	Job		174		174	

Finish Carpentry

Hollow Core Door

	Unit	Material	Labor	Equip.	Total	Specification
Push Plate						
Demolish	Ea.		10.75		10.75	Includes material and labor to install
Install	Ea.	19.40	29		48.40	bronze push-pull plate.
Demolish and Install	Ea.	19.40	39.75		59.15	
Reinstall	Ea.		23.25		23.25	
Clean	Ea.	.31	7.95		8.26	
Minimum Charge	Job		174		174	
Kickplate						
Demolish	Ea.		10.75		10.75	Includes material and labor to install
Install	Ea.	22	23.50		45.50	aluminum kickplate, 10″ x 28″.
Demolish and Install	Ea.	22	34.25		56.25	
Reinstall	Ea.		18.60		18.60	
Clean	Ea.	.31	7.95		8.26	
Minimum Charge	Job		174		174	
Exit Sign						
Demolish	Ea.		19.35		19.35	Includes material and labor to install a
Install	Ea.	53	48		101	wall mounted interior electric exit sign.
Demolish and Install	Ea.	53	67.35		120.35	
Reinstall	Ea.		38.32		38.32	
Clean	Ea.	.02	5.70		5.72	
Minimum Charge	Job		174		174	
Jamb						
Demolish	Ea.		24		24	Includes material and labor to install
Install	Ea.	103	9.45		112.45	flat pine jamb with square cut heads
Demolish and Install	Ea.	103	33.45		136.45	and rabbeted sides for 6′ 8″ high and
Clean	Ea.	.10	6.35		6.45	3-9/16″ door including trim sets for
Paint	Ea.	6	20.50		26.50	both sides.
Minimum Charge	Job		174		174	
Shave & Refit						
Install	Ea.		29		29	Includes labor to shave and rework
Minimum Charge	Job		174		174	door to fit opening at the job site.

Solid Core Door

	Unit	Material	Labor	Equip.	Total	Specification
Pre-hung Wood						
2′ 4″ x 6′ 8″						
Demolish	Ea.		32.50		32.50	Cost includes material and labor to
Install	Ea.	210	17.45		227.45	install 2′ 4″ x 6′ 8″ pre-hung 1-3/4″
Demolish and Install	Ea.	210	49.95		259.95	door, 4-1/2″ split jamb, including
Reinstall	Ea.		17.44		17.44	casing and stop, hinges, aluminum sill,
Clean	Ea.	3.11	10.60		13.71	weatherstripping.
Paint	Ea.	9.30	43.50		52.80	
Minimum Charge	Job		174		174	
2′ 6″ x 6′ 8″						
Demolish	Ea.		32.50		32.50	Cost includes material and labor to
Install	Ea.	204	17.45		221.45	install 2′ 6″ x 6′ 8″ pre-hung 1-3/4″
Demolish and Install	Ea.	204	49.95		253.95	door, 4-1/2″ split jamb, including
Reinstall	Ea.		17.44		17.44	casing and stop, hinges, aluminum sill,
Clean	Ea.	3.11	10.60		13.71	weatherstripping.
Paint	Ea.	9.30	43.50		52.80	
Minimum Charge	Job		174		174	

Finish Carpentry

Solid Core Door		Unit	Material	Labor	Equip.	Total	Specification
2′ 8″ x 6′ 8″							
	Demolish	Ea.		32.50		32.50	Cost includes material and labor to
	Install	Ea.	213	17.45		230.45	install 2′ 8″ x 6′ 8″ pre-hung 1-3/4″
	Demolish and Install	Ea.	213	49.95		262.95	door, 4-1/2″ split jamb, including
	Reinstall	Ea.		17.44		17.44	casing and stop, hinges, aluminum sill,
	Clean	Ea.	3.11	10.60		13.71	weatherstripping.
	Paint	Ea.	9.30	43.50		52.80	
	Minimum Charge	Job		174		174	
3′ x 6′ 8″							
	Demolish	Ea.		32.50		32.50	Cost includes material and labor to
	Install	Ea.	222	18.35		240.35	install 3′ x 6′ 8″ pre-hung 1-3/4″
	Demolish and Install	Ea.	222	50.85		272.85	door, including casing and stop,
	Reinstall	Ea.		18.36		18.36	hinges, jamb, aluminum sill,
	Clean	Ea.	3.11	10.60		13.71	weatherstripped.
	Paint	Ea.	9.30	43.50		52.80	
	Minimum Charge	Job		174		174	
3′ 6″ x 6′ 8″							
	Demolish	Ea.		32.50		32.50	Cost includes material and labor to
	Install	Ea.	375	19.40		394.40	install 3′ 6″ x 6′ 8″ pre-hung 1-3/4″
	Demolish and Install	Ea.	375	51.90		426.90	door, 4-1/2″ split jamb, including
	Reinstall	Ea.		19.38		19.38	casing and stop, hinges, aluminum sill,
	Clean	Ea.	3.11	10.60		13.71	weatherstripping.
	Paint	Ea.	9.30	43.50		52.80	
	Minimum Charge	Job		174		174	
Raised Panel							
	Demolish	Ea.		32.50		32.50	Cost includes material and labor to
	Install	Ea.	435	19.40		454.40	install 3′ x 6′ 8″ pre-hung 1-3/4″
	Demolish and Install	Ea.	435	51.90		486.90	door, jamb, including casing and
	Reinstall	Ea.		19.38		19.38	stop, hinges, aluminum sill,
	Clean	Ea.	3.11	10.60		13.71	weatherstripping.
	Paint	Ea.	9.30	102		111.30	
	Minimum Charge	Job		174		174	
Stain							
	Install	Ea.	4.44	25.50		29.94	Includes labor and material to stain
	Minimum Charge	Job		153		153	single door and trim on both sides.

Bifold Door		Unit	Material	Labor	Equip.	Total	Specification
Wood, Single							
3′							
	Demolish	Ea.		24		24	Cost includes material and labor to
	Install	Ea.	58	53.50		111.50	install 3′ wide solid wood bi-fold door
	Demolish and Install	Ea.	58	77.50		135.50	not including jamb, trim, track, and
	Reinstall	Ea.		53.66		53.66	hardware.
	Clean	Ea.	3.11	10.60		13.71	
	Paint	Ea.	9.25	43.50		52.75	
	Minimum Charge	Job		174		174	
5′							
	Demolish	Ea.		24		24	Includes material and labor to install
	Install	Ea.	107	63.50		170.50	pair of 2′ 6″ wide solid wood bi-fold
	Demolish and Install	Ea.	107	87.50		194.50	doors not including jamb, trim, track,
	Reinstall	Ea.		63.42		63.42	and hardware.
	Clean	Ea.	1.36	18.15		19.51	
	Paint	Ea.	19.60	76.50		96.10	
	Minimum Charge	Job		174		174	

Finish Carpentry

Bifold Door

Bifold Door		Unit	Material	Labor	Equip.	Total	Specification
6'							
	Demolish	Ea.		24		24	Includes material and labor to install
	Install	Ea.	114	70		184	pair of 3' wide solid wood bi-fold
	Demolish and Install	Ea.	114	94		208	doors not including jamb, trim, track,
	Reinstall	Ea.		69.76		69.76	and hardware.
	Clean	Ea.	1.56	21		22.56	
	Paint	Ea.	19.60	76.50		96.10	
	Minimum Charge	Job		174		174	
8'							
	Demolish	Ea.		24		24	Includes material and labor to install
	Install	Ea.	211	39		250	pair of 4' wide solid wood bi-fold
	Demolish and Install	Ea.	211	63		274	doors not including jamb, trim, track,
	Reinstall	Ea.		38.76		38.76	and hardware.
	Clean	Ea.	2.15	25.50		27.65	
	Paint	Ea.	10.20	76.50		86.70	
	Minimum Charge	Job		174		174	

Mirrored

		Unit	Material	Labor	Equip.	Total	Specification
5'							
	Demolish	Ea.		24		24	Includes material and labor to install
	Install	Ea.	445	39		484	pair of 2' 6" wide mirrored bi-fold
	Demolish and Install	Ea.	445	63		508	doors including jamb, trim, track, and
	Reinstall	Ea.		30.76		30.76	hardware.
	Clean	Ea.	1.36	9.05		10.41	
	Minimum Charge	Job		174		174	
6'							
	Demolish	Ea.		24		24	Includes material and labor to install
	Install	Ea.	445	39		484	pair of 3' wide mirrored bi-fold doors
	Demolish and Install	Ea.	445	63		508	including jamb, trim, track, and
	Reinstall	Ea.		38.76		38.76	hardware.
	Clean	Ea.	1.56	10.60		12.16	
	Minimum Charge	Job		174		174	
8'							
	Demolish	Ea.		24		24	Includes material and labor to install
	Install	Ea.	715	58		773	pair of 4' wide mirrored bi-fold doors
	Demolish and Install	Ea.	715	82		797	including jamb, trim, track, and
	Reinstall	Ea.		58.13		58.13	hardware.
	Clean	Ea.	2.15	12.70		14.85	
	Minimum Charge	Job		174		174	

Louvered

		Unit	Material	Labor	Equip.	Total	Specification
3'							
	Demolish	Ea.		24		24	Cost includes material and labor to
	Install	Ea.	177	53.50		230.50	install 3' wide louvered bi-fold door
	Demolish and Install	Ea.	177	77.50		254.50	not including jamb, trim, track, and
	Reinstall	Ea.		53.66		53.66	hardware.
	Clean	Ea.	3.11	10.60		13.71	
	Paint	Ea.	9.25	43.50		52.75	
	Minimum Charge	Job		174		174	
5'							
	Demolish	Ea.		24		24	Includes material and labor to install
	Install	Ea.	243	63.50		306.50	pair of 2' 6" wide louvered wood
	Demolish and Install	Ea.	243	87.50		330.50	bi-fold doors not including jamb, trim,
	Reinstall	Ea.		63.42		63.42	track, and hardware.
	Clean	Ea.	1.36	18.15		19.51	
	Paint	Ea.	19.60	76.50		96.10	
	Minimum Charge	Job		174		174	

Finish Carpentry

Bifold Door	Unit	Material	Labor	Equip.	Total	Specification
6'						
Demolish	Ea.		24		24	Includes material and labor to install
Install	Ea.	270	70		340	pair of 3' wide louvered wood bi-fold
Demolish and Install	Ea.	270	94		364	doors not including jamb, trim, track,
Reinstall	Ea.		69.76		69.76	and hardware.
Clean	Ea.	1.56	21		22.56	
Paint	Ea.	19.60	76.50		96.10	
Minimum Charge	Job		174		174	
8'						
Demolish	Ea.		24		24	Includes material and labor to install
Install	Ea.	430	35		465	pair of 4' wide solid wood bi-fold
Demolish and Install	Ea.	430	59		489	doors not including jamb, trim, track,
Reinstall	Ea.		34.88		34.88	and hardware.
Clean	Ea.	7.35	18.15		25.50	
Paint	Ea.	10.20	76.50		86.70	
Minimum Charge	Job		174		174	
Accordion Type						
Demolish	S.F.		.83		.83	Includes material and labor to install
Install	S.F.	2.01	1.74		3.75	vinyl folding accordion closet doors,
Demolish and Install	S.F.	2.01	2.57		4.58	including track and frame.
Reinstall	S.F.		1.74		1.74	
Clean	S.F.		.30		.30	
Minimum Charge	Job		174		174	
Casing Trim						
Single Width						
Demolish	Ea.		6.15		6.15	Cost includes material and labor to
Install	Opng.	25.50	59		84.50	install 11/16" x 2-1/2" pine ranch
Demolish and Install	Opng.	25.50	65.15		90.65	style casing for one side of a standard
Reinstall	Opng.		47.29		47.29	door opening.
Clean	Opng.	.44	2.54		2.98	
Paint	Ea.	3.22	6.10		9.32	
Minimum Charge	Job		174		174	
Double Width						
Demolish	Ea.		7.30		7.30	Cost includes material and labor to
Install	Opng.	29.50	70		99.50	install 11/16" x 2-1/2" pine ranch
Demolish and Install	Opng.	29.50	77.30		106.80	style door casing for one side of a
Reinstall	Opng.		55.81		55.81	double door opening.
Clean	Opng.	.52	5.10		5.62	
Paint	Ea.	3.22	6.95		10.17	
Minimum Charge	Job		174		174	
Jamb						
Demolish	Ea.		24		24	Includes material and labor to install
Install	Ea.	103	9.45		112.45	flat pine jamb with square cut heads
Demolish and Install	Ea.	103	33.45		136.45	and rabbeted sides for 6' 8" high and
Clean	Ea.	.10	6.35		6.45	3-9/16" door including trim sets for
Paint	Ea.	6	20.50		26.50	both sides.
Minimum Charge	Job		87		87	
Stain						
Install	Ea.	4.44	25.50		29.94	Includes labor and material to stain
Minimum Charge	Job		153		153	single door and trim on both sides.
Shave & Refit						
Install	Ea.		29		29	Includes labor to shave and rework
Minimum Charge	Job		87		87	door to fit opening at the job site.

Finish Carpentry

Louver Door

	Unit	Material	Labor	Equip.	Total	Specification
Full						
2' x 6' 8"						
Demolish	Ea.		6.90		6.90	Cost includes material and labor to
Install	Ea.	186	35		221	install 2' x 6' 8" pine full louver door,
Demolish and Install	Ea.	186	41.90		227.90	jamb, hinges.
Reinstall	Ea.		34.88		34.88	
Clean	Ea.	5.80	32		37.80	
Paint	Ea.	9.25	43.50		52.75	
Minimum Charge	Job		174		174	
2' 8" x 6' 8"						
Demolish	Ea.		6.90		6.90	Cost includes material and labor to
Install	Ea.	207	35		242	install 2' 8" x 6' 8" pine full louver
Demolish and Install	Ea.	207	41.90		248.90	door, jamb, hinges.
Reinstall	Ea.		34.88		34.88	
Clean	Ea.	5.80	32		37.80	
Paint	Ea.	9.25	43.50		52.75	
Minimum Charge	Job		174		174	
3' x 6' 8"						
Demolish	Ea.		6.90		6.90	Cost includes material and labor to
Install	Ea.	218	36.50		254.50	install 3' x 6' 8" pine full louver door,
Demolish and Install	Ea.	218	43.40		261.40	jamb, hinges.
Reinstall	Ea.		36.72		36.72	
Clean	Ea.	5.80	32		37.80	
Paint	Ea.	9.25	43.50		52.75	
Minimum Charge	Job		174		174	
Half						
2' x 6' 8"						
Demolish	Ea.		6.90		6.90	Cost includes material and labor to
Install	Ea.	300	35		335	install 2' x 6' 8" pine half louver door,
Demolish and Install	Ea.	300	41.90		341.90	jamb, hinges.
Reinstall	Ea.		34.88		34.88	
Clean	Ea.	5.80	32		37.80	
Paint	Ea.	9.25	43.50		52.75	
Minimum Charge	Job		174		174	
2' 8" x 6' 8"						
Demolish	Ea.		6.90		6.90	Cost includes material and labor to
Install	Ea.	305	35		340	install 2' 8" x 6' 8" pine half louver
Demolish and Install	Ea.	305	41.90		346.90	door, jamb, hinges.
Reinstall	Ea.		34.88		34.88	
Clean	Ea.	5.80	32		37.80	
Paint	Ea.	9.25	43.50		52.75	
Minimum Charge	Job		174		174	
3' x 6' 8"						
Demolish	Ea.		6.90		6.90	Cost includes material and labor to
Install	Ea.	315	36.50		351.50	install 3' x 6' 8" pine half louver door,
Demolish and Install	Ea.	315	43.40		358.40	jamb, hinges.
Reinstall	Ea.		36.72		36.72	
Clean	Ea.	5.80	32		37.80	
Paint	Ea.	9.25	43.50		52.75	
Minimum Charge	Job		174		174	
Jamb						
Demolish	Ea.		24		24	Includes material and labor to install
Install	Ea.	103	9.45		112.45	flat pine jamb with square cut heads
Demolish and Install	Ea.	103	33.45		136.45	and rabbeted sides for 6' 8" high and
Clean	Ea.	.10	6.35		6.45	3-9/16" door including trim sets for
Paint	Ea.	6	20.50		26.50	both sides.
Minimum Charge	Job		87		87	

Finish Carpentry

Louver Door		Unit	Material	Labor	Equip.	Total	Specification
Stain							
	Install	Ea.	4.44	25.50		29.94	Includes labor and material to stain
	Minimum Charge	Job		153		153	single door and trim on both sides.
Shave & Refit							
	Install	Ea.		29		29	Includes labor to shave and rework
	Minimum Charge	Job		153		153	door to fit opening at the job site.

Bypass Sliding Door		Unit	Material	Labor	Equip.	Total	Specification
Wood							
5'							
	Demolish	Ea.		32.50		32.50	Includes material and labor to install
	Install	Opng.	199	63.50		262.50	pair of 2' 6" wide lauan hollow core
	Demolish and Install	Opng.	199	96		295	wood bi-pass doors with jamb, track
	Reinstall	Opng.		63.42		63.42	hardware.
	Clean	Ea.	1.36	18.15		19.51	
	Paint	Ea.	19.60	76.50		96.10	
	Minimum Charge	Job		174		174	
6'							
	Demolish	Ea.		32.50		32.50	Includes material and labor to install
	Install	Opng.	213	70		283	pair of 3' wide lauan hollow core
	Demolish and Install	Opng.	213	102.50		315.50	wood bi-pass doors with jamb, track
	Reinstall	Opng.		69.76		69.76	and hardware.
	Clean	Ea.	1.56	21		22.56	
	Paint	Ea.	19.60	76.50		96.10	
	Minimum Charge	Job		174		174	
8'							
	Demolish	Ea.		32.50		32.50	Includes material and labor to install
	Install	Opng.	340	87		427	pair of 4' wide lauan hollow core
	Demolish and Install	Opng.	340	119.50		459.50	wood bi-pass doors with jamb, track
	Reinstall	Opng.		87.20		87.20	and hardware.
	Clean	Ea.	2.15	25.50		27.65	
	Paint	Ea.	10.20	76.50		86.70	
	Minimum Charge	Job		174		174	
Mirrored							
5'							
	Demolish	Ea.		32.50		32.50	Includes material and labor to install
	Install	Opng.	224	70		294	pair of 2' 6 " wide mirrored bi-pass
	Demolish and Install	Opng.	224	102.50		326.50	doors with jamb, trim, track hardware.
	Reinstall	Opng.		69.76		69.76	
	Clean	Ea.	1.36	9.05		10.41	
	Minimum Charge	Job		174		174	
6'							
	Demolish	Ea.		32.50		32.50	Includes material and labor to install
	Install	Opng.	256	70		326	pair of 3' wide mirrored bi-pass doors
	Demolish and Install	Opng.	256	102.50		358.50	with jamb, trim, track hardware.
	Reinstall	Opng.		69.76		69.76	
	Clean	Ea.	1.56	10.60		12.16	
	Minimum Charge	Job		174		174	
8'							
	Demolish	Ea.		32.50		32.50	Includes material and labor to install
	Install	Opng.	535	77.50		612.50	pair of 4' wide mirrored bi-pass doors
	Demolish and Install	Opng.	535	110		645	with jamb, trim, track hardware.
	Reinstall	Opng.		77.51		77.51	
	Clean	Ea.	2.15	12.70		14.85	
	Minimum Charge	Job		174		174	

Finish Carpentry

Bypass Sliding Door

	Unit	Material	Labor	Equip.	Total	Specification
Casing Trim						
Single Width						
Demolish	Ea.		6.15		6.15	Cost includes material and labor to
Install	Opng.	25.50	59		84.50	install 11/16" x 2-1/2" pine ranch
Demolish and Install	Opng.	25.50	65.15		90.65	style casing for one side of a standard
Reinstall	Opng.		47.29		47.29	door opening.
Clean	Opng.	.44	2.54		2.98	
Paint	Ea.	3.22	6.10		9.32	
Minimum Charge	Job		174		174	
Double Width						
Demolish	Ea.		7.30		7.30	Cost includes material and labor to
Install	Opng.	29.50	70		99.50	install 11/16" x 2-1/2" pine ranch
Demolish and Install	Opng.	29.50	77.30		106.80	style door casing for one side of a
Reinstall	Opng.		55.81		55.81	double door opening.
Clean	Opng.	.52	5.10		5.62	
Paint	Ea.	3.22	6.95		10.17	
Minimum Charge	Job		174		174	
Jamb						
Demolish	Ea.		24		24	Includes material and labor to install
Install	Ea.	103	9.45		112.45	flat pine jamb with square cut heads
Demolish and Install	Ea.	103	33.45		136.45	and rabbeted sides for 6' 8" high and
Clean	Ea.	.10	6.35		6.45	3-9/16" door including trim sets for
Paint	Ea.	6	20.50		26.50	both sides.
Minimum Charge	Job		87		87	
Stain						
Install	Ea.	4.44	25.50		29.94	Includes labor and material to stain
Minimum Charge	Job		153		153	single door and trim on both sides.
Shave & Refit						
Install	Ea.		29		29	Includes labor to shave and rework
Minimum Charge	Job		87		87	door to fit opening at the job site.

Base Molding

	Unit	Material	Labor	Equip.	Total	Specification
One Piece						
1-5/8"						
Demolish	L.F.		.47		.47	Includes material and labor to install
Install	L.F.	.39	1.03		1.42	pine or birch base molding.
Demolish and Install	L.F.	.39	1.50		1.89	
Reinstall	L.F.		.83		.83	
Clean	L.F.	.02	.15		.17	
Paint	L.F.	.03	.33		.36	
Minimum Charge	Job		174		174	
2-1/2"						
Demolish	L.F.		.47		.47	Cost includes material and labor to
Install	L.F.	.59	1.07		1.66	install 1/2" x 2-1/2" pine or birch
Demolish and Install	L.F.	.59	1.54		2.13	base molding.
Reinstall	L.F.		.86		.86	
Clean	L.F.	.02	.15		.17	
Paint	L.F.	.03	.33		.36	
Minimum Charge	Job		174		174	

Finish Carpentry

Base Molding

	Unit	Material	Labor	Equip.	Total	Specification
3-1/2"						
Demolish	L.F.		.47		.47	Cost includes material and labor to
Install	L.F.	1.70	1.12		2.82	install 9/16" x 3-1/2" pine or birch
Demolish and Install	L.F.	1.70	1.59		3.29	base molding.
Reinstall	L.F.		.89		.89	
Clean	L.F.	.02	.15		.17	
Paint	L.F.	.03	.36		.39	
Minimum Charge	Job		174		174	
6"						
Demolish	L.F.		.47		.47	Cost includes material and labor to
Install	L.F.	1.22	1.22		2.44	install 1" x 6" pine or birch base
Demolish and Install	L.F.	1.22	1.69		2.91	molding.
Reinstall	L.F.		.98		.98	
Clean	L.F.	.02	.21		.23	
Paint	L.F.	.06	.46		.52	
Minimum Charge	Job		174		174	
8"						
Demolish	L.F.		.56		.56	Cost includes material and labor to
Install	L.F.	1.55	1.34		2.89	install 1" x 8" pine or birch base
Demolish and Install	L.F.	1.55	1.90		3.45	molding.
Reinstall	L.F.		1.07		1.07	
Clean	L.F.	.03	.25		.28	
Paint	L.F.	.07	.51		.58	
Minimum Charge	Job		174		174	
Premium / Custom Grade						
Demolish	L.F.		.56		.56	Includes material and labor to install
Install	L.F.	4.20	3.35		7.55	custom 2 or 3 piece oak or other
Demolish and Install	L.F.	4.20	3.91		8.11	hardwood, stain grade trim.
Reinstall	L.F.		2.68		2.68	
Clean	L.F.	.03	.25		.28	
Paint	L.F.	.07	.51		.58	
Minimum Charge	Job		174		174	
One Piece Oak						
1-5/8"						
Demolish	L.F.		.47		.47	Cost includes material and labor to
Install	L.F.	.70	1.03		1.73	install 1/2" x 1-5/8" oak base
Demolish and Install	L.F.	.70	1.50		2.20	molding.
Reinstall	L.F.		.83		.83	
Clean	L.F.	.02	.15		.17	
Paint	L.F.	.08	.45		.53	
Minimum Charge	Job		174		174	
2-1/2"						
Demolish	L.F.		.47		.47	Cost includes material and labor to
Install	L.F.	1.07	1.05		2.12	install 1/2" x 2-1/2" oak base
Demolish and Install	L.F.	1.07	1.52		2.59	molding.
Reinstall	L.F.		.84		.84	
Clean	L.F.	.02	.15		.17	
Paint	L.F.	.08	.46		.54	
Minimum Charge	Job		174		174	
3-1/2"						
Demolish	L.F.		.47		.47	Cost includes material and labor to
Install	L.F.	2.62	1.07		3.69	install 1/2" x 3-1/2" oak base
Demolish and Install	L.F.	2.62	1.54		4.16	molding.
Reinstall	L.F.		.86		.86	
Clean	L.F.	.02	.15		.17	
Paint	L.F.	.09	.47		.56	
Minimum Charge	Job		174		174	

Finish Carpentry

Base Molding		Unit	Material	Labor	Equip.	Total	Specification
6"							
	Demolish	L.F.		.47		.47	Cost includes material and labor to
	Install	L.F.	5.55	1.22		6.77	install 1" x 6" oak base molding.
	Demolish and Install	L.F.	5.55	1.69		7.24	
	Reinstall	L.F.		.98		.98	
	Clean	L.F.	.03	.25		.28	
	Paint	L.F.	.09	.49		.58	
	Minimum Charge	Job		174		174	
8"							
	Demolish	L.F.		.56		.56	Cost includes material and labor to
	Install	L.F.	6.45	1.34		7.79	install 1" x 8" oak base molding.
	Demolish and Install	L.F.	6.45	1.90		8.35	
	Reinstall	L.F.		1.07		1.07	
	Clean	L.F.	.03	.25		.28	
	Paint	L.F.	.09	.51		.60	
	Minimum Charge	Job		174		174	
Vinyl Base Molding							
2-1/2"							
	Demolish	L.F.		.47		.47	Cost includes material and labor to
	Install	L.F.	.72	1.01		1.73	install 2-1/2" vinyl or rubber cove
	Demolish and Install	L.F.	.72	1.48		2.20	base including adhesive.
	Reinstall	L.F.		1.01		1.01	
	Clean	L.F.		.23		.23	
	Minimum Charge	Job		162		162	
4"							
	Demolish	L.F.		.47		.47	Cost includes material and labor to
	Install	L.F.	.77	1.01		1.78	install 4" vinyl or rubber cove base
	Demolish and Install	L.F.	.77	1.48		2.25	including adhesive.
	Reinstall	L.F.		1.01		1.01	
	Clean	L.F.		.23		.23	
	Minimum Charge	Job		162		162	
6"							
	Demolish	L.F.		.47		.47	Cost includes material and labor to
	Install	L.F.	1.21	1.01		2.22	install 6" vinyl or rubber cove base
	Demolish and Install	L.F.	1.21	1.48		2.69	including adhesive.
	Reinstall	L.F.		1.01		1.01	
	Clean	L.F.		.23		.23	
	Minimum Charge	Job		162		162	
Prefinished							
	Demolish	L.F.		.47		.47	Cost includes material and labor to
	Install	L.F.	1.69	1.12		2.81	install 9/16" x 3-5/16" prefinished
	Demolish and Install	L.F.	1.69	1.59		3.28	base molding.
	Reinstall	L.F.		.89		.89	
	Clean	L.F.	.02	.15		.17	
	Paint	L.F.	.03	.36		.39	
	Minimum Charge	Job		174		174	
Shoe							
	Demolish	L.F.		.47		.47	Includes material and labor to install
	Install	L.F.	.39	1.03		1.42	pine or birch base molding.
	Demolish and Install	L.F.	.39	1.50		1.89	
	Reinstall	L.F.		.83		.83	
	Clean	L.F.	.02	.15		.17	
	Paint	L.F.	.13	.49		.62	
	Minimum Charge	Job		174		174	

Finish Carpentry

Base Molding

Base Molding		Unit	Material	Labor	Equip.	Total	Specification
Carpeted							Includes material and labor to install carpet baseboard with adhesive.
	Demolish	L.F.		.19		.19	
	Install	L.F.	1.32	1.94		3.26	
	Demolish and Install	L.F.	1.32	2.13		3.45	
	Reinstall	L.F.		1.55		1.55	
	Clean	L.F.		.10		.10	
	Minimum Charge	Job		160		160	
Ceramic Tile							Includes material and labor to install thin set ceramic tile cove base including grout.
	Demolish	L.F.		1.03		1.03	
	Install	L.F.	3.85	4.42		8.27	
	Demolish and Install	L.F.	3.85	5.45		9.30	
	Clean	L.F.		.48		.48	
	Minimum Charge	Job		160		160	
Quarry Tile							Includes material and labor to install quarry tile base including grout.
	Demolish	L.F.		1.03		1.03	
	Install	L.F.	5.25	5.15		10.40	
	Demolish and Install	L.F.	5.25	6.18		11.43	
	Clean	L.F.		.48		.48	
	Minimum Charge	Job		160		160	
Caulk and Renail							Includes labor and material to renail base and caulk where necessary.
	Install	L.F.	.01	.55		.56	
	Minimum Charge	Job		174		174	

Molding

Molding		Unit	Material	Labor	Equip.	Total	Specification
Oak Base							Cost includes material and labor to install 1/2" x 2-1/2" oak base molding.
	Demolish	L.F.		.47		.47	
	Install	L.F.	1.07	1.05		2.12	
	Demolish and Install	L.F.	1.07	1.52		2.59	
	Reinstall	L.F.		.84		.84	
	Clean	L.F.	.02	.15		.17	
	Paint	L.F.	.08	.46		.54	
	Minimum Charge	Job		174		174	
Shoe							Includes material and labor to install pine or birch base molding.
	Demolish	L.F.		.47		.47	
	Install	L.F.	.39	1.03		1.42	
	Demolish and Install	L.F.	.39	1.50		1.89	
	Reinstall	L.F.		.83		.83	
	Clean	L.F.	.02	.15		.17	
	Paint	L.F.	.13	.49		.62	
	Minimum Charge	Job		174		174	
Quarter Round							Includes material and labor to install pine base quarter round molding.
	Demolish	L.F.		.47		.47	
	Install	L.F.	.70	1.37		2.07	
	Demolish and Install	L.F.	.70	1.84		2.54	
	Reinstall	L.F.		1.09		1.09	
	Clean	L.F.	.02	.15		.17	
	Paint	L.F.	.07	.76		.83	
	Minimum Charge	Job		174		174	
Chair Rail							Cost includes material and labor to install 5/8" x 2-1/2" oak chair rail molding.
	Demolish	L.F.		.47		.47	
	Install	L.F.	1.58	1.29		2.87	
	Demolish and Install	L.F.	1.58	1.76		3.34	
	Reinstall	L.F.		1.03		1.03	
	Clean	L.F.	.02	.15		.17	
	Paint	L.F.	.07	.76		.83	
	Minimum Charge	Job		174		174	

Finish Carpentry

Molding

Molding	Unit	Material	Labor	Equip.	Total	Specification
Crown						
Demolish	L.F.		.47		.47	Includes material and labor to install
Install	L.F.	2.05	1.40		3.45	unfinished pine crown molding.
Demolish and Install	L.F.	2.05	1.87		3.92	
Reinstall	L.F.		1.12		1.12	
Clean	L.F.	.03	.20		.23	
Paint	L.F.	.07	.76		.83	
Minimum Charge	Job		174		174	
Cove						
Demolish	L.F.		.47		.47	Cost includes material and labor to
Install	L.F.	1.43	1.37		2.80	install 1/2" x 2-3/4" pine cove
Demolish and Install	L.F.	1.43	1.84		3.27	molding.
Reinstall	L.F.		1.09		1.09	
Clean	L.F.	.02	.15		.17	
Paint	L.F.	.07	.76		.83	
Minimum Charge	Job		174		174	
Corner						
Demolish	L.F.		.47		.47	Includes material and labor to install
Install	L.F.	.61	.55		1.16	inside or outside pine corner molding.
Demolish and Install	L.F.	.61	1.02		1.63	
Reinstall	L.F.		.44		.44	
Clean	L.F.	.02	.15		.17	
Paint	L.F.	.07	.76		.83	
Minimum Charge	Job		174		174	
Picture						
Demolish	L.F.		.47		.47	Cost includes material and labor to
Install	L.F.	1.49	1.45		2.94	install 9/16" x 2-1/2" pine casing.
Demolish and Install	L.F.	1.49	1.92		3.41	
Reinstall	L.F.		1.16		1.16	
Clean	L.F.	.02	.15		.17	
Paint	L.F.	.07	.76		.83	
Minimum Charge	Job		174		174	
Single Cased Opening						
Demolish	Ea.		24		24	Includes material and labor to install
Install	Ea.	18.80	22		40.80	flat pine jamb with square cut heads
Demolish and Install	Ea.	18.80	46		64.80	and rabbeted sides for 6' 8" high
Clean	Ea.	.22	5.30		5.52	opening including trim sets for both
Paint	Ea.	1.24	7.65		8.89	sides.
Minimum Charge	Job		174		174	
Double Cased Opening						
Demolish	Ea.		24		24	Includes material and labor to install
Install	Ea.	37.50	25		62.50	flat pine jamb with square cut heads
Demolish and Install	Ea.	37.50	49		86.50	and rabbeted sides for 6' 8" high
Clean	Ea.	.25	6.20		6.45	opening including trim sets for both
Paint	Ea.	1.43	9		10.43	sides.
Minimum Charge	Job		174		174	

Window Trim Set

Window Trim Set	Unit	Material	Labor	Equip.	Total	Specification
Single						
Demolish	Ea.		6.25		6.25	Cost includes material and labor to
Install	Opng.	39.50	35		74.50	install 11/16" x 2-1/2" pine ranch
Demolish and Install	Opng.	39.50	41.25		80.75	style window casing.
Reinstall	Opng.		27.90		27.90	
Clean	Ea.	.31	3.63		3.94	
Paint	Ea.	1.16	7.25		8.41	
Minimum Charge	Job		174		174	

Finish Carpentry

Window Trim Set

		Unit	Material	Labor	Equip.	Total	Specification
Double							
	Demolish	Ea.		7.05		7.05	Cost includes material and labor to
	Install	Ea.	54.50	43.50		98	install 11/16" x 2-1/2" pine ranch
	Demolish and Install	Ea.	54.50	50.55		105.05	style trim for casing one side of a
	Reinstall	Ea.		34.88		34.88	double window opening.
	Clean	Ea.	.56	4.23		4.79	
	Paint	Ea.	1.95	12.20		14.15	
	Minimum Charge	Job		174		174	
Triple							
	Demolish	Ea.		8.05		8.05	Cost includes material and labor to
	Install	Opng.	69	58		127	install 11/16" x 2-1/2" pine ranch
	Demolish and Install	Opng.	69	66.05		135.05	style trim for casing one side of a
	Reinstall	Opng.		46.51		46.51	triple window opening.
	Clean	Ea.	.78	5.10		5.88	
	Paint	Ea.	2.73	16.95		19.68	
	Minimum Charge	Job		174		174	
Window Casing Per L.F.							
	Demolish	L.F.		.47		.47	Cost includes material and labor to
	Install	L.F.	1.49	1.45		2.94	install 11/16" x 2-1/2" pine ranch
	Demolish and Install	L.F.	1.49	1.92		3.41	style trim for casing.
	Reinstall	L.F.		1.16		1.16	
	Clean	L.F.	.02	.15		.17	
	Paint	L.F.	.07	.76		.83	
	Minimum Charge	Job		174		174	

Window Sill

		Unit	Material	Labor	Equip.	Total	Specification
Wood							
	Demolish	L.F.		.47		.47	Includes material and labor to install
	Install	L.F.	2.42	1.74		4.16	flat, wood window stool.
	Demolish and Install	L.F.	2.42	2.21		4.63	
	Reinstall	L.F.		1.40		1.40	
	Clean	L.F.	.02	.15		.17	
	Paint	L.F.	.07	.76		.83	
	Minimum Charge	Job		174		174	
Marble							
	Demolish	L.F.		.71		.71	Includes material and labor to install
	Install	L.F.	8.05	7.30		15.35	marble window sills, 6" x 3/4" thick.
	Demolish and Install	L.F.	8.05	8.01		16.06	
	Reinstall	L.F.		5.84		5.84	
	Clean	L.F.	.02	.15		.17	
	Minimum Charge	Job		160		160	

Shelving

		Unit	Material	Labor	Equip.	Total	Specification
Pine							
18"							
	Demolish	L.F.		.81		.81	Cost includes material and labor to
	Install	L.F.	4.41	3.67		8.08	install 1" thick custom shelving board
	Demolish and Install	L.F.	4.41	4.48		8.89	18" wide.
	Reinstall	L.F.		2.94		2.94	
	Clean	L.F.	.04	.42		.46	
	Paint	L.F.	.86	.48		1.34	
	Minimum Charge	Job		174		174	

Finish Carpentry

Shelving

Shelving		Unit	Material	Labor	Equip.	Total	Specification
24"							
	Demolish	L.F.		.81		.81	Cost includes material and labor to
	Install	L.F.	5.90	4.10		10	install 1" thick custom shelving board.
	Demolish and Install	L.F.	5.90	4.91		10.81	
	Reinstall	L.F.		3.28		3.28	
	Clean	L.F.	.07	.64		.71	
	Paint	L.F.	1.18	.64		1.82	
	Minimum Charge	Job		174		174	
Particle Board							
	Demolish	L.F.		.74		.74	Cost includes material and labor to
	Install	L.F.	1.22	1.84		3.06	install 3/4" thick particle shelving
	Demolish and Install	L.F.	1.22	2.58		3.80	board 18" wide.
	Reinstall	L.F.		1.47		1.47	
	Clean	L.F.	.04	.42		.46	
	Paint	S.F.	.12	.71		.83	
	Minimum Charge	Job		174		174	
Plywood							
	Demolish	L.F.		.74		.74	Cost includes material and labor to
	Install	L.F.	.89	5.35		6.24	install 3/4" thick plywood shelving
	Demolish and Install	L.F.	.89	6.09		6.98	board.
	Reinstall	L.F.		4.29		4.29	
	Clean	L.F.	.04	.42		.46	
	Paint	S.F.	.12	.71		.83	
	Minimum Charge	Job		174		174	
Hardwood (oak)							
	Demolish	L.F.		.74		.74	Includes material and labor to install
	Install	L.F.	7.45	2.41		9.86	oak shelving, 1" x 12", incl. cleats and
	Demolish and Install	L.F.	7.45	3.15		10.60	bracing.
	Reinstall	L.F.		1.92		1.92	
	Clean	L.F.	.04	.42		.46	
	Paint	S.F.	.12	.71		.83	
	Minimum Charge	Job		174		174	
Custom Bookcase							
	Demolish	L.F.		9.40		9.40	Includes material and labor to install
	Install	L.F.	25.50	35		60.50	custom modular bookcase unit with
	Demolish and Install	L.F.	25.50	44.40		69.90	clear pine faced frames, shelves 12"
	Reinstall	L.F.		27.90		27.90	O.C., 7' high, 8" deep.
	Clean	L.F.	1.24	9.05		10.29	
	Paint	L.F.	1.01	23.50		24.51	
	Minimum Charge	Job		174		174	
Glass							
	Demolish	L.F.		.74		.74	Includes material and labor to install
	Install	L.F.	28	2.41		30.41	glass shelving, 1" x 12", incl. cleats
	Demolish and Install	L.F.	28	3.15		31.15	and bracing.
	Reinstall	L.F.		1.92		1.92	
	Clean	L.F.	.06	.39		.45	
	Minimum Charge	Job		169		169	

Closet Shelving

Closet Shelving		Unit	Material	Labor	Equip.	Total	Specification
Closet Shelf and Rod							
	Demolish	L.F.		.74		.74	Cost includes material and labor to
	Install	L.F.	2.42	5.80		8.22	install 1" thick custom shelving board
	Demolish and Install	L.F.	2.42	6.54		8.96	18" wide and 1" diameter clothes pole
	Reinstall	L.F.		4.65		4.65	with brackets 3' O.C.
	Clean	L.F.	.03	.32		.35	
	Paint	L.F.	.13	.76		.89	
	Minimum Charge	Job		174		174	

Finish Carpentry

Closet Shelving

Closet Shelving	Unit	Material	Labor	Equip.	Total	Specification
Clothing Rod						
Demolish	L.F.		.49		.49	Includes material and labor to install
Install	L.F.	1.76	1.74		3.50	fir closet pole, 1-5/8″ diameter.
Demolish and Install	L.F.	1.76	2.23		3.99	
Reinstall	L.F.		1.40		1.40	
Clean	L.F.	.03	.44		.47	
Paint	L.F.	.12	.83		.95	
Minimum Charge	Job		174		174	

Stair Assemblies

Stair Assemblies	Unit	Material	Labor	Equip.	Total	Specification
Straight Hardwood						
Demolish	Ea.		335		335	Includes material and labor to install
Install	Flight	1000	233		1233	factory cut and assembled straight
Demolish and Install	Flight	1000	568		1568	closed box stairs with oak treads and
Clean	Flight	3.11	22		25.11	prefinished stair rail with balusters.
Paint	Flight	12.05	40.50		52.55	
Minimum Charge	Job		174		174	
Clear Oak Tread						
Demolish	Ea.		8.80		8.80	Includes material and labor to install
Install	Ea.	65	19.40		84.40	1-1/4″ thick clear oak treads per riser.
Demolish and Install	Ea.	65	28.20		93.20	
Clean	Ea.	.10	2.21		2.31	
Paint	Ea.	.41	2.26		2.67	
Minimum Charge	Job		174		174	
Landing						
Demolish	S.F.		2.94		2.94	Includes material and labor to install
Install	S.F.	7.20	19.40		26.60	clear oak landing.
Demolish and Install	S.F.	7.20	22.34		29.54	
Clean	S.F.		.32		.32	
Paint	S.F.	.29	.75		1.04	
Minimum Charge	Job		174		174	
Refinish / Stain						
Install	S.F.	1.25	1.95		3.20	Includes labor and material to sand
Minimum Charge	Job		153		153	and finish (3 passes) with two coats of
						urethane on a new floor.
Spiral Hardwood						
Demolish	Ea.		335		335	Cost includes material and labor to
Install	Flight	5150	465		5615	install 4′ - 6′ diameter spiral stairs with
Demolish and Install	Flight	5150	800		5950	oak treads, factory cut and assembled
Clean	Flight	3.11	22		25.11	with double handrails.
Paint	Flight	12.05	40.50		52.55	
Minimum Charge	Job		174		174	

Stairs

Stairs	Unit	Material	Labor	Equip.	Total	Specification
Disappearing						
Demolish	Ea.		17.65		17.65	Includes material and labor to install a
Install	Ea.	135	99.50		234.50	folding staircase.
Demolish and Install	Ea.	135	117.15		252.15	
Reinstall	Ea.		79.73		79.73	
Clean	Ea.	.46	15.90		16.36	
Minimum Charge	Job		174		174	

Stair Components	Unit	Material	Labor	Equip.	Total	Specification
Handrail w / Balusters						
Demolish	L.F.		2.35		2.35	Includes material and labor to install
Install	L.F.	23	7.25		30.25	prefinished assembled stair rail with
Demolish and Install	L.F.	23	9.60		32.60	brackets, turned balusters and newel.
Clean	L.F.	.63	1.75		2.38	
Paint	L.F.	1.44	3.59		5.03	
Minimum Charge	Job		174		174	
Bannister						
Demolish	L.F.		2.35		2.35	Includes material and labor to install
Install	L.F.	36	5.80		41.80	built-up oak railings.
Demolish and Install	L.F.	36	8.15		44.15	
Clean	L.F.	.63	1.59		2.22	
Paint	L.F.	1.44	2.77		4.21	
Minimum Charge	Job		174		174	

Wall and Base Cabinets   **Countertop**

Wall Cabinets	Unit	Material	Labor	Equip.	Total	Specification
Good Grade Laminated						
12″ Wide, 1 Door						
Demolish	Ea.		7.05		7.05	Includes material and labor to install
Install	Ea.	116	31.50		147.50	good modular unit with melamine
Demolish and Install	Ea.	116	38.55		154.55	laminated to particle board, in
Reinstall	Ea.		25.37		25.37	textured colors or wood grain print
Clean	Ea.	.46	4.88		5.34	finish, hinges and pulls.
Paint	Ea.	.43	10.90		11.33	
Minimum Charge	Job		174		174	
15″ Wide, 1 Door						
Demolish	Ea.		8.80		8.80	Includes material and labor to install
Install	Ea.	130	32.50		162.50	good modular unit with melamine
Demolish and Install	Ea.	130	41.30		171.30	laminated to particle board, in
Reinstall	Ea.		26.08		26.08	textured colors or wood grain print
Clean	Ea.	.58	6.05		6.63	finish, hinges and pulls.
Paint	Ea.	.54	13.85		14.39	
Minimum Charge	Job		174		174	
18″ Wide, 1 Door						
Demolish	Ea.		10.65		10.65	Includes material and labor to install
Install	Ea.	143	33.50		176.50	good modular unit with melamine
Demolish and Install	Ea.	143	44.15		187.15	laminated to particle board, in
Reinstall	Ea.		26.70		26.70	textured colors or wood grain print
Clean	Ea.	.70	7.25		7.95	finish, hinges and pulls.
Paint	Ea.	.65	16.05		16.70	
Minimum Charge	Job		174		174	
21″ Wide, 1 Door						
Demolish	Ea.		12.25		12.25	Includes material and labor to install
Install	Ea.	161	34		195	good modular unit with melamine
Demolish and Install	Ea.	161	46.25		207.25	laminated to particle board, in
Reinstall	Ea.		27.09		27.09	textured colors or wood grain print
Clean	Ea.	.81	8.45		9.26	finish, hinges and pulls.
Paint	Ea.	.76	19.10		19.86	
Minimum Charge	Job		174		174	
24″ w / Blind Corner						
Demolish	Ea.		14.10		14.10	Includes material and labor to install
Install	Ea.	161	34.50		195.50	good modular unit with melamine
Demolish and Install	Ea.	161	48.60		209.60	laminated to particle board, in
Reinstall	Ea.		27.49		27.49	textured colors or wood grain print
Clean	Ea.	.94	9.75		10.69	finish, hinges and pulls.
Paint	Ea.	.87	22		22.87	
Minimum Charge	Job		174		174	

Cabinets and Countertops

Wall Cabinets	Unit	Material	Labor	Equip.	Total	Specification
27" Wide, 2 Door						
Demolish	Ea.		15.70		15.70	Includes material and labor to install
Install	Ea.	179	35		214	good modular unit with melamine
Demolish and Install	Ea.	179	50.70		229.70	laminated to particle board, in
Reinstall	Ea.		28.19		28.19	textured colors or wood grain print
Clean	Ea.	1.06	11.05		12.11	finish, hinges and pulls.
Paint	Ea.	.98	25.50		26.48	
Minimum Charge	Job		174		174	
30" Wide, 2 Door						
Demolish	Ea.		17.65		17.65	Includes material and labor to install
Install	Ea.	193	36		229	good modular unit with melamine
Demolish and Install	Ea.	193	53.65		246.65	laminated to particle board, in
Reinstall	Ea.		28.92		28.92	textured colors or wood grain print
Clean	Ea.	1.17	12.10		13.27	finish, hinges and pulls.
Paint	Ea.	1.09	28		29.09	
Minimum Charge	Job		174		174	
36" Wide, 2 Door						
Demolish	Ea.		21		21	Includes material and labor to install
Install	Ea.	221	37		258	good modular unit with melamine
Demolish and Install	Ea.	221	58		279	laminated to particle board, in
Reinstall	Ea.		29.69		29.69	textured colors or wood grain print
Clean	Ea.	1.40	14.95		16.35	finish, hinges and pulls.
Paint	Ea.	1.31	34		35.31	
Minimum Charge	Job		174		174	
48" Wide, 2 Door						
Demolish	Ea.		28		28	Includes material and labor to install
Install	Ea.	271	38		309	good modular unit with melamine
Demolish and Install	Ea.	271	66		337	laminated to particle board, in
Reinstall	Ea.		30.33		30.33	textured colors or wood grain print
Clean	Ea.	1.87	19.55		21.42	finish, hinges and pulls.
Paint	Ea.	1.74	43.50		45.24	
Minimum Charge	Job		174		174	
Above-Appliance						
Demolish	Ea.		17.65		17.65	Includes material and labor to install
Install	Ea.	130	28		158	good modular unit with melamine
Demolish and Install	Ea.	130	45.65		175.65	laminated to particle board, in
Reinstall	Ea.		22.50		22.50	textured colors or wood grain print
Clean	Ea.	1.17	12.10		13.27	finish, hinges and pulls.
Paint	Ea.	1.09	28		29.09	
Minimum Charge	Job		174		174	
Better Grade Wood						
12" Wide, 1 Door						
Demolish	Ea.		7.05		7.05	Includes material and labor to install
Install	Ea.	154	31.50		185.50	custom modular unit with solid
Demolish and Install	Ea.	154	38.55		192.55	hardwood faced frames, hardwood
Reinstall	Ea.		25.37		25.37	door frames and drawer fronts,
Clean	Ea.	.46	4.88		5.34	hardwood veneer on raised door
Paint	Ea.	.43	11.75		12.18	panels, hinges and pulls.
Minimum Charge	Job		174		174	
15" Wide, 1 Door						
Demolish	Ea.		8.80		8.80	Includes material and labor to install
Install	Ea.	174	32.50		206.50	custom modular unit with solid
Demolish and Install	Ea.	174	41.30		215.30	hardwood faced frames, hardwood
Reinstall	Ea.		26.08		26.08	door frames and drawer fronts,
Clean	Ea.	.58	6.05		6.63	hardwood veneer on raised door
Paint	Ea.	.54	14.55		15.09	panels, hinges and pulls.
Minimum Charge	Job		174		174	

Cabinets and Countertops

Wall Cabinets

	Unit	Material	Labor	Equip.	Total	Specification
18″ Wide, 1 Door						
Demolish	Ea.		10.65		10.65	Includes material and labor to install
Install	Ea.	190	33.50		223.50	custom modular unit with solid
Demolish and Install	Ea.	190	44.15		234.15	hardwood faced frames, hardwood
Reinstall	Ea.		26.70		26.70	door frames and drawer fronts,
Clean	Ea.	.70	7.25		7.95	hardwood veneer on raised door
Paint	Ea.	.65	17.95		18.60	panels, hinges and pulls.
Minimum Charge	Job		174		174	
21″ Wide, 1 Door						
Demolish	Ea.		12.25		12.25	Includes material and labor to install
Install	Ea.	215	34		249	custom modular unit with solid
Demolish and Install	Ea.	215	46.25		261.25	hardwood faced frames, hardwood
Reinstall	Ea.		27.09		27.09	door frames and drawer fronts,
Clean	Ea.	.81	8.45		9.26	hardwood veneer on raised door
Paint	Ea.	.76	20.50		21.26	panels, hinges and pulls.
Minimum Charge	Job		174		174	
24″ w / Blind Corner						
Demolish	Ea.		14.10		14.10	Includes material and labor to install
Install	Ea.	215	34.50		249.50	custom modular unit with solid
Demolish and Install	Ea.	215	48.60		263.60	hardwood faced frames, hardwood
Reinstall	Ea.		27.49		27.49	door frames and drawer fronts,
Clean	Ea.	.94	9.75		10.69	hardwood veneer on raised door
Paint	Ea.	.87	23.50		24.37	panels, hinges and pulls.
Minimum Charge	Job		174		174	
27″ Wide, 2 Door						
Demolish	Ea.		15.70		15.70	Includes material and labor to install
Install	Ea.	240	35		275	custom modular unit with solid
Demolish and Install	Ea.	240	50.70		290.70	hardwood faced frames, hardwood
Reinstall	Ea.		28.19		28.19	door frames and drawer fronts,
Clean	Ea.	1.06	11.05		12.11	hardwood veneer on raised door
Paint	Ea.	.98	25.50		26.48	panels, hinges and pulls.
Minimum Charge	Job		174		174	
30″ Wide, 2 Door						
Demolish	Ea.		17.65		17.65	Includes material and labor to install
Install	Ea.	257	36		293	custom modular unit with solid
Demolish and Install	Ea.	257	53.65		310.65	hardwood faced frames, hardwood
Reinstall	Ea.		28.92		28.92	door frames and drawer fronts,
Clean	Ea.	1.17	12.10		13.27	hardwood veneer on raised door
Paint	Ea.	1.09	30.50		31.59	panels, hinges and pulls.
Minimum Charge	Job		174		174	
36″ Wide, 2 Door						
Demolish	Ea.		21		21	Includes material and labor to install
Install	Ea.	294	37		331	custom modular unit with solid
Demolish and Install	Ea.	294	58		352	hardwood faced frames, hardwood
Reinstall	Ea.		29.69		29.69	door frames and drawer fronts,
Clean	Ea.	1.40	14.95		16.35	hardwood veneer on raised door
Paint	Ea.	1.31	34		35.31	panels, hinges and pulls.
Minimum Charge	Job		174		174	
48″ Wide, 2 Door						
Demolish	Ea.		28		28	Includes material and labor to install
Install	Ea.	365	38		403	custom modular unit with solid
Demolish and Install	Ea.	365	66		431	hardwood faced frames, hardwood
Reinstall	Ea.		30.33		30.33	door frames and drawer fronts,
Clean	Ea.	1.87	19.55		21.42	hardwood veneer on raised door
Paint	Ea.	1.74	43.50		45.24	panels, hinges and pulls.
Minimum Charge	Job		174		174	

Cabinets and Countertops

Wall Cabinets	Unit	Material	Labor	Equip.	Total	Specification
Above-appliance						
Demolish	Ea.		17.65		17.65	Includes material and labor to install
Install	Ea.	173	28		201	custom modular unit with solid
Demolish and Install	Ea.	173	45.65		218.65	hardwood faced frames, hardwood
Reinstall	Ea.		22.50		22.50	door frames and drawer fronts,
Clean	Ea.	1.17	12.10		13.27	hardwood veneer on raised door
Paint	Ea.	1.09	30.50		31.59	panels, hinges and pulls.
Minimum Charge	Job		174		174	
Premium Grade Wood						
12″ Wide, 1 Door						
Demolish	Ea.		7.05		7.05	Includes material and labor to install
Install	Ea.	185	31.50		216.50	premium modular unit with solid
Demolish and Install	Ea.	185	38.55		223.55	hardwood faced frames, hardwood
Reinstall	Ea.		25.37		25.37	door frames and drawer fronts,
Clean	Ea.	.46	4.88		5.34	hardwood veneer on raised door
Paint	Ea.	.43	12.70		13.13	panels, hinges and pulls.
Minimum Charge	Job		174		174	
15″ Wide, 1 Door						
Demolish	Ea.		8.80		8.80	Includes material and labor to install
Install	Ea.	231	32.50		263.50	premium modular unit with solid
Demolish and Install	Ea.	231	41.30		272.30	hardwood faced frames, hardwood
Reinstall	Ea.		26.08		26.08	door frames and drawer fronts,
Clean	Ea.	.58	6.05		6.63	hardwood veneer on raised door
Paint	Ea.	.54	16.05		16.59	panels, hinges and pulls.
Minimum Charge	Job		174		174	
18″ Wide, 1 Door						
Demolish	Ea.		10.65		10.65	Includes material and labor to install
Install	Ea.	229	33.50		262.50	premium modular unit with solid
Demolish and Install	Ea.	229	44.15		273.15	hardwood faced frames, hardwood
Reinstall	Ea.		26.70		26.70	door frames and drawer fronts,
Clean	Ea.	.70	7.25		7.95	hardwood veneer on raised door
Paint	Ea.	.65	19.10		19.75	panels, hinges and pulls.
Minimum Charge	Job		174		174	
21″ Wide, 1 Door						
Demolish	Ea.		12.25		12.25	Includes material and labor to install
Install	Ea.	256	34		290	premium modular unit with solid
Demolish and Install	Ea.	256	46.25		302.25	hardwood faced frames, hardwood
Reinstall	Ea.		27.09		27.09	door frames and drawer fronts,
Clean	Ea.	.81	8.45		9.26	hardwood veneer on raised door
Paint	Ea.	.76	22		22.76	panels, hinges and pulls.
Minimum Charge	Job		174		174	
24″ w / Blind Corner						
Demolish	Ea.		14.10		14.10	Includes material and labor to install
Install	Ea.	257	34.50		291.50	premium modular unit with solid
Demolish and Install	Ea.	257	48.60		305.60	hardwood faced frames, hardwood
Reinstall	Ea.		27.49		27.49	door frames and drawer fronts,
Clean	Ea.	.94	9.75		10.69	hardwood veneer on raised door
Paint	Ea.	.87	25.50		26.37	panels, hinges and pulls.
Minimum Charge	Job		174		174	
27″ Wide, 2 Door						
Demolish	Ea.		15.70		15.70	Includes material and labor to install
Install	Ea.	257	35		292	premium modular unit with solid
Demolish and Install	Ea.	257	50.70		307.70	hardwood faced frames, hardwood
Reinstall	Ea.		28.19		28.19	door frames and drawer fronts,
Clean	Ea.	1.06	11.05		12.11	hardwood veneer on raised door
Paint	Ea.	.98	28		28.98	panels, hinges and pulls.
Minimum Charge	Job		174		174	

Cabinets and Countertops

Wall Cabinets		Unit	Material	Labor	Equip.	Total	Specification
30" Wide, 2 Door							
	Demolish	Ea.		17.65		17.65	Includes material and labor to install
	Install	Ea.	310	36		346	premium modular unit with solid
	Demolish and Install	Ea.	310	53.65		363.65	hardwood faced frames, hardwood
	Reinstall	Ea.		28.92		28.92	door frames and drawer fronts,
	Clean	Ea.	1.17	12.10		13.27	hardwood veneer on raised door
	Paint	Ea.	1.09	30.50		31.59	panels, hinges and pulls.
	Minimum Charge	Job		174		174	
36" Wide, 2 Door							
	Demolish	Ea.		21		21	Includes material and labor to install
	Install	Ea.	350	37		387	premium modular unit with solid
	Demolish and Install	Ea.	350	58		408	hardwood faced frames, hardwood
	Reinstall	Ea.		29.69		29.69	door frames and drawer fronts,
	Clean	Ea.	1.40	14.95		16.35	hardwood veneer on raised door
	Paint	Ea.	1.31	38		39.31	panels, hinges and pulls.
	Minimum Charge	Job		174		174	
48" Wide, 2 Door							
	Demolish	Ea.		28		28	Includes material and labor to install
	Install	Ea.	435	38		473	premium modular unit with solid
	Demolish and Install	Ea.	435	66		501	hardwood faced frames, hardwood
	Reinstall	Ea.		30.33		30.33	door frames and drawer fronts,
	Clean	Ea.	1.87	19.55		21.42	hardwood veneer on raised door
	Paint	Ea.	1.74	51		52.74	panels, hinges and pulls.
	Minimum Charge	Job		174		174	
Above-appliance							
	Demolish	Ea.		17.65		17.65	Includes material and labor to install
	Install	Ea.	207	28		235	premium modular unit with solid
	Demolish and Install	Ea.	207	45.65		252.65	hardwood faced frames, hardwood
	Reinstall	Ea.		22.50		22.50	door frames and drawer fronts,
	Clean	Ea.	1.17	12.10		13.27	hardwood veneer on raised door
	Paint	Ea.	1.09	30.50		31.59	panels, hinges, pulls.
	Minimum Charge	Job		174		174	
Premium Hardwood							
12" Wide, 1 Door							
	Demolish	Ea.		7.05		7.05	Includes material and labor to install
	Install	Ea.	201	31.50		232.50	premium hardwood modular unit with
	Demolish and Install	Ea.	201	38.55		239.55	solid hardwood faced frames, drawer
	Reinstall	Ea.		25.37		25.37	fronts, and door panels, steel drawer
	Clean	Ea.	.46	4.88		5.34	guides, hinges, pulls, laminated
	Paint	Ea.	.43	13.85		14.28	interior.
	Minimum Charge	Job		174		174	
15" Wide, 1 Door							
	Demolish	Ea.		8.80		8.80	Includes material and labor to install
	Install	Ea.	305	32.50		337.50	premium hardwood modular unit with
	Demolish and Install	Ea.	305	41.30		346.30	solid hardwood faced frames, drawer
	Reinstall	Ea.		26.08		26.08	fronts, and door panels, steel drawer
	Clean	Ea.	.58	6.05		6.63	guides, hinges, pulls, laminated
	Paint	Ea.	.54	16.95		17.49	interior.
	Minimum Charge	Job		174		174	
18" Wide, 1 Door							
	Demolish	Ea.		10.65		10.65	Includes material and labor to install
	Install	Ea.	250	33.50		283.50	premium hardwood modular unit with
	Demolish and Install	Ea.	250	44.15		294.15	solid hardwood faced frames, drawer
	Reinstall	Ea.		26.70		26.70	fronts, and door panels, steel drawer
	Clean	Ea.	.70	7.25		7.95	guides, hinges, pulls, laminated
	Paint	Ea.	.65	20.50		21.15	interior.
	Minimum Charge	Job		174		174	

Cabinets and Countertops

Wall Cabinets

	Unit	Material	Labor	Equip.	Total	Specification
21″ Wide, 1 Door						
Demolish	Ea.		12.25		12.25	Includes material and labor to install
Install	Ea.	281	34		315	premium hardwood modular unit with
Demolish and Install	Ea.	281	46.25		327.25	solid hardwood faced frames, drawer
Reinstall	Ea.		27.09		27.09	fronts, and door panels, steel drawer
Clean	Ea.	.81	8.45		9.26	guides, hinges, pulls, laminated
Paint	Ea.	.76	23.50		24.26	interior.
Minimum Charge	Job		174		174	
24″ w / Blind Corner						
Demolish	Ea.		14.10		14.10	Includes material and labor to install
Install	Ea.	282	34.50		316.50	premium hardwood modular unit with
Demolish and Install	Ea.	282	48.60		330.60	solid hardwood faced frames, drawer
Reinstall	Ea.		27.49		27.49	fronts, and door panels, steel drawer
Clean	Ea.	.94	9.75		10.69	guides, hinges, pulls, laminated
Paint	Ea.	.87	28		28.87	interior.
Minimum Charge	Job		174		174	
27″ Wide, 2 Door						
Demolish	Ea.		15.70		15.70	Includes material and labor to install
Install	Ea.	315	35		350	premium hardwood modular unit with
Demolish and Install	Ea.	315	50.70		365.70	solid hardwood faced frames, drawer
Reinstall	Ea.		28.19		28.19	fronts, and door panels, steel drawer
Clean	Ea.	1.06	11.05		12.11	guides, hinges, pulls, laminated
Paint	Ea.	.98	30.50		31.48	interior.
Minimum Charge	Job		174		174	
30″ Wide, 2 Door						
Demolish	Ea.		17.65		17.65	Includes material and labor to install
Install	Ea.	335	36		371	premium hardwood modular unit with
Demolish and Install	Ea.	335	53.65		388.65	solid hardwood faced frames, drawer
Reinstall	Ea.		28.92		28.92	fronts, and door panels, steel drawer
Clean	Ea.	1.17	12.10		13.27	guides, hinges, pulls, laminated
Paint	Ea.	1.09	34		35.09	interior.
Minimum Charge	Job		174		174	
36″ Wide, 2 Door						
Demolish	Ea.		21		21	Includes material and labor to install
Install	Ea.	385	37		422	premium hardwood modular unit with
Demolish and Install	Ea.	385	58		443	solid hardwood faced frames, drawer
Reinstall	Ea.		29.69		29.69	fronts, and door panels, steel drawer
Clean	Ea.	1.40	14.95		16.35	guides, hinges, pulls, laminated
Paint	Ea.	1.31	43.50		44.81	interior.
Minimum Charge	Job		174		174	
48″ Wide, 2 Door						
Demolish	Ea.		28		28	Includes material and labor to install
Install	Ea.	475	38		513	premium hardwood modular unit with
Demolish and Install	Ea.	475	66		541	solid hardwood faced frames, drawer
Reinstall	Ea.		30.33		30.33	fronts, and door panels, steel drawer
Clean	Ea.	1.87	19.55		21.42	guides, hinges, pulls, laminated
Paint	Ea.	1.74	51		52.74	interior.
Minimum Charge	Job		174		174	
Above-appliance						
Demolish	Ea.		17.65		17.65	Includes material and labor to install
Install	Ea.	226	28		254	premium hardwood modular unit with
Demolish and Install	Ea.	226	45.65		271.65	solid hardwood faced frames, drawer
Reinstall	Ea.		22.50		22.50	fronts, and door panels, steel drawer
Clean	Ea.	1.17	12.10		13.27	guides, hinges, pulls, laminated
Paint	Ea.	1.09	34		35.09	interior.
Minimum Charge	Job		174		174	

Cabinets and Countertops

Wall Cabinets		Unit	Material	Labor	Equip.	Total	Specification
Medicine Cabinet							
Good Grade							
	Demolish	Ea.		11.75		11.75	Cost includes material and labor to
	Install	Ea.	134	25		159	install 14" x 18" recessed medicine
	Demolish and Install	Ea.	134	36.75		170.75	cabinet with stainless steel frame,
	Reinstall	Ea.		19.93		19.93	mirror. Light not included.
	Clean	Ea.	.02	11.40		11.42	
	Paint	Ea.	1.18	19.10		20.28	
	Minimum Charge	Job		174		174	
Good Grade Lighted							
	Demolish	Ea.		11.75		11.75	Cost includes material and labor to
	Install	Ea.	350	29		379	install 14" x 18" recessed medicine
	Demolish and Install	Ea.	350	40.75		390.75	cabinet with stainless steel frame,
	Reinstall	Ea.		23.25		23.25	mirror and light.
	Clean	Ea.	.02	11.40		11.42	
	Paint	Ea.	1.18	19.10		20.28	
	Minimum Charge	Job		174		174	
Better Grade							
	Demolish	Ea.		11.75		11.75	Includes material and labor to install
	Install	Ea.	113	25		138	recessed medicine cabinet with
	Demolish and Install	Ea.	113	36.75		149.75	polished chrome frame and mirror.
	Reinstall	Ea.		19.93		19.93	Light not included.
	Clean	Ea.	.02	11.40		11.42	
	Paint	Ea.	1.18	19.10		20.28	
	Minimum Charge	Job		174		174	
Better Grade Lighted							
	Demolish	Ea.		11.75		11.75	Includes material and labor to install
	Install	Ea.	335	29		364	recessed medicine cabinet with
	Demolish and Install	Ea.	335	40.75		375.75	polished chrome frame, mirror and
	Reinstall	Ea.		23.25		23.25	light.
	Clean	Ea.	.02	11.40		11.42	
	Paint	Ea.	1.18	19.10		20.28	
	Minimum Charge	Job		174		174	
Premium Grade							
	Demolish	Ea.		11.75		11.75	Cost includes material and labor to
	Install	Ea.	263	25		288	install 14" x 24" recessed medicine
	Demolish and Install	Ea.	263	36.75		299.75	cabinet with beveled mirror, frameless
	Reinstall	Ea.		19.93		19.93	swing door. Light not included.
	Clean	Ea.	.02	11.40		11.42	
	Paint	Ea.	1.18	19.10		20.28	
	Minimum Charge	Job		174		174	
Premium Grade Light							
	Demolish	Ea.		11.75		11.75	Cost includes material and labor to
	Install	Ea.	485	29		514	install 36" x 30" surface mounted
	Demolish and Install	Ea.	485	40.75		525.75	medicine cabinet with beveled tri-view
	Reinstall	Ea.		23.25		23.25	mirror, frameless swing door and light.
	Clean	Ea.	.02	11.40		11.42	
	Paint	Ea.	1.18	19.10		20.28	
	Minimum Charge	Job		174		174	
Bath Mirror							
	Demolish	Ea.		8.80		8.80	Cost includes material and labor to
	Install	Ea.	59	17.45		76.45	install 18" x 24" surface mount mirror
	Demolish and Install	Ea.	59	26.25		85.25	with stainless steel frame.
	Reinstall	Ea.		13.95		13.95	
	Clean	Ea.	.02	.76		.78	
	Minimum Charge	Job		174		174	

Cabinets and Countertops

Wall Cabinets

Wall Cabinets	Unit	Material	Labor	Equip.	Total	Specification
Garage						
Demolish	L.F.		7.05		7.05	Includes material and labor to install
Install	L.F.	56.50	17.45		73.95	good modular unit with melamine
Demolish and Install	L.F.	56.50	24.50		81	laminated to particle board, in
Reinstall	L.F.		13.95		13.95	textured colors or wood grain print
Clean	L.F.	.46	4.88		5.34	finish, hinges and pulls.
Paint	L.F.	.43	10.90		11.33	
Minimum Charge	Job		174		174	
Strip and Refinish						
Install	L.F.	1.08	5.55		6.63	Includes labor and material to strip,
Minimum Charge	Job		153		153	prep and refinish exterior of cabinets.

Tall Cabinets

Tall Cabinets	Unit	Material	Labor	Equip.	Total	Specification
Good Grade Laminated						
Demolish	L.F.		12.55		12.55	Includes material and labor to install
Install	L.F.	231	197		428	custom modular unit with solid
Demolish and Install	L.F.	231	209.55		440.55	hardwood faced frames, hardwood
Reinstall	L.F.		157.21		157.21	door frames and drawer fronts,
Clean	L.F.	1.24	9.05		10.29	hardwood veneer on raised door
Paint	L.F.	1.01	22		23.01	panels, hinges and pulls.
Minimum Charge	Job		174		174	
Better Grade Wood						
Demolish	L.F.		12.55		12.55	Includes material and labor to install
Install	L.F.	310	197		507	custom modular unit with solid
Demolish and Install	L.F.	310	209.55		519.55	hardwood faced frames, hardwood
Reinstall	L.F.		157.21		157.21	door frames and drawer fronts,
Clean	L.F.	1.24	9.05		10.29	hardwood veneer on raised door
Paint	L.F.	1.01	23.50		24.51	panels, hinges and pulls.
Minimum Charge	Job		174		174	
Premium Grade Wood						
Demolish	L.F.		12.55		12.55	Includes material and labor to install
Install	L.F.	415	197		612	premium modular unit with solid
Demolish and Install	L.F.	415	209.55		624.55	hardwood faced frames, hardwood
Reinstall	L.F.		157.21		157.21	door frames and drawer fronts,
Clean	L.F.	1.24	9.05		10.29	hardwood veneer on raised door
Paint	L.F.	1.01	25.50		26.51	panels, hinges and pulls.
Minimum Charge	Job		174		174	
Premium Hardwood						
Demolish	L.F.		12.55		12.55	Includes material and labor to install
Install	L.F.	540	197		737	premium hardwood modular unit with
Demolish and Install	L.F.	540	209.55		749.55	solid hardwood faced frames, drawer
Reinstall	L.F.		157.21		157.21	fronts, and door panels, steel drawer
Clean	L.F.	1.24	9.05		10.29	guides, hinges and pulls, laminated
Paint	L.F.	1.01	28		29.01	interior.
Minimum Charge	Job		174		174	

Cabinets and Countertops

Tall Cabinets	Unit	Material	Labor	Equip.	Total	Specification
Single Oven, 27″ Wide						
Good Grade Laminated						
Demolish	Ea.		28		28	Includes material and labor to install
Install	Ea.	525	87		612	good modular unit with melamine
Demolish and Install	Ea.	525	115		640	laminated to particle board, in
Reinstall	Ea.		69.76		69.76	textured colors or wood grain print
Clean	Ea.	3.89	15.90		19.79	finish, hinges and pulls.
Paint	Ea.	1.71	49		50.71	
Minimum Charge	Job		174		174	
Better Grade Veneer						
Demolish	Ea.		28		28	Includes material and labor to install
Install	Ea.	925	87		1012	custom modular unit with solid
Demolish and Install	Ea.	925	115		1040	hardwood faced frames, hardwood
Reinstall	Ea.		69.76		69.76	door frames and drawer fronts,
Clean	Ea.	3.89	15.90		19.79	hardwood veneer on raised door
Paint	Ea.	1.71	52.50		54.21	panels, hinges and pulls.
Minimum Charge	Job		174		174	
Premium Grade Veneer						
Demolish	Ea.		28		28	Includes material and labor to install
Install	Ea.	830	87		917	premium modular unit with solid
Demolish and Install	Ea.	830	115		945	hardwood faced frames, hardwood
Reinstall	Ea.		69.76		69.76	door frames and drawer fronts,
Clean	Ea.	3.89	15.90		19.79	hardwood veneer on raised door
Paint	Ea.	1.71	57.50		59.21	panels, hinges and pulls.
Minimum Charge	Job		174		174	
Premium Hardwood						
Demolish	Ea.		28		28	Includes material and labor to install
Install	Ea.	910	87		997	premium hardwood modular unit with
Demolish and Install	Ea.	910	115		1025	solid hardwood faced frames, drawer
Reinstall	Ea.		69.76		69.76	fronts, and door panels, steel drawer
Clean	Ea.	3.89	15.90		19.79	guides, hinges and pulls, laminated
Paint	Ea.	1.71	62.50		64.21	interior.
Minimum Charge	Job		174		174	
Single Oven, 30″ Wide						
Good Grade Laminated						
Demolish	Ea.		28		28	Includes material and labor to install
Install	Ea.	690	87		777	good modular unit with melamine
Demolish and Install	Ea.	690	115		805	laminated to particle board, in
Reinstall	Ea.		69.76		69.76	textured colors or wood grain print
Clean	Ea.	3.89	15.90		19.79	finish, hinges and pulls.
Paint	Ea.	1.71	49		50.71	
Minimum Charge	Job		174		174	
Better Grade Veneer						
Demolish	Ea.		28		28	Includes material and labor to install
Install	Ea.	815	87		902	custom modular unit with solid
Demolish and Install	Ea.	815	115		930	hardwood faced frames, hardwood
Reinstall	Ea.		69.76		69.76	door frames and drawer fronts,
Clean	Ea.	3.89	15.90		19.79	hardwood veneer on raised door
Paint	Ea.	1.71	49		50.71	panels, hinges and pulls.
Minimum Charge	Job		174		174	

Tall Cabinets	Unit	Material	Labor	Equip.	Total	Specification
Premium Grade Veneer						Includes material and labor to install
Demolish	Ea.		28		28	premium modular unit with solid
Install	Ea.	885	87		972	hardwood faced frames, hardwood
Demolish and Install	Ea.	885	115		1000	door frames and drawer fronts,
Reinstall	Ea.		69.76		69.76	hardwood veneer on raised door
Clean	Ea.	3.89	15.90		19.79	panels, hinges and pulls.
Paint	Ea.	1.71	49		50.71	
Minimum Charge	Job		174		174	
Premium Hardwood						Includes material and labor to install
Demolish	Ea.		28		28	premium hardwood modular unit with
Install	Ea.	1075	87		1162	solid hardwood faced frames, drawer
Demolish and Install	Ea.	1075	115		1190	fronts, and door panels, steel drawer
Reinstall	Ea.		69.76		69.76	guides, hinges and pulls, laminated
Clean	Ea.	3.89	15.90		19.79	interior.
Paint	Ea.	1.71	49		50.71	
Minimum Charge	Job		174		174	

Utility 18″ W x 12″ D

	Unit	Material	Labor	Equip.	Total	Specification
Good Grade Laminated						Includes material and labor to install
Demolish	Ea.		28		28	good modular unit with melamine
Install	Ea.	345	70		415	laminated to particle board, in
Demolish and Install	Ea.	345	98		443	textured colors or wood grain print
Reinstall	Ea.		55.81		55.81	finish, hinges and pulls.
Clean	Ea.	.91	7.70		8.61	
Paint	Ea.	.65	19.10		19.75	
Minimum Charge	Job		174		174	
Better Grade Veneer						Includes material and labor to install
Demolish	Ea.		28		28	custom modular unit with solid
Install	Ea.	460	70		530	hardwood faced frames, hardwood
Demolish and Install	Ea.	460	98		558	door frames and drawer fronts,
Reinstall	Ea.		55.81		55.81	hardwood veneer on raised door
Clean	Ea.	.91	7.70		8.61	panels, hinges and pulls.
Paint	Ea.	.65	20.50		21.15	
Minimum Charge	Job		174		174	
Premium Grade Veneer						Includes material and labor to install
Demolish	Ea.		28		28	premium modular unit with solid
Install	Ea.	550	70		620	hardwood faced frames, hardwood
Demolish and Install	Ea.	550	98		648	door frames and drawer fronts,
Reinstall	Ea.		55.81		55.81	hardwood veneer on raised door
Clean	Ea.	.91	7.70		8.61	panels, hinges and pulls.
Paint	Ea.	.65	23.50		24.15	
Minimum Charge	Job		174		174	
Premium Hardwood						Includes material and labor to install
Demolish	Ea.		28		28	premium hardwood modular unit with
Install	Ea.	600	70		670	solid hardwood faced frames, drawer
Demolish and Install	Ea.	600	98		698	fronts, and door panels, steel drawer
Reinstall	Ea.		55.81		55.81	guides, hinges and pulls, laminated
Clean	Ea.	.91	7.70		8.61	interior.
Paint	Ea.	.65	25.50		26.15	
Minimum Charge	Job		174		174	

Cabinets and Countertops

Tall Cabinets	Unit	Material	Labor	Equip.	Total	Specification
Utility 24″ W x 24″ D						
Good Grade Laminated						
Demolish	Ea.		28		28	Includes material and labor to install
Install	Ea.	470	87		557	good modular unit with melamine
Demolish and Install	Ea.	470	115		585	laminated to particle board, in
Reinstall	Ea.		69.76		69.76	textured colors or wood grain print
Clean	Ea.	1.20	10.15		11.35	finish, hinges and pulls.
Paint	Ea.	.87	25.50		26.37	
Minimum Charge	Job		174		174	
Better Grade Veneer						
Demolish	Ea.		28		28	Includes material and labor to install
Install	Ea.	620	87		707	custom modular unit with solid
Demolish and Install	Ea.	620	115		735	hardwood faced frames, hardwood
Reinstall	Ea.		69.76		69.76	door frames and drawer fronts,
Clean	Ea.	1.20	10.15		11.35	hardwood veneer on raised door
Paint	Ea.	.87	28		28.87	panels, hinges and pulls.
Minimum Charge	Job		174		174	
Premium Grade Veneer						
Demolish	Ea.		28		28	Includes material and labor to install
Install	Ea.	750	87		837	premium modular unit with solid
Demolish and Install	Ea.	750	115		865	hardwood faced frames, hardwood
Reinstall	Ea.		69.76		69.76	door frames and drawer fronts,
Clean	Ea.	1.20	10.15		11.35	hardwood veneer on raised door
Paint	Ea.	.87	30.50		31.37	panels, hinges and pulls.
Minimum Charge	Job		174		174	
Premium Hardwood						
Demolish	Ea.		28		28	Includes material and labor to install
Install	Ea.	815	87		902	premium hardwood modular unit with
Demolish and Install	Ea.	815	115		930	solid hardwood faced frames, drawer
Reinstall	Ea.		69.76		69.76	fronts, and door panels, steel drawer
Clean	Ea.	1.20	10.15		11.35	guides, hinges and pulls, laminated
Paint	Ea.	.87	34		34.87	interior.
Minimum Charge	Job		174		174	
Plain Shelves						
Install	Ea.	25.50	22		47.50	Includes labor and material to install
Rotating Shelves						tall cabinet shelving, per set.
Install	Ea.	121	17.45		138.45	Includes labor and material to install
Strip and Refinish						lazy susan shelving for wall cabinet.
Install	L.F.	1.62	13.85		15.47	Includes labor and material to strip,
Minimum Charge	Job		153		153	prep and refinish exterior of cabinets.
Stain						
Install	L.F.	.18	2.77		2.95	Includes labor and material to replace
Minimum Charge	Job		153		153	normal prep and stain on doors and
						exterior cabinets.

Cabinets and Countertops

Base Cabinets	Unit	Material	Labor	Equip.	Total	Specification
Good Grade Laminated						
12" w, 1 Door, 1 Drawer						
Demolish	Ea.		7.05		7.05	Includes material and labor to install
Install	Ea.	161	28		189	good modular unit with melamine
Demolish and Install	Ea.	161	35.05		196.05	laminate, textured colors or wood
Reinstall	Ea.		22.50		22.50	grain print finish including hinges and
Clean	Ea.	.63	5.10		5.73	pulls.
Paint	Ea.	.43	12.70		13.13	
Minimum Charge	Job		174		174	
15" w, 1 Door, 1 Drawer						
Demolish	Ea.		8.80		8.80	Includes material and labor to install
Install	Ea.	168	29		197	good modular unit with melamine
Demolish and Install	Ea.	168	37.80		205.80	laminate, textured colors or wood
Reinstall	Ea.		23.25		23.25	grain print finish including hinges and
Clean	Ea.	.78	6.35		7.13	pulls.
Paint	Ea.	.54	16.05		16.59	
Minimum Charge	Job		174		174	
18" w, 1 Door, 1 Drawer						
Demolish	Ea.		10.45		10.45	Includes material and labor to install
Install	Ea.	184	30		214	good modular unit with melamine
Demolish and Install	Ea.	184	40.45		224.45	laminate, textured colors or wood
Reinstall	Ea.		23.95		23.95	grain print finish including hinges and
Clean	Ea.	.91	7.70		8.61	pulls.
Paint	Ea.	.65	19.10		19.75	
Minimum Charge	Job		174		174	
21" w, 1 Door, 1 Drawer						
Demolish	Ea.		12.25		12.25	Includes material and labor to install
Install	Ea.	200	30.50		230.50	good modular unit with melamine
Demolish and Install	Ea.	200	42.75		242.75	laminate, textured colors or wood
Reinstall	Ea.		24.59		24.59	grain print finish including hinges and
Clean	Ea.	1.11	9.05		10.16	pulls.
Paint	Ea.	.76	22		22.76	
Minimum Charge	Job		174		174	
24" w, 1 Door, 1 Drawer						
Demolish	Ea.		14.10		14.10	Includes material and labor to install
Install	Ea.	219	31.50		250.50	good modular unit with melamine
Demolish and Install	Ea.	219	45.60		264.60	laminate, textured colors or wood
Reinstall	Ea.		25.03		25.03	grain print finish including hinges and
Clean	Ea.	1.20	10.15		11.35	pulls.
Paint	Ea.	.87	25.50		26.37	
Minimum Charge	Job		174		174	
30" Wide Sink Base						
Demolish	Ea.		17.65		17.65	Includes material and labor to install
Install	Ea.	239	32.50		271.50	good modular unit with melamine
Demolish and Install	Ea.	239	50.15		289.15	laminate, textured colors or wood
Reinstall	Ea.		26.08		26.08	grain print finish including hinges and
Clean	Ea.	1.56	12.70		14.26	pulls.
Paint	Ea.	1.09	30.50		31.59	
Minimum Charge	Job		174		174	
36" Wide Sink Base						
Demolish	Ea.		21.50		21.50	Includes material and labor to install
Install	Ea.	264	34.50		298.50	good modular unit with melamine
Demolish and Install	Ea.	264	56		320	laminate, textured colors or wood
Reinstall	Ea.		27.49		27.49	grain print finish including hinges and
Clean	Ea.	1.73	14.95		16.68	pulls.
Paint	Ea.	1.31	38		39.31	
Minimum Charge	Job		174		174	

Cabinets and Countertops

Base Cabinets	Unit	Material	Labor	Equip.	Total	Specification
36" Blind Corner						
Demolish	Ea.		21.50		21.50	Includes material and labor to install
Install	Ea.	345	39		384	good modular unit with melamine
Demolish and Install	Ea.	345	60.50		405.50	laminate, textured colors or wood
Reinstall	Ea.		31		31	grain print finish including hinges and
Clean	Ea.	1.73	14.95		16.68	pulls.
Paint	Ea.	1.31	38		39.31	
Minimum Charge	Job		174		174	
42" w, 2 Door, 2 Drawer						
Demolish	Ea.		25.50		25.50	Includes material and labor to install
Install	Ea.	320	35		355	good modular unit with melamine
Demolish and Install	Ea.	320	60.50		380.50	laminate, textured colors or wood
Reinstall	Ea.		28.19		28.19	grain print finish including hinges and
Clean	Ea.	2.22	18.15		20.37	pulls.
Paint	Ea.	1.51	43.50		45.01	
Minimum Charge	Job		174		174	
Clean Interior						
Clean	L.F.	.63	3.85		4.48	Includes labor and materials to clean
Minimum Charge	Job		127		127	the interior of a 27" cabinet.
Better Grade Wood						
12" w, 1 Door, 1 Drawer						
Demolish	Ea.		7.05		7.05	Includes material and labor to install
Install	Ea.	215	28		243	custom modular unit including solid
Demolish and Install	Ea.	215	35.05		250.05	hardwood faced frames, hardwood
Reinstall	Ea.		22.50		22.50	door frames and drawer fronts,
Clean	Ea.	.63	5.10		5.73	hardwood veneer on raised door
Paint	Ea.	.43	13.85		14.28	panels, hinges and pulls.
Minimum Charge	Job		174		174	
15" w, 1 Door, 1 Drawer						
Demolish	Ea.		8.80		8.80	Includes material and labor to install
Install	Ea.	224	29		253	custom modular unit including solid
Demolish and Install	Ea.	224	37.80		261.80	hardwood faced frames, hardwood
Reinstall	Ea.		23.25		23.25	door frames and drawer fronts,
Clean	Ea.	.78	6.35		7.13	hardwood veneer on raised door
Paint	Ea.	.54	17.95		18.49	panels, hinges and pulls.
Minimum Charge	Job		174		174	
18" w, 1 Door, 1 Drawer						
Demolish	Ea.		10.45		10.45	Includes material and labor to install
Install	Ea.	245	30		275	custom modular unit including solid
Demolish and Install	Ea.	245	40.45		285.45	hardwood faced frames, hardwood
Reinstall	Ea.		23.95		23.95	door frames and drawer fronts,
Clean	Ea.	.91	7.70		8.61	hardwood veneer on raised door
Paint	Ea.	.65	20.50		21.15	panels, hinges and pulls.
Minimum Charge	Job		174		174	
21" w, 1 Door, 1 Drawer						
Demolish	Ea.		12.25		12.25	Includes material and labor to install
Install	Ea.	266	30.50		296.50	custom modular unit including solid
Demolish and Install	Ea.	266	42.75		308.75	hardwood faced frames, hardwood
Reinstall	Ea.		24.59		24.59	door frames and drawer fronts,
Clean	Ea.	1.11	9.05		10.16	hardwood veneer on raised door
Paint	Ea.	.76	23.50		24.26	panels, hinges and pulls.
Minimum Charge	Job		174		174	

Cabinets and Countertops

Base Cabinets	Unit	Material	Labor	Equip.	Total	Specification
24" w, 1 Door, 1 Drawer						
Demolish	Ea.		14.10		14.10	Includes material and labor to install
Install	Ea.	292	31.50		323.50	custom modular unit including solid
Demolish and Install	Ea.	292	45.60		337.60	hardwood faced frames, hardwood
Reinstall	Ea.		25.03		25.03	door frames and drawer fronts,
Clean	Ea.	1.20	10.15		11.35	hardwood veneer on raised door
Paint	Ea.	.87	28		28.87	panels, hinges and pulls.
Minimum Charge	Job		174		174	
30" Wide Sink Base						
Demolish	Ea.		17.65		17.65	Includes material and labor to install
Install	Ea.	320	32.50		352.50	custom modular unit including solid
Demolish and Install	Ea.	320	50.15		370.15	hardwood faced frames, hardwood
Reinstall	Ea.		26.08		26.08	door frames and drawer fronts,
Clean	Ea.	1.56	12.70		14.26	hardwood veneer on raised door
Paint	Ea.	1.09	34		35.09	panels, hinges and pulls.
Minimum Charge	Job		174		174	
36" Wide Sink Base						
Demolish	Ea.		21.50		21.50	Includes material and labor to install
Install	Ea.	350	34.50		384.50	custom modular unit including solid
Demolish and Install	Ea.	350	56		406	hardwood faced frames, hardwood
Reinstall	Ea.		27.49		27.49	door frames and drawer fronts,
Clean	Ea.	1.73	14.95		16.68	hardwood veneer on raised door
Paint	Ea.	1.31	43.50		44.81	panels, hinges and pulls.
Minimum Charge	Job		174		174	
36" Blind Corner						
Demolish	Ea.		21.50		21.50	Includes material and labor to install
Install	Ea.	460	39		499	custom modular unit including solid
Demolish and Install	Ea.	460	60.50		520.50	hardwood faced frames, hardwood
Reinstall	Ea.		31		31	door frames and drawer fronts,
Clean	Ea.	1.73	14.95		16.68	hardwood veneer on raised door
Paint	Ea.	1.31	43.50		44.81	panels, hinges and pulls.
Minimum Charge	Job		174		174	
42" w, 2 Door, 2 Drawer						
Demolish	Ea.		25.50		25.50	Includes material and labor to install
Install	Ea.	430	35		465	modular unit including solid hardwood
Demolish and Install	Ea.	430	60.50		490.50	faced frames, hardwood door frames
Reinstall	Ea.		28.19		28.19	and drawer fronts, hardwood veneer
Clean	Ea.	2.22	18.15		20.37	on raised door panels, hinges and
Paint	Ea.	1.51	51		52.51	pulls.
Minimum Charge	Job		174		174	
Clean Interior						
Clean	L.F.	.63	3.85		4.48	Includes labor and materials to clean
Minimum Charge	Job		127		127	the interior of a 27" cabinet.
Premium Grade Wood						
12" w, 1 Door, 1 Drawer						
Demolish	Ea.		7.05		7.05	Includes material and labor to install
Install	Ea.	257	14.05		271.05	premium hardwood modular unit with
Demolish and Install	Ea.	257	21.10		278.10	solid hardwood faced frames, drawer
Reinstall	Ea.		11.25		11.25	fronts, and door panels, steel drawer
Clean	Ea.	.63	5.10		5.73	guides, hinges, pulls and laminated
Paint	Ea.	.43	15.25		15.68	interior.
Minimum Charge	Job		174		174	

Cabinets and Countertops

Base Cabinets	Unit	Material	Labor	Equip.	Total	Specification
15" w, 1 Door, 1 Drawer						
Demolish	Ea.		8.80		8.80	Includes material and labor to install
Install	Ea.	270	29		299	premium modular unit including solid
Demolish and Install	Ea.	270	37.80		307.80	hardwood faced frames, hardwood
Reinstall	Ea.		23.25		23.25	door frames and drawer fronts,
Clean	Ea.	.78	6.35		7.13	hardwood veneer on raised door
Paint	Ea.	.54	19.10		19.64	panels, hinges and pulls.
Minimum Charge	Job		174		174	
18" w, 1 Door, 1 Drawer						
Demolish	Ea.		10.45		10.45	Includes material and labor to install
Install	Ea.	295	30		325	premium modular unit including solid
Demolish and Install	Ea.	295	40.45		335.45	hardwood faced frames, hardwood
Reinstall	Ea.		23.95		23.95	door frames and drawer fronts,
Clean	Ea.	.91	7.70		8.61	hardwood veneer on raised door
Paint	Ea.	.65	23.50		24.15	panels, hinges and pulls.
Minimum Charge	Job		174		174	
21" w, 1 Door, 1 Drawer						
Demolish	Ea.		12.25		12.25	Includes material and labor to install
Install	Ea.	320	30.50		350.50	premium modular unit including solid
Demolish and Install	Ea.	320	42.75		362.75	hardwood faced frames, hardwood
Reinstall	Ea.		24.59		24.59	door frames and drawer fronts,
Clean	Ea.	1.11	9.05		10.16	hardwood veneer on raised door
Paint	Ea.	.76	25.50		26.26	panels, hinges and pulls.
Minimum Charge	Job		174		174	
24" w, 1 Door, 1 Drawer						
Demolish	Ea.		14.10		14.10	Includes material and labor to install
Install	Ea.	350	31.50		381.50	premium modular unit including solid
Demolish and Install	Ea.	350	45.60		395.60	hardwood faced frames, hardwood
Reinstall	Ea.		25.03		25.03	door frames and drawer fronts,
Clean	Ea.	1.20	10.15		11.35	hardwood veneer on raised door
Paint	Ea.	.87	30.50		31.37	panels, hinges and pulls.
Minimum Charge	Job		174		174	
30" Wide Sink Base						
Demolish	Ea.		17.65		17.65	Includes material and labor to install
Install	Ea.	380	32.50		412.50	premium modular unit including solid
Demolish and Install	Ea.	380	50.15		430.15	hardwood faced frames, hardwood
Reinstall	Ea.		26.08		26.08	door frames and drawer fronts,
Clean	Ea.	1.56	12.70		14.26	hardwood veneer on raised door
Paint	Ea.	1.09	38		39.09	panels, hinges and pulls.
Minimum Charge	Job		174		174	
36" Wide Sink Base						
Demolish	Ea.		21.50		21.50	Includes material and labor to install
Install	Ea.	425	34.50		459.50	premium modular unit including solid
Demolish and Install	Ea.	425	56		481	hardwood faced frames, hardwood
Reinstall	Ea.		27.49		27.49	door frames and drawer fronts,
Clean	Ea.	1.73	14.95		16.68	hardwood veneer on raised door
Paint	Ea.	1.31	43.50		44.81	panels, hinges and pulls.
Minimum Charge	Job		174		174	
36" Blind Corner						
Demolish	Ea.		21.50		21.50	Includes material and labor to install
Install	Ea.	555	39		594	premium modular unit including solid
Demolish and Install	Ea.	555	60.50		615.50	hardwood faced frames, hardwood
Reinstall	Ea.		31		31	door frames and drawer fronts,
Clean	Ea.	1.73	14.95		16.68	hardwood veneer on raised door
Paint	Ea.	1.31	43.50		44.81	panels, hinges and pulls.
Minimum Charge	Job		174		174	

Cabinets and Countertops

Base Cabinets		Unit	Material	Labor	Equip.	Total	Specification
42″ w, 2 Door, 2 Drawer							
	Demolish	Ea.		25.50		25.50	Includes material and labor to install
	Install	Ea.	570	35		605	premium modular unit including solid
	Demolish and Install	Ea.	570	60.50		630.50	hardwood faced frames, hardwood
	Reinstall	Ea.		28.19		28.19	door frames and drawer fronts,
	Clean	Ea.	2.22	18.15		20.37	hardwood veneer on raised door
	Paint	Ea.	1.51	51		52.51	panels, hinges and pulls.
	Minimum Charge	Job		174		174	
Clean Interior							
	Clean	L.F.	.63	3.85		4.48	Includes labor and materials to clean
	Minimum Charge	Job		127		127	the interior of a 27″ cabinet.

Premium Hardwood

		Unit	Material	Labor	Equip.	Total	Specification
12″ w, 1 Door, 1 Drawer							
	Demolish	Ea.		7.05		7.05	Includes material and labor to install
	Install	Ea.	281	28		309	premium hardwood modular unit with
	Demolish and Install	Ea.	281	35.05		316.05	solid hardwood faced frames, drawer
	Reinstall	Ea.		22.50		22.50	fronts, and door panels, steel drawer
	Clean	Ea.	.63	5.10		5.73	guides, hinges, pulls and laminated
	Paint	Ea.	.43	16.95		17.38	interior.
	Minimum Charge	Job		174		174	
15″ w, 1 Door, 1 Drawer							
	Demolish	Ea.		8.80		8.80	Includes material and labor to install
	Install	Ea.	295	29		324	premium hardwood modular unit with
	Demolish and Install	Ea.	295	37.80		332.80	solid hardwood faced frames, drawer
	Reinstall	Ea.		23.25		23.25	fronts, and door panels, steel drawer
	Clean	Ea.	.78	6.35		7.13	guides, hinges, pulls and laminated
	Paint	Ea.	.54	22		22.54	interior.
	Minimum Charge	Job		174		174	
18″ w, 1 Door, 1 Drawer							
	Demolish	Ea.		10.45		10.45	Includes material and labor to install
	Install	Ea.	320	30		350	premium hardwood modular unit with
	Demolish and Install	Ea.	320	40.45		360.45	solid hardwood faced frames, drawer
	Reinstall	Ea.		23.95		23.95	fronts, and door panels, steel drawer
	Clean	Ea.	.91	7.70		8.61	guides, hinges, pulls and laminated
	Paint	Ea.	.65	25.50		26.15	interior.
	Minimum Charge	Job		174		174	
21″ w, 1 Door, 1 Drawer							
	Demolish	Ea.		12.25		12.25	Includes material and labor to install
	Install	Ea.	350	30.50		380.50	premium hardwood modular unit with
	Demolish and Install	Ea.	350	42.75		392.75	solid hardwood faced frames, drawer
	Reinstall	Ea.		24.59		24.59	fronts, and door panels, steel drawer
	Clean	Ea.	1.11	9.05		10.16	guides, hinges, pulls and laminated
	Paint	Ea.	.76	28		28.76	interior.
	Minimum Charge	Job		174		174	
24″ w, 1 Door, 1 Drawer							
	Demolish	Ea.		14.10		14.10	Includes material and labor to install
	Install	Ea.	385	31.50		416.50	premium hardwood modular unit with
	Demolish and Install	Ea.	385	45.60		430.60	solid hardwood faced frames, drawer
	Reinstall	Ea.		25.03		25.03	fronts, and door panels, steel drawer
	Clean	Ea.	1.20	10.15		11.35	guides, hinges and pulls, laminated
	Paint	Ea.	.87	34		34.87	interior.
	Minimum Charge	Job		174		174	

Cabinets and Countertops

Base Cabinets	Unit	Material	Labor	Equip.	Total	Specification
30″ Wide Sink Base						
Demolish	Ea.		17.65		17.65	Includes material and labor to install
Install	Ea.	420	32.50		452.50	premium hardwood modular unit with
Demolish and Install	Ea.	420	50.15		470.15	solid hardwood faced frames, drawer
Reinstall	Ea.		26.08		26.08	fronts, and door panels, steel drawer
Clean	Ea.	1.56	12.70		14.26	guides, hinges, pulls and laminated
Paint	Ea.	1.09	38		39.09	interior.
Minimum Charge	Job		174		174	
36″ Wide Sink Base						
Demolish	Ea.		21.50		21.50	Includes material and labor to install
Install	Ea.	460	34.50		494.50	premium hardwood modular unit with
Demolish and Install	Ea.	460	56		516	solid hardwood faced frames, drawer
Reinstall	Ea.		27.49		27.49	fronts, and door panels, steel drawer
Clean	Ea.	1.73	14.95		16.68	guides, hinges, pulls and laminated
Paint	Ea.	1.31	51		52.31	interior.
Minimum Charge	Job		174		174	
36″ Blind Corner						
Demolish	Ea.		21.50		21.50	Includes material and labor to install
Install	Ea.	605	39		644	premium hardwood modular unit with
Demolish and Install	Ea.	605	60.50		665.50	solid hardwood faced frames, drawer
Reinstall	Ea.		31		31	fronts, and door panels, steel drawer
Clean	Ea.	1.73	14.95		16.68	guides, hinges, pulls and laminated
Paint	Ea.	1.31	51		52.31	interior.
Minimum Charge	Job		174		174	
42″ w, 2 Door, 2 Drawer						
Demolish	Ea.		25.50		25.50	Includes material and labor to install
Install	Ea.	750	35		785	premium hardwood modular unit with
Demolish and Install	Ea.	750	60.50		810.50	solid hardwood faced frames, drawer
Reinstall	Ea.		28.19		28.19	fronts, and door panels, steel drawer
Clean	Ea.	2.22	18.15		20.37	guides, hinges, pulls and laminated
Paint	Ea.	1.51	61		62.51	interior.
Minimum Charge	Job		174		174	
Clean Interior						
Clean	L.F.	.63	3.85		4.48	Includes labor and materials to clean
Minimum Charge	Job		127		127	the interior of a 27″ cabinet.

Island

24″ Wide	Unit	Material	Labor	Equip.	Total	Specification
24″ Wide						
Demolish	Ea.		14.10		14.10	Includes material and labor to install
Install	Ea.	425	29		454	custom modular unit with solid
Demolish and Install	Ea.	425	43.10		468.10	hardwood faced frames, hardwood
Reinstall	Ea.		23.25		23.25	door frames and drawer fronts,
Clean	Ea.	1.73	14.95		16.68	hardwood veneer on raised door
Paint	Ea.	1.30	43.50		44.80	panels, hinges and pulls.
Minimum Charge	Job		174		174	
30″ Wide						
Demolish	Ea.		17.65		17.65	Includes material and labor to install
Install	Ea.	515	36.50		551.50	custom modular unit with solid
Demolish and Install	Ea.	515	54.15		569.15	hardwood faced frames, hardwood
Reinstall	Ea.		29.07		29.07	door frames and drawer fronts,
Clean	Ea.	2.22	18.15		20.37	hardwood veneer on raised door
Paint	Ea.	1.63	51		52.63	panels, hinges and pulls.
Minimum Charge	Job		174		174	

Cabinets and Countertops

Base Cabinets	Unit	Material	Labor	Equip.	Total	Specification
36″ Wide						
Demolish	Ea.		21.50		21.50	Includes material and labor to install
Install	Ea.	590	43.50		633.50	custom modular unit with solid
Demolish and Install	Ea.	590	65		655	hardwood faced frames, hardwood
Reinstall	Ea.		34.88		34.88	door frames and drawer fronts,
Clean	Ea.	2.60	23		25.60	hardwood veneer on raised door
Paint	Ea.	1.97	61		62.97	panels, hinges and pulls.
Minimum Charge	Job		174		174	
48″ Wide						
Demolish	Ea.		28		28	Includes material and labor to install
Install	Ea.	845	58		903	custom modular unit with solid
Demolish and Install	Ea.	845	86		931	hardwood faced frames, hardwood
Reinstall	Ea.		46.51		46.51	door frames and drawer fronts,
Clean	Ea.	3.11	28		31.11	hardwood veneer on raised door
Paint	Ea.	2.26	76.50		78.76	panels, hinges and pulls.
Minimum Charge	Job		174		174	
60″ Wide						
Demolish	Ea.		31.50		31.50	Includes material and labor to install
Install	Ea.	1025	72.50		1097.50	custom modular unit with solid
Demolish and Install	Ea.	1025	104		1129	hardwood faced frames, hardwood
Reinstall	Ea.		58.13		58.13	door frames and drawer fronts,
Clean	Ea.	3.89	36.50		40.39	hardwood veneer on raised door
Paint	Ea.	2.26	102		104.26	panels, hinges and pulls.
Minimum Charge	Job		174		174	
Clean Interior						
Clean	L.F.	.63	3.85		4.48	Includes labor and materials to clean
Minimum Charge	Job		127		127	the interior of a 27″ cabinet.
Liquor Bar						
Demolish	L.F.		7.05		7.05	Includes material and labor to install
Install	L.F.	162	35		197	custom modular unit with solid
Demolish and Install	L.F.	162	42.05		204.05	hardwood faced frames, hardwood
Reinstall	L.F.		27.90		27.90	door frames and drawer fronts,
Clean	L.F.	.63	5.30		5.93	hardwood veneer on raised door
Paint	L.F.	.43	13.85		14.28	panels, hinges and pulls.
Minimum Charge	Job		174		174	
Clean Interior						
Clean	L.F.	.63	3.85		4.48	Includes labor and materials to clean
Minimum Charge	Job		127		127	the interior of a 27″ cabinet.
Seal Interior						
Install	L.F.	.34	5		5.34	Includes labor and materials to seal a
Minimum Charge	Job		153		153	cabinet interior.
Back Bar						
Demolish	L.F.		7.05		7.05	Includes material and labor to install
Install	L.F.	162	35		197	custom modular unit with solid
Demolish and Install	L.F.	162	42.05		204.05	hardwood faced frames, hardwood
Reinstall	L.F.		27.90		27.90	door frames and drawer fronts,
Clean	L.F.	.63	5.30		5.93	hardwood veneer on raised door
Paint	L.F.	.43	13.85		14.28	panels, hinges and pulls.
Minimum Charge	Job		174		174	
Clean Interior						
Clean	L.F.	.63	3.85		4.48	Includes labor and materials to clean
Minimum Charge	Job		127		127	the interior of a 27″ cabinet.

Cabinets and Countertops

Base Cabinets		Unit	Material	Labor	Equip.	Total	Specification
Seal Interior							
	Install	L.F.	.34	5		5.34	Includes labor and materials to seal a
	Minimum Charge	Job		153		153	cabinet interior.
Storage (garage)							
	Demolish	L.F.		7.05		7.05	Includes material and labor to install
	Install	L.F.	121	35		156	good modular unit with melamine
	Demolish and Install	L.F.	121	42.05		163.05	laminate, textured colors or wood
	Reinstall	L.F.		27.90		27.90	grain print finish including hinges and
	Clean	L.F.	.63	5.30		5.93	pulls.
	Paint	L.F.	.43	13.85		14.28	
	Minimum Charge	Job		174		174	
Clean Interior							
	Clean	L.F.	.63	3.85		4.48	Includes labor and materials to clean
	Minimum Charge	Job		127		127	the interior of a 27" cabinet.
Seal Interior							
	Install	L.F.	.34	5		5.34	Includes labor and materials to seal a
	Minimum Charge	Job		153		153	cabinet interior.
Built-in Desk							
	Demolish	L.F.		7.05		7.05	Includes material and labor to install
	Install	L.F.	162	35		197	custom modular unit with solid
	Demolish and Install	L.F.	162	42.05		204.05	hardwood faced frames, hardwood
	Reinstall	L.F.		27.90		27.90	door frames and drawer fronts,
	Clean	L.F.	.63	5.30		5.93	hardwood veneer on raised door
	Paint	L.F.	.43	13.85		14.28	panels, hinges and pulls.
	Minimum Charge	Job		174		174	
Built-in Bookcase							
	Demolish	L.F.		9.40		9.40	Includes material and labor to install
	Install	L.F.	25.50	35		60.50	custom modular bookcase unit with
	Demolish and Install	L.F.	25.50	44.40		69.90	clear pine faced frames, shelves 12"
	Reinstall	L.F.		27.90		27.90	O.C., 7' high, 8" deep.
	Clean	L.F.	1.24	9.05		10.29	
	Paint	L.F.	1.01	23.50		24.51	
	Minimum Charge	Job		174		174	
Strip and Refinish							
	Install	L.F.	1.08	5.55		6.63	Includes labor and material to strip,
	Minimum Charge	Job		153		153	prep and refinish exterior of cabinets.
Stain							
	Install	L.F.	.39	1.42		1.81	Includes labor and material to stain
	Minimum Charge	Job		174		174	door and exterior of cabinets.

Countertop		Unit	Material	Labor	Equip.	Total	Specification
Laminated							
With Splash							
	Demolish	L.F.		4.70		4.70	Includes material and labor to install
	Install	L.F.	30.50	12.45		42.95	one piece laminated top with 4"
	Demolish and Install	L.F.	30.50	17.15		47.65	backsplash.
	Reinstall	L.F.		9.97		9.97	
	Clean	L.F.	.08	.64		.72	
	Minimum Charge	Job		174		174	

Cabinets and Countertops

Countertop		Unit	Material	Labor	Equip.	Total	Specification
Roll Top							
	Demolish	L.F.		4.70		4.70	Includes material and labor to install
	Install	L.F.	10.80	11.65		22.45	one piece laminated top with rolled
	Demolish and Install	L.F.	10.80	16.35		27.15	drip edge (post formed) and
	Reinstall	L.F.		9.30		9.30	backsplash.
	Clean	L.F.	.11	.96		1.07	
	Minimum Charge	Job		174		174	
Ceramic Tile							
	Demolish	S.F.		2.35		2.35	Cost includes material and labor to
	Install	S.F.	15.60	7		22.60	install 4-1/4" x 4-1/4" to 6" x 6"
	Demolish and Install	S.F.	15.60	9.35		24.95	glazed tile set in mortar bed and grout
	Clean	S.F.	.06	.48		.54	on particle board substrate.
	Minimum Charge	Job		160		160	
Cultured Marble							
	Demolish	L.F.		4.70		4.70	Includes material and labor to install
	Install	L.F.	30.50	35.50		66	marble countertop, 24" wide, no
	Demolish and Install	L.F.	30.50	40.20		70.70	backsplash.
	Reinstall	L.F.		28.35		28.35	
	Clean	L.F.	.11	.96		1.07	
	Minimum Charge	Job		160		160	
With Splash							
	Demolish	L.F.		4.70		4.70	Includes material and labor to install
	Install	L.F.	34	35.50		69.50	marble countertop, 24" wide, with
	Demolish and Install	L.F.	34	40.20		74.20	backsplash.
	Reinstall	L.F.		28.35		28.35	
	Clean	L.F.	.11	.96		1.07	
	Minimum Charge	Job		160		160	
Quarry Tile							
	Demolish	S.F.		2.35		2.35	Includes material and labor to install
	Install	S.F.	8.25	7		15.25	quarry tile countertop with no
	Demolish and Install	S.F.	8.25	9.35		17.60	backsplash.
	Clean	S.F.	.06	.48		.54	
	Minimum Charge	Job		160		160	
With Splash							
	Demolish	S.F.		2.35		2.35	Includes material and labor to install
	Install	S.F.	10.90	7.75		18.65	quarry tile countertop with backsplash.
	Demolish and Install	S.F.	10.90	10.10		21	
	Clean	S.F.	.06	.48		.54	
	Minimum Charge	Job		160		160	
Butcher Block							
	Demolish	S.F.		2.35		2.35	Includes material and labor to install
	Install	S.F.	31.50	6.25		37.75	solid laminated maple countertop with
	Demolish and Install	S.F.	31.50	8.60		40.10	no backsplash.
	Reinstall	S.F.		4.98		4.98	
	Clean	S.F.	.02	.20		.22	
	Minimum Charge	Job		174		174	
Solid Surface							
	Demolish	L.F.		4.70		4.70	Includes material and labor to install
	Install	L.F.	49	25		74	solid surface counter top 22" deep.
	Demolish and Install	L.F.	49	29.70		78.70	
	Reinstall	L.F.		19.93		19.93	
	Clean	L.F.	.11	.96		1.07	
	Minimum Charge	Job		174		174	

Cabinets and Countertops

Vanity Cabinets	Unit	Material	Labor	Equip.	Total	Specification
Good Grade Laminated						
24" Wide, 2 Door						
Demolish	Ea.		14.10		14.10	Includes material and labor to install
Install	Ea.	163	35		198	modular unit with melamine laminated
Demolish and Install	Ea.	163	49.10		212.10	to particle board, textured colors or
Reinstall	Ea.		27.90		27.90	wood grain print finish, hinges and
Clean	Ea.	1.20	10.15		11.35	pulls.
Paint	Ea.	.87	25.50		26.37	
Minimum Charge	Job		174		174	
30" Wide, 2 Door						
Demolish	Ea.		17.65		17.65	Includes material and labor to install
Install	Ea.	186	43.50		229.50	modular unit with melamine laminated
Demolish and Install	Ea.	186	61.15		247.15	to particle board, textured colors or
Reinstall	Ea.		34.88		34.88	wood grain print finish, hinges and
Clean	Ea.	1.56	12.70		14.26	pulls.
Paint	Ea.	1.09	30.50		31.59	
Minimum Charge	Job		174		174	
36" Wide, 2 Door						
Demolish	Ea.		21.50		21.50	Includes material and labor to install
Install	Ea.	249	52.50		301.50	modular unit with melamine laminated
Demolish and Install	Ea.	249	74		323	to particle board, textured colors or
Reinstall	Ea.		41.87		41.87	wood grain print finish, hinges and
Clean	Ea.	1.73	14.95		16.68	pulls.
Paint	Ea.	1.31	38		39.31	
Minimum Charge	Job		174		174	
42" w, 3 Doors / Drawer						
Demolish	Ea.		25.50		25.50	Includes material and labor to install
Install	Ea.	360	56.50		416.50	modular unit with melamine laminated
Demolish and Install	Ea.	360	82		442	to particle board, textured colors or
Reinstall	Ea.		45.08		45.08	wood grain print finish, hinges and
Clean	Ea.	2.22	18.15		20.37	pulls.
Paint	Ea.	1.51	43.50		45.01	
Minimum Charge	Job		174		174	
48" w, 3 Doors / Drawer						
Demolish	Ea.		28		28	Includes material and labor to install
Install	Ea.	297	61		358	modular unit with melamine laminated
Demolish and Install	Ea.	297	89		386	to particle board, textured colors or
Reinstall	Ea.		48.83		48.83	wood grain print finish, hinges and
Clean	Ea.	3.11	28		31.11	pulls.
Paint	Ea.	2.26	76.50		78.76	
Minimum Charge	Job		174		174	
Clean Interior						
Install	L.F.	.63	5.30		5.93	Includes labor and materials to clean
Minimum Charge	Job		127		127	base cabinetry.
Better Grade Wood						
24" Wide, 2 Door						
Demolish	Ea.		14.10		14.10	Includes material and labor to install
Install	Ea.	217	35		252	custom modular unit with solid
Demolish and Install	Ea.	217	49.10		266.10	hardwood faced frames, hardwood
Reinstall	Ea.		27.90		27.90	door frames and drawer fronts,
Clean	Ea.	1.20	10.15		11.35	hardwood veneer on raised door
Paint	Ea.	.87	28		28.87	panels, hinges and pulls.
Minimum Charge	Job		174		174	

Cabinets and Countertops

Vanity Cabinets

	Unit	Material	Labor	Equip.	Total	Specification
30" Wide, 2 Door						
Demolish	Ea.		17.65		17.65	Includes material and labor to install
Install	Ea.	249	43.50		292.50	custom modular unit with solid
Demolish and Install	Ea.	249	61.15		310.15	hardwood faced frames, hardwood
Reinstall	Ea.		34.88		34.88	door frames and drawer fronts,
Clean	Ea.	1.56	12.70		14.26	hardwood veneer on raised door
Paint	Ea.	1.09	34		35.09	panels, hinges and pulls.
Minimum Charge	Job		174		174	
36" Wide, 2 Door						
Demolish	Ea.		21.50		21.50	Includes material and labor to install
Install	Ea.	330	52.50		382.50	custom modular unit with solid
Demolish and Install	Ea.	330	74		404	hardwood faced frames, hardwood
Reinstall	Ea.		41.87		41.87	door frames and drawer fronts,
Clean	Ea.	1.73	14.95		16.68	hardwood veneer on raised door
Paint	Ea.	1.31	43.50		44.81	panels, hinges and pulls.
Minimum Charge	Job		174		174	
42" w, 3 Doors / Drawer						
Demolish	Ea.		25.50		25.50	Includes material and labor to install
Install	Ea.	360	56.50		416.50	custom modular unit with solid
Demolish and Install	Ea.	360	82		442	hardwood faced frames, hardwood
Reinstall	Ea.		45.08		45.08	door frames and drawer fronts,
Clean	Ea.	2.22	18.15		20.37	hardwood veneer on raised door
Paint	Ea.	1.51	51		52.51	panels, hinges and pulls.
Minimum Charge	Job		174		174	
48" w, 3 Doors / Drawer						
Demolish	Ea.		28		28	Includes material and labor to install
Install	Ea.	395	61		456	custom modular unit with solid
Demolish and Install	Ea.	395	89		484	hardwood faced frames, hardwood
Reinstall	Ea.		48.83		48.83	door frames and drawer fronts,
Clean	Ea.	3.11	28		31.11	hardwood veneer on raised door
Paint	Ea.	2.26	76.50		78.76	panels, hinges and pulls.
Minimum Charge	Job		174		174	
Clean Interior						
Install	L.F.	.63	5.30		5.93	Includes labor and materials to clean
Minimum Charge	Job		127		127	base cabinetry.
Premium Grade Wood						
24" Wide, 2 Door						
Demolish	Ea.		14.10		14.10	Includes material and labor to install
Install	Ea.	256	35		291	premium modular unit with solid
Demolish and Install	Ea.	256	49.10		305.10	hardwood faced frames, hardwood
Reinstall	Ea.		27.90		27.90	door frames and drawer fronts,
Clean	Ea.	1.20	10.15		11.35	hardwood veneer on raised door
Paint	Ea.	.87	30.50		31.37	panels, hinges and pulls.
Minimum Charge	Job		174		174	
30" Wide, 2 Door						
Demolish	Ea.		17.65		17.65	Includes material and labor to install
Install	Ea.	296	43.50		339.50	premium modular unit with solid
Demolish and Install	Ea.	296	61.15		357.15	hardwood faced frames, hardwood
Reinstall	Ea.		34.88		34.88	door frames and drawer fronts,
Clean	Ea.	1.56	12.70		14.26	hardwood veneer on raised door
Paint	Ea.	1.09	38		39.09	panels, hinges and pulls.
Minimum Charge	Job		174		174	

Cabinets and Countertops

Vanity Cabinets

	Unit	Material	Labor	Equip.	Total	Specification
36″ Wide, 2 Door						
Demolish	Ea.		21.50		21.50	Includes material and labor to install
Install	Ea.	460	52.50		512.50	premium modular unit with solid
Demolish and Install	Ea.	460	74		534	hardwood faced frames, hardwood
Reinstall	Ea.		41.87		41.87	door frames and drawer fronts,
Clean	Ea.	1.73	14.95		16.68	hardwood veneer on raised door
Paint	Ea.	1.31	43.50		44.81	panels, hinges and pulls.
Minimum Charge	Job		174		174	
42″ w, 3 Doors / Drawer						
Demolish	Ea.		25.50		25.50	Includes material and labor to install
Install	Ea.	440	56.50		496.50	premium modular unit with solid
Demolish and Install	Ea.	440	82		522	hardwood faced frames, hardwood
Reinstall	Ea.		45.08		45.08	door frames and drawer fronts,
Clean	Ea.	2.22	18.15		20.37	hardwood veneer on raised door
Paint	Ea.	1.51	51		52.51	panels, hinges and pulls.
Minimum Charge	Job		174		174	
48″ w, 3 Doors / Drawer						
Demolish	Ea.		28		28	Includes material and labor to install
Install	Ea.	540	61		601	premium modular unit with solid
Demolish and Install	Ea.	540	89		629	hardwood faced frames, hardwood
Reinstall	Ea.		48.83		48.83	door frames and drawer fronts,
Clean	Ea.	3.11	28		31.11	hardwood veneer on raised door
Paint	Ea.	2.26	76.50		78.76	panels, hinges and pulls.
Minimum Charge	Job		174		174	
Clean Interior						
Install	L.F.	.63	5.30		5.93	Includes labor and materials to clean
Minimum Charge	Job		127		127	base cabinetry.

Premium Hardwood

	Unit	Material	Labor	Equip.	Total	Specification
24″ Wide, 2 Door						
Demolish	Ea.		14.10		14.10	Includes material and labor to install
Install	Ea.	320	35		355	premium modular unit with solid
Demolish and Install	Ea.	320	49.10		369.10	hardwood faced frames, hardwood
Reinstall	Ea.		27.90		27.90	door frames and drawer fronts,
Clean	Ea.	1.20	10.15		11.35	hardwood veneer on raised door
Paint	Ea.	.87	34		34.87	panels, hinges and pulls.
Minimum Charge	Job		174		174	
30″ Wide, 2 Door						
Demolish	Ea.		17.65		17.65	Includes material and labor to install
Install	Ea.	365	43.50		408.50	premium modular unit with solid
Demolish and Install	Ea.	365	61.15		426.15	hardwood faced frames, hardwood
Reinstall	Ea.		34.88		34.88	door frames and drawer fronts,
Clean	Ea.	1.56	12.70		14.26	hardwood veneer on raised door
Paint	Ea.	1.09	38		39.09	panels, hinges and pulls.
Minimum Charge	Job		174		174	
36″ Wide, 2 Door						
Demolish	Ea.		21.50		21.50	Includes material and labor to install
Install	Ea.	530	52.50		582.50	premium modular unit with solid
Demolish and Install	Ea.	530	74		604	hardwood faced frames, hardwood
Reinstall	Ea.		41.87		41.87	door frames and drawer fronts,
Clean	Ea.	1.73	14.95		16.68	hardwood veneer on raised door
Paint	Ea.	1.31	51		52.31	panels, hinges and pulls.
Minimum Charge	Job		174		174	

Cabinets and Countertops

Vanity Cabinets

Vanity Cabinets	Unit	Material	Labor	Equip.	Total	Specification
42″ w, 3 Doors / Drawer						
Demolish	Ea.		25.50		25.50	Includes material and labor to install
Install	Ea.	490	56.50		546.50	premium modular unit with solid
Demolish and Install	Ea.	490	82		572	hardwood faced frames, hardwood
Reinstall	Ea.		45.08		45.08	door frames and drawer fronts,
Clean	Ea.	2.22	18.15		20.37	hardwood veneer on raised door
Paint	Ea.	1.51	61		62.51	panels, hinges and pulls.
Minimum Charge	Job		174		174	
48″ w, 3 Doors / Drawer						
Demolish	Ea.		28		28	Includes material and labor to install
Install	Ea.	605	61		666	premium modular unit with solid
Demolish and Install	Ea.	605	89		694	hardwood faced frames, hardwood
Reinstall	Ea.		48.83		48.83	door frames and drawer fronts,
Clean	Ea.	3.11	28		31.11	hardwood veneer on raised door
Paint	Ea.	2.26	76.50		78.76	panels, hinges and pulls.
Minimum Charge	Job		174		174	
Clean Interior						
Install	L.F.	.63	5.30		5.93	Includes labor and materials to clean
Minimum Charge	Job		127		127	base cabinetry.
Strip and Refinish						
Install	L.F.	1.08	5.55		6.63	Includes labor and material to strip,
Minimum Charge	Job		153		153	prep and refinish exterior of cabinets.

Vanity Cabinet Tops

Vanity Cabinet Tops	Unit	Material	Labor	Equip.	Total	Specification
Cultured Marble Top						
Demolish	L.F.		4.70		4.70	Includes material and labor to install
Install	L.F.	38.50	21		59.50	cultured marble vanity top with
Demolish and Install	L.F.	38.50	25.70		64.20	integral sink, 22″ deep.
Reinstall	L.F.		16.68		16.68	
Clean	L.F.	.11	.96		1.07	
Minimum Charge	Job		160		160	
Laminated Top						
Demolish	L.F.		2.82		2.82	Includes material and labor to install
Install	L.F.	30.50	12.45		42.95	one piece laminated top with 4″
Demolish and Install	L.F.	30.50	15.27		45.77	backsplash.
Reinstall	L.F.		9.97		9.97	
Clean	L.F.	.08	.64		.72	
Minimum Charge	Job		174		174	
Solid Surface						
Demolish	L.F.		4.70		4.70	Includes material and labor to install
Install	L.F.	68	26		94	solid surface vanity top with integral
Demolish and Install	L.F.	68	30.70		98.70	sink installed, 22″ deep.
Reinstall	L.F.		20.67		20.67	
Clean	L.F.	.11	.96		1.07	
Minimum Charge	Job		174		174	

Painting

Roller Handle

Paint Brush

Paint Preparation, Walls		Unit	Material	Labor	Equip.	Total	Specification
Cover / Protect Floors							
	Install	S.F.	.54	.12		.66	Includes material and labor to install plastic masking sheet on large uninterrupted areas.
	Minimum Charge	Job		154		154	
Cover / Protect Walls							
	Install	S.F.	.54	.12		.66	Includes material and labor to install plastic masking sheet on walls.
	Minimum Charge	Job		154		154	
Clean Walls							
Light							
	Clean	S.F.		.10		.10	Includes labor to wash gypsum drywall or plaster wall surfaces.
	Minimum Charge	Job		154		154	
Heavy							
	Clean	S.F.		.12		.12	Includes labor and material for heavy cleaning (multiple applications) with detergent and solvent.
	Minimum Charge	Job		154		154	
Prep Walls							
	Paint	S.F.	.03	.15		.18	Includes labor and material to prepare for painting including scraping, patching and puttying.
	Minimum Charge	Job		154		154	
Sand Walls							
	Paint	S.F.		.10		.10	Includes labor to sand gypsum drywall wall surfaces.
	Minimum Charge	Job		154		154	
Seal Walls							
Large Area							
	Paint	S.F.	.06	.23		.29	Includes labor and material to paint with primer / sealer by roller.
	Minimum Charge	Job		154		154	
Spot							
	Paint	S.F.	.06	.27		.33	Includes labor and material to seal existing drywall with shellac-based material.
	Minimum Charge	Job		154		154	

Paint Preparation, Ceilings		Unit	Material	Labor	Equip.	Total	Specification
Cover / Protect Floors							
	Install	S.F.	.54	.12		.66	Includes material and labor to install plastic masking sheet on large uninterrupted areas.
	Minimum Charge	Job		154		154	

Painting

Paint Preparation, Ceilings

Paint Preparation, Ceilings		Unit	Material	Labor	Equip.	Total	Specification
Cover / Protect Walls							
	Install	S.F.	.54	.12		.66	Includes material and labor to install
	Minimum Charge	Job		154		154	plastic masking sheet on walls.
Prep Ceiling							
	Paint	S.F.	.06	.57		.63	Includes labor and material to fill in
	Minimum Charge	Job		154		154	hairline cracks up to 1/8" with filler, sand and prep for paint.
Sand Ceiling							
	Paint	S.F.		.14		.14	Includes labor to sand gypsum drywall
	Minimum Charge	Job		154		154	ceiling surfaces.
Seal Ceiling							
Large Area							
	Paint	S.F.	.06	.23		.29	Includes labor and material to paint
	Minimum Charge	Job		153		153	ceiling, one coat flat latex.
Spot							
	Paint	S.F.	.06	.27		.33	Includes labor and material to seal
	Minimum Charge	Job		153		153	existing drywall with shellac-based material.

Paint / Texture, Walls

Paint / Texture, Walls		Unit	Material	Labor	Equip.	Total	Specification
Texture Walls							
Spray							
	Install	S.F.	.06	.38		.44	Includes labor and material to install
	Minimum Charge	Job		153		153	by spray, texture finish.
Trowel							
	Install	S.F.	.01	.20		.21	Includes labor and material to apply
	Minimum Charge	Job		153		153	hand troweled texture.
Paint Walls 1 Coat							
	Paint	S.F.	.06	.23		.29	Includes labor and material to paint,
	Minimum Charge	Job		153		153	one coat flat latex.
Paint Walls 2 Coats							
	Paint	S.F.	.12	.38		.50	Includes labor and material to paint,
	Minimum Charge	Job		153		153	two coats flat latex.
Paint Walls 3 Coats							
	Paint	S.F.	.18	.47		.65	Includes labor and material to paint,
	Minimum Charge	Job		153		153	three coats flat latex.
Clean and Seal							
	Paint	S.F.	.24	.81		1.05	Includes labor and material to do light
	Minimum Charge	Job		153		153	cleaning, sealing with 2 coats of latex paint.

Paint / Texture, Ceilings

Paint / Texture, Ceilings		Unit	Material	Labor	Equip.	Total	Specification
Texture Ceiling							
Spray							
	Install	S.F.	.06	.38		.44	Includes labor and material to install
	Minimum Charge	Job		153		153	by spray, texture finish.
Trowel							
	Install	S.F.	.01	.20		.21	Includes labor and material to apply
	Minimum Charge	Job		153		153	hand troweled texture.

Painting

Paint / Texture, Ceilings		Unit	Material	Labor	Equip.	Total	Specification
Paint Ceiling, 1 Coat							
	Paint	S.F.	.06	.23		.29	Includes labor and material to paint
	Minimum Charge	Job		153		153	ceiling, one coat flat latex.
Paint Ceiling, 2 Coats							
	Paint	S.F.	.12	.38		.50	Includes labor and material to paint
	Minimum Charge	Job		153		153	two coats flat latex paint.
Paint Ceiling, 3 Coats							
	Paint	S.F.	.18	.45		.63	Includes labor and material to paint
	Minimum Charge	Job		153		153	ceiling, three coats flat latex.
Clean and Seal							
	Paint	S.F.	.18	.64		.82	Includes labor and material for light
	Minimum Charge	Job		153		153	cleaning, sealing and paint 3 coats of flat latex paint.

Plank Flooring	Wood Strip Flooring	Tile Flooring

Wood Plank Flooring	Unit	Material	Labor	Equip.	Total	Specification
Maple						
2-1/4"						
Demolish	S.F.		1.13		1.13	Includes material and labor to install
Install	S.F.	4.65	1.74		6.39	unfinished T&G maple flooring,
Demolish and Install	S.F.	4.65	2.87		7.52	25/32" thick, 2-1/4" wide, random
Reinstall	S.F.		1.39		1.39	3' to 16' lengths including felt
Clean	S.F.		.32		.32	underlayment, nailed in place over
Paint	S.F.	1.25	1.95		3.20	prepared subfloor.
Minimum Charge	Job		174		174	
3-1/4"						
Demolish	S.F.		1.13		1.13	Includes material and labor to install
Install	S.F.	3.60	1.74		5.34	unfinished T&G maple flooring,
Demolish and Install	S.F.	3.60	2.87		6.47	33/32" thick, 3-1/4" wide, random
Reinstall	S.F.		1.39		1.39	3' to 16' lengths including felt
Clean	S.F.		.32		.32	underlayment, nailed in place over
Paint	S.F.	1.25	1.95		3.20	prepared subfloor.
Minimum Charge	Job		174		174	
Sand and Finish						
Install	S.F.	.84	.64		1.48	Includes labor and material to sand (3
Minimum Charge	Job		174		174	passes) and finish and two coats of
						urethane existing floor.
Oak						
2-1/4"						
Demolish	S.F.		1.13		1.13	Includes material and labor to install
Install	S.F.	4.18	1.74		5.92	unfinished T&G oak flooring, 25/32"
Demolish and Install	S.F.	4.18	2.87		7.05	thick, 2-1/4" wide, random 3' to 16'
Reinstall	S.F.		1.39		1.39	lengths including felt underlayment,
Clean	S.F.		.32		.32	nailed in place over prepared
Paint	S.F.	1.25	1.95		3.20	subfloor.
Minimum Charge	Job		174		174	
3-1/4"						
Demolish	S.F.		1.13		1.13	Includes material and labor to install
Install	S.F.	5.70	1.23		6.93	unfinished T&G oak flooring, 25/32"
Demolish and Install	S.F.	5.70	2.36		8.06	thick, 3-1/4" wide, random 3' to 16'
Reinstall	S.F.		.99		.99	lengths including felt underlayment,
Clean	S.F.		.32		.32	nailed in place over prepared
Paint	S.F.	1.25	1.95		3.20	subfloor.
Minimum Charge	Job		174		174	

Flooring

Wood Plank Flooring	Unit	Material	Labor	Equip.	Total	Specification
Sand and Finish						
Install	S.F.	1.25	1.95		3.20	Includes labor and material to sand
Minimum Charge	Job		174		174	and finish (3 passes) with two coats of urethane on a new floor.
Walnut						
3″ to 7″						
Demolish	S.F.		1.13		1.13	Includes material and labor to install
Install	S.F.	9.05	1.23		10.28	unfinished T&G walnut flooring,
Demolish and Install	S.F.	9.05	2.36		11.41	25/32″ thick, random 3′ to 16′
Reinstall	S.F.		.99		.99	lengths including felt underlayment,
Clean	S.F.		.32		.32	nailed in place over prepared
Paint	S.F.	1.25	1.95		3.20	subfloor.
Minimum Charge	Job		174		174	
Sand and Finish						
Install	S.F.	1.25	1.95		3.20	Includes labor and material to sand
Minimum Charge	Job		174		174	and finish (3 passes) with two coats of urethane on a new floor.
Pine						
2-1/4″						
Demolish	S.F.		1.13		1.13	Includes material and labor to install
Install	S.F.	7.35	1.31		8.66	unfinished T&G pine flooring, 25/32″
Demolish and Install	S.F.	7.35	2.44		9.79	thick, 2-1/4″ wide, random 3′ to 16′
Reinstall	S.F.		1.05		1.05	lengths including felt underlayment,
Clean	S.F.		.32		.32	nailed in place over prepared
Paint	S.F.	.84	.64		1.48	subfloor.
Minimum Charge	Job		174		174	
Sand and Finish						
Install	S.F.	1.25	1.95		3.20	Includes labor and material to sand
Minimum Charge	Job		174		174	and finish (3 passes) with two coats of urethane on a new floor.
Prefinished Oak						
2-1/4″						
Demolish	S.F.		1.13		1.13	Includes material and labor to install
Install	S.F.	5.75	1.74		7.49	prefinished T&G oak flooring, 2-1/4″
Demolish and Install	S.F.	5.75	2.87		8.62	wide, random 3′ to 16′ lengths
Reinstall	S.F.		1.39		1.39	including felt underlayment, nailed in
Clean	S.F.		.32		.32	place over prepared subfloor.
Minimum Charge	Job		174		174	
3-1/4″						
Demolish	S.F.		1.13		1.13	Includes material and labor to install
Install	S.F.	7.40	1.60		9	prefinished T&G oak flooring, 25/32″
Demolish and Install	S.F.	7.40	2.73		10.13	thick, 3″ to 7″ wide, random 3′ to 16′
Reinstall	S.F.		1.28		1.28	lengths including felt underlayment,
Clean	S.F.		.32		.32	nailed in place over prepared
Minimum Charge	Job		174		174	subfloor.
Sand and Finish New						
Install	S.F.	.84	.86		1.70	Includes labor and material to sand (3
Minimum Charge	Job		174		174	passes) and finish, two coats of urethane, new floor.
Refinish Existing						
Install	S.F.	.84	.64		1.48	Includes labor and material to sand (3
Minimum Charge	Job		174		174	passes) and finish and two coats of urethane existing floor.

Flooring

Wood Plank Flooring

Wood Plank Flooring		Unit	Material	Labor	Equip.	Total	Specification
Urethane Coat							
	Install	S.F.	.15	.16		.31	Includes labor and material to install
	Minimum Charge	Job		174		174	two coats of urethane, on existing
							floor.
Clean and Wax							
	Clean	S.F.	.02	.20		.22	Includes labor and material to wash
	Minimum Charge	Job		174		174	and wax a hardwood floor.
Underlayment Per S.F.							
	Demolish	S.F.		.51		.51	Cost includes material and labor to
	Install	S.F.	.77	.47		1.24	install 1/4" lauan subfloor, standard
	Demolish and Install	S.F.	.77	.98		1.75	interior grade, nailed every 6".
	Clean	S.F.		.25		.25	
	Minimum Charge	Job		174		174	

Wood Parquet Tile Floor

Wood Parquet Tile Floor		Unit	Material	Labor	Equip.	Total	Specification
Oak							
9" x 9"							
	Demolish	S.F.		.86		.86	Includes material and labor to install
	Install	S.F.	4.76	2.79		7.55	prefinished oak 9" x 9" parquet block
	Demolish and Install	S.F.	4.76	3.65		8.41	flooring, 5/16" thick, installed in
	Reinstall	S.F.		2.23		2.23	mastic.
	Clean	S.F.		.32		.32	
	Paint	S.F.	1.25	1.95		3.20	
	Minimum Charge	Job		174		174	
13" x 13"							
	Demolish	S.F.		.86		.86	Includes material and labor to install
	Install	S.F.	9	2.18		11.18	prefinished oak 13" x 13" parquet
	Demolish and Install	S.F.	9	3.04		12.04	block flooring, 5/16" thick, installed in
	Reinstall	S.F.		1.74		1.74	mastic.
	Clean	S.F.		.32		.32	
	Paint	S.F.	1.25	1.95		3.20	
	Minimum Charge	Job		174		174	
Cherry							
9" x 9"							
	Demolish	S.F.		.86		.86	Includes material and labor to install
	Install	S.F.	8.15	2.79		10.94	prefinished cherry 9" x 9" parquet
	Demolish and Install	S.F.	8.15	3.65		11.80	block flooring, 5/16" thick, installed in
	Reinstall	S.F.		2.79		2.79	mastic.
	Clean	S.F.		.32		.32	
	Paint	S.F.	1.25	1.95		3.20	
	Minimum Charge	Job		174		174	
13" x 13"							
	Demolish	S.F.		.86		.86	Includes material and labor to install
	Install	S.F.	9.30	2.18		11.48	prefinished cherry 13" x 13" parquet
	Demolish and Install	S.F.	9.30	3.04		12.34	block flooring, 5/16" thick, installed in
	Reinstall	S.F.		2.18		2.18	mastic.
	Clean	S.F.		.32		.32	
	Paint	S.F.	1.25	1.95		3.20	
	Minimum Charge	Job		174		174	

Flooring

Wood Parquet Tile Floor	Unit	Material	Labor	Equip.	Total	Specification
Walnut						
9″ x 9″						
Demolish	S.F.		.86		.86	Includes material and labor to install
Install	S.F.	5.15	2.79		7.94	prefinished walnut 9″ x 9″ parquet
Demolish and Install	S.F.	5.15	3.65		8.80	flooring in mastic.
Reinstall	S.F.		2.23		2.23	
Clean	S.F.		.32		.32	
Paint	S.F.	1.25	1.95		3.20	
Minimum Charge	Job		174		174	
13″ x 13″						
Demolish	S.F.		.86		.86	Includes material and labor to install
Install	S.F.	9.65	2.18		11.83	prefinished walnut 13″ x 13″ parquet
Demolish and Install	S.F.	9.65	3.04		12.69	block flooring, 5/16″ thick, installed in
Reinstall	S.F.		2.18		2.18	mastic.
Clean	S.F.		.32		.32	
Paint	S.F.	1.25	1.95		3.20	
Minimum Charge	Job		174		174	
Acrylic Impregnated						
Oak 12″ x 12″						
Demolish	S.F.		.86		.86	Includes material and labor to install
Install	S.F.	8.30	2.49		10.79	red oak 12″ x 12″ acrylic
Demolish and Install	S.F.	8.30	3.35		11.65	impregnated parquet block flooring,
Reinstall	S.F.		2.49		2.49	5/16″ thick, installed in mastic.
Clean	S.F.		.32		.32	
Minimum Charge	Job		174		174	
Cherry 12″ x 12″						
Demolish	S.F.		.86		.86	Includes material and labor to install
Install	S.F.	11.40	2.49		13.89	prefinished cherry 12″ x 12″ acrylic
Demolish and Install	S.F.	11.40	3.35		14.75	impregnated parquet block flooring,
Reinstall	S.F.		2.49		2.49	5/16″ thick, installed in mastic.
Clean	S.F.		.32		.32	
Minimum Charge	Job		174		174	
Ash 12″ x 12″						
Demolish	S.F.		.86		.86	Includes material and labor to install
Install	S.F.	10.25	2.49		12.74	ash 12″ x 12″ acrylic impregnated
Demolish and Install	S.F.	10.25	3.35		13.60	parquet block flooring, 5/16″ thick in
Reinstall	S.F.		2.49		2.49	mastic.
Clean	S.F.		.32		.32	
Minimum Charge	Job		174		174	
Teak						
Demolish	S.F.		.86		.86	Includes material and labor to install
Install	S.F.	7.35	2.18		9.53	prefinished teak parquet block
Demolish and Install	S.F.	7.35	3.04		10.39	flooring, 5/16″ thick, and installed in
Reinstall	S.F.		2.18		2.18	mastic.
Clean	S.F.		.32		.32	
Paint	S.F.	1.25	1.95		3.20	
Minimum Charge	Job		174		174	
Clean and Wax						
Clean	S.F.	.02	.20		.22	Includes labor and material to wash
Minimum Charge	Job		174		174	and wax a hardwood floor.

Flooring

Ceramic Tile Flooring

Ceramic Tile Flooring		Unit	Material	Labor	Equip.	Total	Specification
Economy Grade							
	Demolish	S.F.		.84		.84	Cost includes material and labor to
	Install	S.F.	4.59	6.25		10.84	install 4-1/4" x 4-1/4" ceramic tile in
	Demolish and Install	S.F.	4.59	7.09		11.68	mortar bed including grout.
	Clean	S.F.		.48		.48	
	Minimum Charge	Job		160		160	
Average Grade							
	Demolish	S.F.		.84		.84	Cost includes material and labor to
	Install	S.F.	6.10	6.40		12.50	install 4-1/4" x 4-1/4" ceramic tile in
	Demolish and Install	S.F.	6.10	7.24		13.34	mortar bed including grout.
	Clean	S.F.		.48		.48	
	Minimum Charge	Job		160		160	
Premium Grade							
	Demolish	S.F.		.84		.84	Cost includes material and labor to
	Install	S.F.	9.65	6.40		16.05	install 4-1/4" x 4-1/4" ceramic tile in
	Demolish and Install	S.F.	9.65	7.24		16.89	mortar bed including grout.
	Clean	S.F.		.48		.48	
	Minimum Charge	Job		160		160	
Re-grout							
	Install	S.F.	.15	2.56		2.71	Includes material and labor to regrout
	Minimum Charge	Job		160		160	tile floors.

Hard Tile Flooring

Hard Tile Flooring		Unit	Material	Labor	Equip.	Total	Specification
Thick Set Paver							
Brick							
	Demolish	S.F.		1.76		1.76	Cost includes material and labor to
	Install	S.F.	.90	14.80		15.70	install 6" x 12" adobe brick paver with
	Demolish and Install	S.F.	.90	16.56		17.46	1/2" mortar joints.
	Clean	S.F.		.48		.48	
	Minimum Charge	Job		160		160	
Mexican Red							
	Demolish	S.F.		1.76		1.76	Cost includes material and labor to
	Install	S.F.	1.61	6.65		8.26	install 12" x 12" red Mexican paver
	Demolish and Install	S.F.	1.61	8.41		10.02	tile in mortar bed with grout.
	Clean	S.F.		.48		.48	
	Minimum Charge	Job		160		160	
Saltillo							
	Demolish	S.F.		.84		.84	Cost includes material and labor to
	Install	S.F.	1.13	6.65		7.78	install 12" x 12"saltillo tile in mortar
	Demolish and Install	S.F.	1.13	7.49		8.62	bed with grout.
	Clean	S.F.		.48		.48	
	Minimum Charge	Job		160		160	
Marble							
Premium Grade							
	Demolish	S.F.		.84		.84	Cost includes material and labor to
	Install	S.F.	20	13.90		33.90	install 3/8" x 12" x 12" marble, in
	Demolish and Install	S.F.	20	14.74		34.74	mortar bed with grout.
	Clean	S.F.		.48		.48	
	Minimum Charge	Job		160		160	

Flooring

Hard Tile Flooring

Hard Tile Flooring	Unit	Material	Labor	Equip.	Total	Specification
Terrazzo						
Gray Cement						
Demolish	S.F.		1.12		1.12	Cost includes material and labor to
Install	S.F.	3.08	4.39	2.16	9.63	install 1-3/4″ terrazzo , #1 and #2
Demolish and Install	S.F.	3.08	5.51	2.16	10.75	chips in gray Portland cement.
Clean	S.F.		.48		.48	
Minimum Charge	Job		286	140	426	
White Cement						
Demolish	S.F.		1.12		1.12	Cost includes material and labor to
Install	S.F.	3.51	4.39	2.16	10.06	install 1-3/4″ terrazzo, #1 and #2
Demolish and Install	S.F.	3.51	5.51	2.16	11.18	chips in white Portland cement.
Clean	S.F.		.48		.48	
Minimum Charge	Job		286	140	426	
Non-skid Gray						
Demolish	S.F.		1.12		1.12	Cost includes material and labor to
Install	S.F.	5.60	9.50	4.67	19.77	install 1-3/4″ terrazzo, #1 and #2
Demolish and Install	S.F.	5.60	10.62	4.67	20.89	chips in gray Portland cement with
Clean	S.F.		.48		.48	light non-skid abrasive.
Minimum Charge	Job		286	140	426	
Non-skid White						
Demolish	S.F.		1.12		1.12	Cost includes material and labor to
Install	S.F.	6.15	9.50	4.67	20.32	install 1-3/4″ terrazzo, #1 and #2
Demolish and Install	S.F.	6.15	10.62	4.67	21.44	chips in white Portland cement with
Clean	S.F.		.48		.48	light non-skid abrasive.
Minimum Charge	Job		286	140	426	
Gray w / Brass Divider						
Demolish	S.F.		1.13		1.13	Cost includes material and labor to
Install	S.F.	7.70	10.40	5.10	23.20	install 1-3/4″ terrazzo, #1 and #2
Demolish and Install	S.F.	7.70	11.53	5.10	24.33	chips in gray Portland cement with
Clean	S.F.		.48		.48	brass strips 2′ O.C. each way.
Minimum Charge	Job		286	140	426	
Clean						
Clean	S.F.	.13	.32		.45	Includes labor and material to clean
Minimum Charge	Job		286	140	426	terrazzo flooring.
Slate						
Demolish	S.F.		1.01		1.01	Includes material and labor to install
Install	S.F.	5.50	3.14		8.64	slate tile flooring in thin set with grout.
Demolish and Install	S.F.	5.50	4.15		9.65	
Clean	S.F.		.48		.48	
Minimum Charge	Job		160		160	
Granite						
Demolish	S.F.		.83		.83	Includes material and labor to install
Install	S.F.	20	3.20		23.20	granite tile flooring in thin set with
Demolish and Install	S.F.	20	4.03		24.03	grout.
Reinstall	S.F.		3.20		3.20	
Clean	S.F.		.48		.48	
Minimum Charge	Job		160		160	
Re-grout						
Install	S.F.	.15	2.56		2.71	Includes material and labor to regrout
Minimum Charge	Job		160		160	tile floors.

Flooring

Sheet Vinyl Per S.F.

Sheet Vinyl Per S.F.	Unit	Material	Labor	Equip.	Total	Specification
Economy Grade						
Demolish	S.F.		.40		.40	Includes material and labor to install
Install	S.F.	1.84	1		2.84	resilient sheet vinyl flooring.
Demolish and Install	S.F.	1.84	1.40		3.24	
Clean	S.F.		.32		.32	
Minimum Charge	Job		160		160	
Average Grade						
Demolish	S.F.		.40		.40	Includes material and labor to install
Install	S.F.	2.85	1		3.85	resilient sheet vinyl flooring.
Demolish and Install	S.F.	2.85	1.40		4.25	
Clean	S.F.		.32		.32	
Minimum Charge	Job		160		160	
Premium Grade						
Demolish	S.F.		.40		.40	Includes material and labor to install
Install	S.F.	4.84	1		5.84	resilient sheet vinyl flooring.
Demolish and Install	S.F.	4.84	1.40		6.24	
Clean	S.F.		.32		.32	
Minimum Charge	Job		160		160	
Clean and Wax						
Clean	S.F.	.02	.16		.18	Includes labor and material to wash
Minimum Charge	Job		160		160	and wax a vinyl tile floor.

Sheet Vinyl

Sheet Vinyl	Unit	Material	Labor	Equip.	Total	Specification
Economy Grade						
Demolish	S.Y.		3.64		3.64	Includes material and labor to install
Install	S.Y.	16.50	9		25.50	no wax sheet vinyl flooring.
Demolish and Install	S.Y.	16.50	12.64		29.14	
Clean	S.Y.		2.21		2.21	
Minimum Charge	Job		160		160	
Average Grade						
Demolish	S.Y.		3.64		3.64	Includes material and labor to install
Install	S.Y.	25.50	9		34.50	no wax sheet vinyl flooring.
Demolish and Install	S.Y.	25.50	12.64		38.14	
Clean	S.Y.		2.21		2.21	
Minimum Charge	Job		160		160	
Premium Grade						
Demolish	S.Y.		3.64		3.64	Includes material and labor to install
Install	S.Y.	43.50	9		52.50	no wax sheet vinyl flooring.
Demolish and Install	S.Y.	43.50	12.64		56.14	
Clean	S.Y.		2.21		2.21	
Minimum Charge	Job		160		160	
Clean and Wax						
Clean	S.F.	.02	.16		.18	Includes labor and material to wash
Minimum Charge	Job		160		160	and wax a vinyl tile floor.

Vinyl Tile

Vinyl Tile	Unit	Material	Labor	Equip.	Total	Specification
Economy Grade						
Demolish	S.F.		.56		.56	Includes material and labor to install
Install	S.F.	2.42	.64		3.06	no wax 12″ x 12″ vinyl tile.
Demolish and Install	S.F.	2.42	1.20		3.62	
Clean	S.F.		.32		.32	
Minimum Charge	Job		160		160	

Flooring

Vinyl Tile

	Unit	Material	Labor	Equip.	Total	Specification
Average Grade						
Demolish	S.F.		.56		.56	Includes material and labor to install
Install	S.F.	5.90	.64		6.54	no wax 12″ x 12″ vinyl tile.
Demolish and Install	S.F.	5.90	1.20		7.10	
Clean	S.F.		.32		.32	
Minimum Charge	Job		160		160	
Premium Grade						
Demolish	S.F.		.56		.56	Includes material and labor to install
Install	S.F.	12.10	.64		12.74	no wax 12″ x 12″ vinyl tile.
Demolish and Install	S.F.	12.10	1.20		13.30	
Clean	S.F.		.32		.32	
Minimum Charge	Job		160		160	
Clean and Wax						
Clean	S.F.	.02	.16		.18	Includes labor and material to wash
Minimum Charge	Job		160		160	and wax a vinyl tile floor.

Carpeting Per S.F.

	Unit	Material	Labor	Equip.	Total	Specification
Economy Grade						
Demolish	S.F.		.06		.06	Includes material and labor to install
Install	S.F.	1.20	.62		1.82	carpet including tack strips and hot
Demolish and Install	S.F.	1.20	.68		1.88	melt tape on seams.
Clean	S.F.		.44		.44	
Minimum Charge	Job		160		160	
Average Grade						
Demolish	S.F.		.06		.06	Includes material and labor to install
Install	S.F.	1.40	.62		2.02	carpet including tack strips and hot
Demolish and Install	S.F.	1.40	.68		2.08	melt tape on seams.
Clean	S.F.		.44		.44	
Minimum Charge	Job		160		160	
Premium Grade						
Demolish	S.F.		.06		.06	Includes material and labor to install
Install	S.F.	2.48	.62		3.10	carpet including tack strips and hot
Demolish and Install	S.F.	2.48	.68		3.16	melt tape on seams.
Clean	S.F.		.44		.44	
Minimum Charge	Job		160		160	
Indoor/Outdoor						
Demolish	S.F.		.06		.06	Includes material and labor to install
Install	S.F.	1.29	.76		2.05	indoor-outdoor carpet.
Demolish and Install	S.F.	1.29	.82		2.11	
Clean	S.F.		.44		.44	
Minimum Charge	Job		160		160	
100% Wool						
Average Grade						
Demolish	S.F.		.06		.06	Includes material and labor to install
Install	S.F.	8.20	.63		8.83	carpet including tack strips and hot
Demolish and Install	S.F.	8.20	.69		8.89	melt tape on seams.
Clean	S.F.		.44		.44	
Minimum Charge	Job		160		160	
Premium Grade						
Demolish	S.F.		.06		.06	Includes material and labor to install
Install	S.F.	9.20	.63		9.83	carpet including tack strips and hot
Demolish and Install	S.F.	9.20	.69		9.89	melt tape on seams.
Clean	S.F.		.44		.44	
Minimum Charge	Job		160		160	

Flooring

Carpeting Per S.F.

Carpeting Per S.F.		Unit	Material	Labor	Equip.	Total	Specification
Luxury Grade							
	Demolish	S.F.		.06		.06	Includes material and labor to install
	Install	S.F.	12.80	.63		13.43	carpet including tack strips and hot
	Demolish and Install	S.F.	12.80	.69		13.49	melt tape on seams.
	Clean	S.F.		.44		.44	
	Minimum Charge	Job		160		160	
Berber							
Average Grade							
	Demolish	S.F.		.06		.06	Includes material and labor to install
	Install	S.F.	3.34	.76		4.10	carpet including tack strips and hot
	Demolish and Install	S.F.	3.34	.82		4.16	melt tape on seams.
	Clean	S.F.		.44		.44	
	Minimum Charge	Job		160		160	
Premium Grade							
	Demolish	S.F.		.06		.06	Includes material and labor to install
	Install	S.F.	5.95	.76		6.71	carpet including tack strips and hot
	Demolish and Install	S.F.	5.95	.82		6.77	melt tape on seams.
	Clean	S.F.		.44		.44	
	Minimum Charge	Job		160		160	

Carpeting

Carpeting		Unit	Material	Labor	Equip.	Total	Specification
Indoor/Outdoor							
	Demolish	S.Y.		.94		.94	Includes material and labor to install
	Install	S.Y.	11.55	6.80		18.35	indoor-outdoor carpet.
	Demolish and Install	S.Y.	11.55	7.74		19.29	
	Reinstall	S.Y.		5.44		5.44	
	Clean	S.Y.		3.97		3.97	
	Minimum Charge	Job		160		160	
Economy Grade							
	Demolish	S.Y.		.94		.94	Includes material and labor to install
	Install	S.Y.	10.80	5.60		16.40	carpet including tack strips and hot
	Demolish and Install	S.Y.	10.80	6.54		17.34	melt tape on seams.
	Reinstall	S.Y.		4.49		4.49	
	Clean	S.Y.		3.97		3.97	
	Minimum Charge	Job		160		160	
Average Grade							
	Demolish	S.Y.		.94		.94	Includes material and labor to install
	Install	S.Y.	12.55	5.60		18.15	carpet including tack strips and hot
	Demolish and Install	S.Y.	12.55	6.54		19.09	melt tape on seams.
	Reinstall	S.Y.		4.49		4.49	
	Clean	S.Y.		3.97		3.97	
	Minimum Charge	Job		160		160	
Premium Grade							
	Demolish	S.Y.		.94		.94	Includes material and labor to install
	Install	S.Y.	22	5.60		27.60	carpet including tack strips and hot
	Demolish and Install	S.Y.	22	6.54		28.54	melt tape on seams.
	Reinstall	S.Y.		4.49		4.49	
	Clean	S.Y.		3.97		3.97	
	Minimum Charge	Job		160		160	

Flooring

Carpeting	Unit	Material	Labor	Equip.	Total	Specification
100% Wool						
Average Grade						
Demolish	S.Y.		.94		.94	Includes material and labor to install
Install	S.Y.	73.50	5.70		79.20	wool carpet including tack strips and
Demolish and Install	S.Y.	73.50	6.64		80.14	hot melt tape on seams.
Reinstall	S.Y.		4.57		4.57	
Clean	S.Y.		3.97		3.97	
Minimum Charge	Job		160		160	
Premium Grade						
Demolish	S.Y.		.94		.94	Includes material and labor to install
Install	S.Y.	82.50	5.70		88.20	carpet including tack strips and hot
Demolish and Install	S.Y.	82.50	6.64		89.14	melt tape on seams.
Reinstall	S.Y.		4.57		4.57	
Clean	S.Y.		3.97		3.97	
Minimum Charge	Job		160		160	
Luxury Grade						
Demolish	S.Y.		.94		.94	Includes material and labor to install
Install	S.Y.	116	5.70		121.70	carpet including tack strips and hot
Demolish and Install	S.Y.	116	6.64		122.64	melt tape on seams.
Reinstall	S.Y.		4.57		4.57	
Clean	S.Y.		3.97		3.97	
Minimum Charge	Job		160		160	
Berber						
Average Grade						
Demolish	S.Y.		.94		.94	Includes material and labor to install
Install	S.Y.	30.50	6.80		37.30	berber carpet including tack strips and
Demolish and Install	S.Y.	30.50	7.74		38.24	hot melt tape on seams.
Reinstall	S.Y.		5.44		5.44	
Clean	S.Y.		3.97		3.97	
Minimum Charge	Job		160		160	
Premium Grade						
Demolish	S.Y.		.94		.94	Includes material and labor to install
Install	S.Y.	54	6.80		60.80	carpet including tack strips and hot
Demolish and Install	S.Y.	54	7.74		61.74	melt tape on seams.
Reinstall	S.Y.		5.44		5.44	
Clean	S.Y.		3.97		3.97	
Minimum Charge	Job		160		160	

Carpet Pad Per S.F.	Unit	Material	Labor	Equip.	Total	Specification
Urethane						
Demolish	S.F.		.02		.02	Includes material and labor to install
Install	S.F.	.43	.24		.67	urethane carpet pad.
Demolish and Install	S.F.	.43	.26		.69	
Minimum Charge	Job		160		160	
Foam Rubber Slab						
Demolish	S.F.		.02		.02	Includes material and labor to install
Install	S.F.	.59	.24		.83	carpet pad, 3/8" thick foam rubber.
Demolish and Install	S.F.	.59	.26		.85	
Minimum Charge	Job		160		160	
Waffle						
Demolish	S.F.		.02		.02	Includes material and labor to install
Install	S.F.	.37	.03		.40	rubber waffle carpet pad.
Demolish and Install	S.F.	.37	.05		.42	
Minimum Charge	Job		160		160	

Flooring

Carpet Pad Per S.F.

Carpet Pad Per S.F.	Unit	Material	Labor	Equip.	Total	Specification
Jute						
Demolish	S.F.		.02		.02	Includes material and labor to install
Install	S.F.	.67	.21		.88	jute hair carpet pad.
Demolish and Install	S.F.	.67	.23		.90	
Minimum Charge	Job		160		160	
Rebound						
Demolish	S.F.		.02		.02	Includes material and labor to install
Install	S.F.	.31	.18		.49	rebound carpet pad.
Demolish and Install	S.F.	.31	.20		.51	
Minimum Charge	Job		160		160	

Carpet Pad

Carpet Pad	Unit	Material	Labor	Equip.	Total	Specification
Urethane						
Demolish	S.Y.		.16		.16	Includes material and labor to install
Install	S.Y.	3.85	2.13		5.98	urethane carpet pad.
Demolish and Install	S.Y.	3.85	2.29		6.14	
Reinstall	S.Y.		1.70		1.70	
Minimum Charge	Job		160		160	
Foam Rubber Slab						
Demolish	S.Y.		.16		.16	Includes material and labor to install
Install	S.Y.	5.35	2.13		7.48	rubber slab carpet pad.
Demolish and Install	S.Y.	5.35	2.29		7.64	
Reinstall	S.Y.		1.70		1.70	
Minimum Charge	Job		160		160	
Waffle						
Demolish	S.Y.		.16		.16	Includes material and labor to install
Install	S.Y.	3.32	.24		3.56	rubber waffle carpet pad.
Demolish and Install	S.Y.	3.32	.40		3.72	
Reinstall	S.Y.		.19		.19	
Minimum Charge	Job		160		160	
Jute						
Demolish	S.Y.		.16		.16	Includes material and labor to install
Install	S.Y.	6.05	1.88		7.93	jute hair carpet pad.
Demolish and Install	S.Y.	6.05	2.04		8.09	
Reinstall	S.Y.		1.50		1.50	
Minimum Charge	Job		160		160	
Rebound						
Demolish	S.Y.		.16		.16	Includes material and labor to install
Install	S.Y.	2.83	1.59		4.42	rebound carpet pad.
Demolish and Install	S.Y.	2.83	1.75		4.58	
Reinstall	S.Y.		1.27		1.27	
Minimum Charge	Job		160		160	

Stair Components

Stair Components	Unit	Material	Labor	Equip.	Total	Specification
Carpeting						
Demolish	Ea.		1.71		1.71	Includes material and labor to install
Install	Riser	41.50	8.40		49.90	carpet and pad on stairs.
Demolish and Install	Riser	41.50	10.11		51.61	
Reinstall	Riser		6.73		6.73	
Clean	Ea.		3.97		3.97	
Minimum Charge	Job		160		160	

Finish Mechanical

Tub/Shower

Stainless Steel Double Sink

Water Closet

Commode	Unit	Material	Labor	Equip.	Total	Specification
Demolish	Ea.		39		39	Includes material and labor to install a toilet with valve, seat and cover.
Install	Ea.	201	132		333	
Demolish and Install	Ea.	201	171		372	
Reinstall	Ea.		105.66		105.66	
Clean	Ea.	.31	12.70		13.01	
Minimum Charge	Job		194		194	
Wall w / Flush Valve						
Demolish	Ea.		39		39	Includes material and labor to install a wall hung toilet with flush valve, seat and cover.
Install	Ea.	345	121		466	
Demolish and Install	Ea.	345	160		505	
Reinstall	Ea.		96.55		96.55	
Clean	Ea.	.31	12.70		13.01	
Minimum Charge	Job		194		194	
Designer Floor Mount						
Demolish	Ea.		39		39	Includes material and labor to install a toilet with valve, seat and cover.
Install	Ea.	585	132		717	
Demolish and Install	Ea.	585	171		756	
Reinstall	Ea.		105.66		105.66	
Clean	Ea.	.31	12.70		13.01	
Minimum Charge	Job		194		194	
Urinal						
Demolish	Ea.		39		39	Includes material and labor to install a wall hung urinal with flush valve.
Install	Ea.	360	233		593	
Demolish and Install	Ea.	360	272		632	
Reinstall	Ea.		186.67		186.67	
Clean	Ea.	.31	12.70		13.01	
Minimum Charge	Job		194		194	
Bidet						
Vitreous China						
Demolish	Ea.		39		39	Includes material and labor to install a vitreous china bidet complete with trim.
Install	Ea.	695	140		835	
Demolish and Install	Ea.	695	179		874	
Reinstall	Ea.		112		112	
Clean	Ea.	.31	12.70		13.01	
Minimum Charge	Job		194		194	

Finish Mechanical

Commode		Unit	Material	Labor	Equip.	Total	Specification
Chrome Fittings							
	Install	Ea.	169	48.50		217.50	Includes labor and materials to install chrome fittings for a bidet.
Brass Fittings							
	Install	Ea.	186	48.50		234.50	Includes labor and materials to install brass fittings for a bidet.
Seat							
	Demolish	Ea.		16.20		16.20	Includes material and labor to install a
	Install	Ea.	30.50	16.20		46.70	toilet seat and cover.
	Demolish and Install	Ea.	30.50	32.40		62.90	
	Reinstall	Ea.		12.96		12.96	
	Clean	Ea.	.15	5.30		5.45	
	Minimum Charge	Job		194		194	
Rough-in							
	Install	Ea.	273	246		519	Includes labor and materials to install water supply pipe with valves and drain, waste and vent pipe with all couplings, hangers and fasteners necessary.

Sink (assembly)		Unit	Material	Labor	Equip.	Total	Specification
Single							
Porcelain							
	Demolish	Ea.		24.50		24.50	Includes material and labor to install a
	Install	Ea.	261	125		386	single bowl enamel finished cast iron
	Demolish and Install	Ea.	261	149.50		410.50	kitchen sink with faucet and drain.
	Reinstall	Ea.		100		100	
	Clean	Ea.	.20	9.05		9.25	
	Minimum Charge	Job		194		194	
Stainless Steel							
	Demolish	Ea.		24.50		24.50	Includes material and labor to install a
	Install	Ea.	470	125		595	single bowl stainless steel kitchen sink
	Demolish and Install	Ea.	470	149.50		619.50	with faucet and drain.
	Reinstall	Ea.		100		100	
	Clean	Ea.	.20	9.05		9.25	
	Minimum Charge	Job		194		194	
Double							
Porcelain							
	Demolish	Ea.		28		28	Includes material and labor to install a
	Install	Ea.	550	146		696	double bowl enamel finished cast iron
	Demolish and Install	Ea.	550	174		724	kitchen sink with faucet and drain.
	Reinstall	Ea.		116.67		116.67	
	Clean	Ea.	.31	12.70		13.01	
	Minimum Charge	Job		194		194	
Stainless Steel							
	Demolish	Ea.		28		28	Includes material and labor to install a
	Install	Ea.	295	146		441	double bowl stainless steel kitchen sink
	Demolish and Install	Ea.	295	174		469	with faucet and drain.
	Reinstall	Ea.		116.67		116.67	
	Clean	Ea.	.31	12.70		13.01	
	Minimum Charge	Job		194		194	
Bar							
	Demolish	Ea.		24.50		24.50	Includes material and labor to install a
	Install	Ea.	88	97		185	small single bowl stainless steel bar /
	Demolish and Install	Ea.	88	121.50		209.50	vegetable / kitchen sink with faucet
	Reinstall	Ea.		77.76		77.76	and drain.
	Clean	Ea.	.20	9.05		9.25	
	Minimum Charge	Job		194		194	

Finish Mechanical

Sink (assembly)

	Unit	Material	Labor	Equip.	Total	Specification
Floor						
Demolish	Ea.		78		78	Includes material and labor to install
Install	Ea.	665	159		824	an enameled cast iron floor mounted
Demolish and Install	Ea.	665	237		902	corner service sink with faucet and
Reinstall	Ea.		127.27		127.27	drain.
Clean	Ea.	.31	6.35		6.66	
Minimum Charge	Job		194		194	
Pedestal						
Demolish	Ea.		24.50		24.50	Includes material and labor to install a
Install	Ea.	375	109		484	vitreous china pedestal lavatory with
Demolish and Install	Ea.	375	133.50		508.50	faucet set and pop-up drain.
Reinstall	Ea.		87.50		87.50	
Clean	Ea.	.20	9.05		9.25	
Minimum Charge	Job		194		194	
Vanity Lavatory						
Demolish	Ea.		24.50		24.50	Includes material and labor to install a
Install	Ea.	272	130		402	vitreous china lavatory with faucet set
Demolish and Install	Ea.	272	154.50		426.50	and pop-up drain.
Reinstall	Ea.		103.70		103.70	
Clean	Ea.	.20	9.05		9.25	
Minimum Charge	Job		194		194	
Rough-in						
Install	Ea.	223	325		548	Includes labor and materials to install copper supply pipe with valves and drain, waste and vent pipe with all couplings, hangers and fasteners necessary.
Laundry						
Demolish	Ea.		25.50		25.50	Includes material and labor to install a
Install	Ea.	107	108		215	laundry sink (wall mounted or with
Demolish and Install	Ea.	107	133.50		240.50	legs) including faucet and pop up
Clean	Ea.	.20	9.05		9.25	drain.
Minimum Charge	Job		194		194	

Sink Only

	Unit	Material	Labor	Equip.	Total	Specification
Single						
Porcelain						
Demolish	Ea.		24.50		24.50	Includes material and labor to install a
Install	Ea.	187	139		326	single bowl enamel finished cast iron
Demolish and Install	Ea.	187	163.50		350.50	kitchen sink.
Reinstall	Ea.		111.09		111.09	
Clean	Ea.	.20	9.05		9.25	
Minimum Charge	Job		194		194	
Stainless Steel						
Demolish	Ea.		24.50		24.50	Includes material and labor to install a
Install	Ea.	390	97		487	single bowl stainless steel kitchen sink.
Demolish and Install	Ea.	390	121.50		511.50	
Reinstall	Ea.		77.76		77.76	
Clean	Ea.	.20	9.05		9.25	
Minimum Charge	Job		194		194	

Finish Mechanical

Sink Only

	Unit	Material	Labor	Equip.	Total	Specification
Double						
Porcelain						
Demolish	Ea.		28		28	Includes material and labor to install a
Install	Ea.	271	162		433	double bowl enamel finished cast iron
Demolish and Install	Ea.	271	190		461	kitchen sink.
Reinstall	Ea.		129.60		129.60	
Clean	Ea.	.31	12.70		13.01	
Minimum Charge	Job		194		194	
Stainless Steel						
Demolish	Ea.		28		28	Includes material and labor to install a
Install	Ea.	590	125		715	double bowl stainless steel kitchen
Demolish and Install	Ea.	590	153		743	sink.
Reinstall	Ea.		100.34		100.34	
Clean	Ea.	.31	12.70		13.01	
Minimum Charge	Job		194		194	
Bar						
Demolish	Ea.		24.50		24.50	Includes material and labor to install a
Install	Ea.	46	97		143	small single bowl stainless steel bar /
Demolish and Install	Ea.	46	121.50		167.50	vegetable / kitchen sink.
Reinstall	Ea.		77.76		77.76	
Clean	Ea.	.20	9.05		9.25	
Minimum Charge	Job		194		194	
Floor						
Demolish	Ea.		78		78	Includes material and labor to install
Install	Ea.	565	177		742	an enameled cast iron floor mounted
Demolish and Install	Ea.	565	255		820	corner service sink.
Reinstall	Ea.		141.38		141.38	
Clean	Ea.	.31	6.35		6.66	
Minimum Charge	Job		194		194	
Pedestal						
Demolish	Ea.		24.50		24.50	Includes material and labor to install a
Install	Ea.	268	122		390	vitreous china pedestal lavatory.
Demolish and Install	Ea.	268	146.50		414.50	
Reinstall	Ea.		97.20		97.20	
Clean	Ea.	.20	9.05		9.25	
Minimum Charge	Job		194		194	
Vanity Lavatory						
Demolish	Ea.		24.50		24.50	Includes material and labor to install a
Install	Ea.	186	122		308	vitreous china lavatory.
Demolish and Install	Ea.	186	146.50		332.50	
Reinstall	Ea.		97.20		97.20	
Clean	Ea.	.20	9.05		9.25	
Minimum Charge	Job		194		194	
Rough-in						
Install	Ea.	186	305		491	Includes labor and materials to install copper supply pipe with valves and drain, waste and vent pipe with all couplings, hangers and fasteners necessary.
Faucet Sink						
Good Grade						
Demolish	Ea.		19.45		19.45	Includes material and labor to install a
Install	Ea.	60.50	27		87.50	sink faucet and fittings.
Demolish and Install	Ea.	60.50	46.45		106.95	
Reinstall	Ea.		21.47		21.47	
Clean	Ea.	.12	7.95		8.07	
Minimum Charge	Job		194		194	

Finish Mechanical

Sink Only

Sink Only	Unit	Material	Labor	Equip.	Total	Specification
Better Grade						
Demolish	Ea.		19.45		19.45	Includes material and labor to install a
Install	Ea.	102	27		129	sink faucet and fittings.
Demolish and Install	Ea.	102	46.45		148.45	
Reinstall	Ea.		21.47		21.47	
Clean	Ea.	.12	7.95		8.07	
Minimum Charge	Job		194		194	
Premium Grade						
Demolish	Ea.		19.45		19.45	Includes material and labor to install a
Install	Ea.	198	33.50		231.50	sink faucet and fittings.
Demolish and Install	Ea.	198	52.95		250.95	
Reinstall	Ea.		26.77		26.77	
Clean	Ea.	.12	7.95		8.07	
Minimum Charge	Job		194		194	
Drain & Basket						
Demolish	Ea.		12.15		12.15	Includes material and labor to install a
Install	Ea.	13.35	24.50		37.85	sink drain and basket.
Demolish and Install	Ea.	13.35	36.65		50	
Reinstall	Ea.		19.44		19.44	
Clean	Ea.	.12	7.95		8.07	
Minimum Charge	Job		194		194	

Bathtub

Bathtub	Unit	Material	Labor	Equip.	Total	Specification
Enameled						
Steel						
Demolish	Ea.		70		70	
Install	Ea.	340	127		467	
Demolish and Install	Ea.	340	197		537	
Reinstall	Ea.		101.82		101.82	
Clean	Ea.	.63	21		21.63	
Minimum Charge	Job		194		194	
Cast Iron						
Demolish	Ea.		70		70	Includes material and labor to install a
Install	Ea.	830	159		989	cast iron bath tub with spout, mixing
Demolish and Install	Ea.	830	229		1059	valve, shower head and pop-up drain.
Reinstall	Ea.		127.27		127.27	
Clean	Ea.	.63	21		21.63	
Minimum Charge	Job		194		194	
Fiberglass						
Demolish	Ea.		70		70	Includes material and labor to install
Install	Ea.	985	127		1112	an acrylic soaking tub with spout,
Demolish and Install	Ea.	985	197		1182	mixing valve, shower head and
Reinstall	Ea.		101.82		101.82	pop-up drain.
Clean	Ea.	.63	21		21.63	
Minimum Charge	Job		194		194	
Institutional						
Demolish	Ea.		70		70	Includes material and labor to install a
Install	Ea.	990	233		1223	hospital / institutional type bathtub
Demolish and Install	Ea.	990	303		1293	with spout, mixing valve, shower head
Reinstall	Ea.		186.67		186.67	and pop-up drain.
Clean	Ea.	.63	21		21.63	
Minimum Charge	Job		194		194	

Finish Mechanical

Bathtub

	Unit	Material	Labor	Equip.	Total	Specification
Whirlpool (Acrylic)						
Demolish	Ea.		216		216	Includes material and labor to install a
Install	Ea.	2675	700		3375	molded fiberglass whirlpool tub.
Demolish and Install	Ea.	2675	916		3591	
Reinstall	Ea.		560		560	
Clean	Ea.	.63	21		21.63	
Minimum Charge	Job		194		194	
Pump / Motor						
Demolish	Ea.		54		54	Includes material and labor to install a
Install	Ea.	201	117		318	whirlpool tub pump / motor.
Demolish and Install	Ea.	201	171		372	
Reinstall	Ea.		93.41		93.41	
Clean	Ea.	.15	15.90		16.05	
Minimum Charge	Job		194		194	
Heater / Motor						
Demolish	Ea.		54		54	Includes material and labor to install a
Install	Ea.	287	173		460	whirlpool tub heater.
Demolish and Install	Ea.	287	227		514	
Reinstall	Ea.		138.24		138.24	
Clean	Ea.	.15	15.90		16.05	
Minimum Charge	Job		194		194	
Thermal Cover						
Demolish	Ea.		8.65		8.65	Includes material and labor to install a
Install	Ea.	187	24.50		211.50	whirlpool tub thermal cover.
Demolish and Install	Ea.	187	33.15		220.15	
Reinstall	Ea.		19.44		19.44	
Clean	Ea.	1.56	15.90		17.46	
Minimum Charge	Job		194		194	
Tub / Shower Combination						
Demolish	Ea.		43		43	Includes material and labor to install a
Install	Ea.	600	175		775	fiberglass module tub with shower
Demolish and Install	Ea.	600	218		818	surround with spout, mixing valve,
Reinstall	Ea.		140		140	shower head and pop-up drain.
Clean	Ea.	.78	32		32.78	
Minimum Charge	Job		194		194	
Accessories						
Sliding Door						
Demolish	Ea.		17.65		17.65	Includes material and labor to install a
Install	Ea.	430	81.50		511.50	48" aluminum framed tempered glass
Demolish and Install	Ea.	430	99.15		529.15	shower door.
Reinstall	Ea.		65.15		65.15	
Clean	Ea.	1.30	10.60		11.90	
Minimum Charge	Job		194		194	
Shower Head						
Demolish	Ea.		4.86		4.86	Includes material and labor to install a
Install	Ea.	71.50	16.20		87.70	water saving shower head.
Demolish and Install	Ea.	71.50	21.06		92.56	
Reinstall	Ea.		12.96		12.96	
Clean	Ea.	.01	3.18		3.19	
Minimum Charge	Job		194		194	
Faucet Set						
Demolish	Ea.		24.50		24.50	Includes material and labor to install a
Install	Ea.	114	48.50		162.50	combination spout / diverter for a
Demolish and Install	Ea.	114	73		187	bathtub.
Reinstall	Ea.		38.88		38.88	
Clean	Ea.	.12	7.95		8.07	
Minimum Charge	Job		194		194	

Finish Mechanical

Bathtub

	Unit	Material	Labor	Equip.	Total	Specification
Shower Rod						Includes material and labor to install a chrome plated shower rod.
Demolish	Ea.		2.94		2.94	
Install	Ea.	15.75	27		42.75	
Demolish and Install	Ea.	15.75	29.94		45.69	
Reinstall	Ea.		21.46		21.46	
Clean	Ea.		3.18		3.18	
Minimum Charge	Job		194		194	
Rough-in						Includes labor and materials to install copper supply pipe with valves and drain, waste and vent pipe with all couplings, hangers and fasteners necessary.
Install	Ea.	241	340		581	

Shower

	Unit	Material	Labor	Equip.	Total	Specification
Fiberglass						
32″ x 32″						Includes material and labor to install a fiberglass shower stall with door, mixing valve and shower head and drain fitting.
Demolish	Ea.		146		146	
Install	Ea.	500	127		627	
Demolish and Install	Ea.	500	273		773	
Reinstall	Ea.		101.82		101.82	
Clean	Ea.	.63	21		21.63	
Minimum Charge	Job		194		194	
36″ x 36″						Includes material and labor to install a fiberglass shower stall with door, mixing valve and shower head and drain fitting.
Demolish	Ea.		146		146	
Install	Ea.	580	127		707	
Demolish and Install	Ea.	580	273		853	
Reinstall	Ea.		101.82		101.82	
Clean	Ea.	.63	21		21.63	
Minimum Charge	Job		194		194	
35″ x 60″						Includes material and labor to install a handicap fiberglass shower stall with door & seat, mixing valve and shower head and drain fitting.
Demolish	Ea.		146		146	
Install	Ea.	645	175		820	
Demolish and Install	Ea.	645	321		966	
Reinstall	Ea.		140		140	
Clean	Ea.	.63	21		21.63	
Minimum Charge	Job		194		194	
Accessories						
Single Door						Includes material and labor to install a shower door.
Demolish	Ea.		17.65		17.65	
Install	Ea.	121	43		164	
Demolish and Install	Ea.	121	60.65		181.65	
Reinstall	Ea.		34.38		34.38	
Clean	Ea.	1.30	10.60		11.90	
Minimum Charge	Job		194		194	
Shower Head						Includes material and labor to install a water saving shower head.
Demolish	Ea.		4.86		4.86	
Install	Ea.	71.50	16.20		87.70	
Demolish and Install	Ea.	71.50	21.06		92.56	
Reinstall	Ea.		12.96		12.96	
Clean	Ea.	.01	3.18		3.19	
Minimum Charge	Job		194		194	

Finish Mechanical

Shower

Shower		Unit	Material	Labor	Equip.	Total	Specification
Faucet Set							
	Demolish	Ea.		24.50		24.50	Includes material and labor to install a
	Install	Ea.	114	48.50		162.50	combination spout / diverter for a
	Demolish and Install	Ea.	114	73		187	bathtub.
	Reinstall	Ea.		38.88		38.88	
	Clean	Ea.	.12	7.95		8.07	
	Minimum Charge	Job		194		194	
Shower Pan							
	Demolish	Ea.		18		18	Includes material and labor to install
	Install	Ea.	206	48.50		254.50	the base portion of a fiberglass
	Demolish and Install	Ea.	206	66.50		272.50	shower unit.
	Reinstall	Ea.		38.88		38.88	
	Clean	Ea.	.63	21		21.63	
	Minimum Charge	Job		194		194	
Rough-in							
	Install	Ea.	340	340		680	Includes labor and materials to install copper supply pipe with valves and drain, waste and vent pipe with all couplings, hangers and fasteners necessary.

Ductwork

Ductwork		Unit	Material	Labor	Equip.	Total	Specification
Return Grill							
Under 10″ Wide							
	Demolish	Ea.		8.60		8.60	Includes material and labor to install a
	Install	Ea.	37.50	21.50		59	10″ x 10″ air return register.
	Demolish and Install	Ea.	37.50	30.10		67.60	
	Reinstall	Ea.		17.19		17.19	
	Clean	Ea.	.15	7.05		7.20	
	Minimum Charge	Job		193		193	
12″ to 20″ Wide							
	Demolish	Ea.		8.60		8.60	Includes material and labor to install a
	Install	Ea.	68	23		91	16″ x 16″ air return register.
	Demolish and Install	Ea.	68	31.60		99.60	
	Reinstall	Ea.		18.20		18.20	
	Clean	Ea.	.15	7.05		7.20	
	Minimum Charge	Job		193		193	
Over 20″ Wide							
	Demolish	Ea.		13		13	Includes material and labor to install a
	Install	Ea.	132	35		167	24″ x 24″ air return register.
	Demolish and Install	Ea.	132	48		180	
	Reinstall	Ea.		28.13		28.13	
	Clean	Ea.	.15	7.95		8.10	
	Minimum Charge	Job		193		193	
Supply Grill							
Baseboard Type							
	Demolish	Ea.		6.15		6.15	Includes material and labor to install a
	Install	Ea.	19.95	19.35		39.30	14″ x 6″ baseboard air supply
	Demolish and Install	Ea.	19.95	25.50		45.45	register.
	Reinstall	Ea.		15.47		15.47	
	Clean	Ea.	.15	7.05		7.20	
	Minimum Charge	Job		193		193	

Finish Mechanical

Ductwork	Unit	Material	Labor	Equip.	Total	Specification
10″ Wide						
Demolish	Ea.		7.15		7.15	Includes material and labor to install a
Install	Ea.	29.50	19.35		48.85	10″ x 6″ air supply register.
Demolish and Install	Ea.	29.50	26.50		56	
Reinstall	Ea.		15.47		15.47	
Clean	Ea.	.15	7.05		7.20	
Minimum Charge	Job		193		193	
12″ to 15″ Wide						
Demolish	Ea.		7.15		7.15	Includes material and labor to install a
Install	Ea.	41.50	23		64.50	14″ x 8″ air supply register.
Demolish and Install	Ea.	41.50	30.15		71.65	
Reinstall	Ea.		18.20		18.20	
Clean	Ea.	.15	7.05		7.20	
Minimum Charge	Job		193		193	
18″ to 24″ Wide						
Demolish	Ea.		8.60		8.60	Includes material and labor to install a
Install	Ea.	57	30		87	24″ x 8″ air supply register.
Demolish and Install	Ea.	57	38.60		95.60	
Reinstall	Ea.		23.80		23.80	
Clean	Ea.	.15	7.95		8.10	
Minimum Charge	Job		193		193	
30″ to 36″ Wide						
Demolish	Ea.		10.75		10.75	Includes material and labor to install a
Install	Ea.	75	27.50		102.50	30″ x 8″ air supply register.
Demolish and Install	Ea.	75	38.25		113.25	
Reinstall	Ea.		22.10		22.10	
Clean	Ea.	.15	15.90		16.05	
Minimum Charge	Job		193		193	
Diffuser						
12″ Louver						
Demolish	Ea.		8.60		8.60	Includes material and labor to install a
Install	Ea.	38.50	27.50		66	12″ aluminum louvered diffuser.
Demolish and Install	Ea.	38.50	36.10		74.60	
Reinstall	Ea.		22.10		22.10	
Clean	Ea.	.15	7.05		7.20	
Minimum Charge	Job		193		193	
14″ to 20″ Louver						
Demolish	Ea.		8.60		8.60	Includes material and labor to install a
Install	Ea.	43	38.50		81.50	14″ to 20″ aluminum louvered diffuser.
Demolish and Install	Ea.	43	47.10		90.10	
Reinstall	Ea.		30.94		30.94	
Clean	Ea.	.15	7.95		8.10	
Minimum Charge	Job		193		193	
25″ to 32″ Louver						
Demolish	Ea.		13		13	Includes material and labor to install a
Install	Ea.	56	55.50		111.50	25″ to 32″ aluminum louvered diffuser.
Demolish and Install	Ea.	56	68.50		124.50	
Reinstall	Ea.		44.21		44.21	
Clean	Ea.	.15	10.60		10.75	
Minimum Charge	Job		193		193	
12″ Round						
Demolish	Ea.		11.95		11.95	Includes material and labor to install a
Install	Ea.	36.50	32		68.50	12″ diameter aluminum diffuser with a
Demolish and Install	Ea.	36.50	43.95		80.45	butterfly damper.
Reinstall	Ea.		25.79		25.79	
Clean	Ea.	.15	7.05		7.20	
Minimum Charge	Job		193		193	

Finish Mechanical

Ductwork	Unit	Material	Labor	Equip.	Total	Specification
14″ to 20″ Round						
Demolish	Ea.		15.90		15.90	Includes material and labor to install a
Install	Ea.	135	43		178	20″ diameter aluminum diffuser with a
Demolish and Install	Ea.	135	58.90		193.90	butterfly damper.
Reinstall	Ea.		34.38		34.38	
Clean	Ea.	.15	7.95		8.10	
Minimum Charge	Job		193		193	
Flush-out / Sanitize						
Install	L.F.	.63	3.18		3.81	Includes labor and material costs to
Minimum Charge	Job		193		193	clean ductwork.

Lighting Fixture   **Duplex Receptacle**

Electrical Per S.F.	Unit	Material	Labor	Equip.	Total	Specification
Residential						
Light Fixtures						
Install	S.F.	.24	.19		.43	Includes labor and material to replace house light fixtures. Apply S.F. cost to floor area including garages but not basements.

Circuits	Unit	Material	Labor	Equip.	Total	Specification
Outlet with Wiring						
Demolish	Ea.		21.50		21.50	Includes material and labor to install
Install	Ea.	12.10	26.50		38.60	general purpose outlet with cover, up
Demolish and Install	Ea.	12.10	48		60.10	to 20 feet of #12/2 wire.
Reinstall	Ea.		21.07		21.07	
Clean	Ea.		1.59		1.59	
Minimum Charge	Job		192		192	
GFCI Outlet w / Wiring						
Demolish	Ea.		21.50		21.50	Cost includes material and labor to
Install	Ea.	43.50	31		74.50	install 120V GFCI receptacle and wall
Demolish and Install	Ea.	43.50	52.50		96	plate, with up to 20 feet of #12/2
Reinstall	Ea.		24.90		24.90	wire.
Clean	Ea.		1.59		1.59	
Minimum Charge	Job		192		192	
Exterior Outlet w / Wiring						
Demolish	Ea.		21.50		21.50	Cost includes material and labor to
Install	Ea.	37.50	32		69.50	install 110V waterproof external
Demolish and Install	Ea.	37.50	53.50		91	receptacle and cover plate with up to
Reinstall	Ea.		25.55		25.55	20' of #12/2 wire.
Clean	Ea.		1.59		1.59	
Minimum Charge	Job		192		192	
240V Outlet w / Wiring						
Demolish	Ea.		21.50		21.50	Cost includes material and labor to
Install	Ea.	86.50	60		146.50	install 240V receptacle and cover
Demolish and Install	Ea.	86.50	81.50		168	plate with up to 20' of #12/2 wire.
Reinstall	Ea.		47.83		47.83	
Clean	Ea.		1.59		1.59	
Minimum Charge	Job		192		192	

Finish Electrical

Circuits	Unit	Material	Labor	Equip.	Total	Specification
TV / Cable Jack w / Wiring						
Demolish	Ea.		21.50		21.50	Includes material and labor to install
Install	Ea.	17.90	24		41.90	TV/cable outlet box and cover plate
Demolish and Install	Ea.	17.90	45.50		63.40	including 20' of wiring.
Reinstall	Ea.		19.16		19.16	
Clean	Ea.		1.59		1.59	
Minimum Charge	Job		192		192	
Phone Jack with Wiring						
Demolish	Ea.		21.50		21.50	Includes material and labor to install
Install	Ea.	10.30	14.75		25.05	phone jack and cover plate including
Demolish and Install	Ea.	10.30	36.25		46.55	up to 20' of telephone cable.
Reinstall	Ea.		11.79		11.79	
Clean	Ea.		.99		.99	
Minimum Charge	Job		192		192	

Light Switch	Unit	Material	Labor	Equip.	Total	Specification
Single						
Switch						
Demolish	Ea.		4.26		4.26	Includes material and labor to install
Install	Ea.	5.45	18.25		23.70	single switch and wall plate. Wiring is
Demolish and Install	Ea.	5.45	22.51		27.96	not included.
Reinstall	Ea.		14.60		14.60	
Clean	Ea.		1.59		1.59	
Minimum Charge	Job		192		192	
Plate						
Demolish	Ea.		2.13		2.13	Includes material and labor to install a
Install	Ea.	.34	4.79		5.13	one gang brown wall switch plate.
Demolish and Install	Ea.	.34	6.92		7.26	
Reinstall	Ea.		3.83		3.83	
Clean	Ea.	.07	.32		.39	
Minimum Charge	Job		192		192	
Double						
Switch						
Demolish	Ea.		8.50		8.50	Includes material and labor to install
Install	Ea.	8.10	27.50		35.60	double switch and wall plate. Wiring
Demolish and Install	Ea.	8.10	36		44.10	is not included.
Reinstall	Ea.		21.90		21.90	
Clean	Ea.	.07	5.10		5.17	
Minimum Charge	Job		192		192	
Plate						
Demolish	Ea.		2.13		2.13	Includes material and labor to install a
Install	Ea.	.90	7.25		8.15	two gang brown wall switch plate.
Demolish and Install	Ea.	.90	9.38		10.28	
Reinstall	Ea.		5.78		5.78	
Clean	Ea.	.07	.32		.39	
Minimum Charge	Job		192		192	
Triple						
Switch						
Demolish	Ea.		12.90		12.90	Includes material and labor to install
Install	Ea.	25	32		57	triple switch and wall plate. Wiring is
Demolish and Install	Ea.	25	44.90		69.90	not included.
Reinstall	Ea.		25.55		25.55	
Clean	Ea.	.07	7.60		7.67	
Minimum Charge	Job		192		192	

Finish Electrical

Light Switch	Unit	Material	Labor	Equip.	Total	Specification
Plate						
Demolish	Ea.		2.13		2.13	Includes material and labor to install a
Install	Ea.	1.32	12		13.32	three gang brown wall switch plate.
Demolish and Install	Ea.	1.32	14.13		15.45	
Reinstall	Ea.		9.58		9.58	
Clean	Ea.	.07	.32		.39	
Minimum Charge	Job		192		192	
Dimmer						
Switch						
Demolish	Ea.		4.26		4.26	Includes material and labor to install
Install	Ea.	19.20	22.50		41.70	dimmer switch. Wiring is not included.
Demolish and Install	Ea.	19.20	26.76		45.96	
Reinstall	Ea.		18.03		18.03	
Clean	Ea.		1.59		1.59	
Minimum Charge	Job		192		192	
Plate						
Demolish	Ea.		2.13		2.13	Includes material and labor to install a
Install	Ea.	.34	4.79		5.13	one gang brown wall switch plate.
Demolish and Install	Ea.	.34	6.92		7.26	
Reinstall	Ea.		3.83		3.83	
Clean	Ea.	.07	.32		.39	
Minimum Charge	Job		192		192	
With Wiring						
Single						
Demolish	Ea.		21.50		21.50	Includes material and labor to install
Install	Ea.	13.90	22.50		36.40	single switch and plate including up to
Demolish and Install	Ea.	13.90	44		57.90	20' of romex cable.
Reinstall	Ea.		17.93		17.93	
Clean	Ea.		1.59		1.59	
Minimum Charge	Job		192		192	
Double						
Demolish	Ea.		21.50		21.50	Includes material and labor to install
Install	Ea.	32.50	38.50		71	double switch and plate including up
Demolish and Install	Ea.	32.50	60		92.50	to 20' of romex cable.
Reinstall	Ea.		30.66		30.66	
Clean	Ea.	.07	5.10		5.17	
Minimum Charge	Job		192		192	
Triple						
Demolish	Ea.		32		32	Includes material and labor to install
Install	Ea.	36.50	43		79.50	triple switch and plate including up to
Demolish and Install	Ea.	36.50	75		111.50	20' of romex cable.
Reinstall	Ea.		34.48		34.48	
Clean	Ea.	.07	7.60		7.67	
Minimum Charge	Job		192		192	
Dimmer						
Demolish	Ea.		21.50		21.50	Includes material and labor to install
Install	Ea.	27.50	26.50		54	dimmer switch and cover plate
Demolish and Install	Ea.	27.50	48		75.50	including up to 20' of romex cable.
Reinstall	Ea.		21.07		21.07	
Clean	Ea.		1.59		1.59	
Minimum Charge	Job		192		192	

Finish Electrical

Cover for Outlet / Switch

	Unit	Material	Labor	Equip.	Total	Specification
High Grade						
Demolish	Ea.		2.13		2.13	Includes material and labor to install
Install	Ea.	2.27	4.79		7.06	wall plate, stainless steel, 1 gang.
Demolish and Install	Ea.	2.27	6.92		9.19	
Reinstall	Ea.		3.83		3.83	
Clean	Ea.	.07	.32		.39	
Minimum Charge	Job		192		192	
Deluxe Grade						
Demolish	Ea.		2.13		2.13	Includes material and labor to install
Install	Ea.	4.84	4.79		9.63	wall plate, brushed brass, 1 gang.
Demolish and Install	Ea.	4.84	6.92		11.76	
Reinstall	Ea.		3.83		3.83	
Clean	Ea.	.07	.32		.39	
Minimum Charge	Job		192		192	

Residential Light Fixture

	Unit	Material	Labor	Equip.	Total	Specification
Ceiling						
Good Quality						
Demolish	Ea.		8.85		8.85	Includes material and labor to install
Install	Ea.	29.50	9.60		39.10	standard quality incandescent ceiling
Demolish and Install	Ea.	29.50	18.45		47.95	light fixture. Wiring is not included.
Reinstall	Ea.		7.66		7.66	
Clean	Ea.	.15	10.60		10.75	
Minimum Charge	Job		192		192	
Better Quality						
Demolish	Ea.		8.85		8.85	Includes material and labor to install
Install	Ea.	89	20		109	custom quality incandescent ceiling
Demolish and Install	Ea.	89	28.85		117.85	light fixture. Wiring is not included.
Reinstall	Ea.		16.13		16.13	
Clean	Ea.	.15	10.60		10.75	
Minimum Charge	Job		192		192	
Premium Quality						
Demolish	Ea.		8.85		8.85	Includes material and labor to install
Install	Ea.	263	20		283	deluxe quality incandescent ceiling
Demolish and Install	Ea.	263	28.85		291.85	light fixture. Wiring is not included.
Reinstall	Ea.		16.13		16.13	
Clean	Ea.	.15	10.60		10.75	
Minimum Charge	Job		192		192	
Recessed Ceiling						
Demolish	Ea.		17.75		17.75	Includes material and labor to install
Install	Ea.	40.50	12.75		53.25	standard quality recessed eyeball
Demolish and Install	Ea.	40.50	30.50		71	spotlight with housing. Wiring is not
Reinstall	Ea.		10.22		10.22	included.
Clean	Ea.	.03	2.65		2.68	
Minimum Charge	Job		192		192	
Recessed Eyeball						
Demolish	Ea.		17.75		17.75	Includes material and labor to install
Install	Ea.	60.50	13.70		74.20	custom quality recessed eyeball
Demolish and Install	Ea.	60.50	31.45		91.95	spotlight with housing. Wiring is not
Reinstall	Ea.		10.95		10.95	included.
Clean	Ea.	.03	2.65		2.68	
Minimum Charge	Job		192		192	

Finish Electrical

Residential Light Fixture

	Unit	Material	Labor	Equip.	Total	Specification
Recessed Shower Type						
Demolish	Ea.		17.75		17.75	Includes material and labor to install
Install	Ea.	50.50	12.75		63.25	deluxe quality recessed eyeball
Demolish and Install	Ea.	50.50	30.50		81	spotlight with housing. Wiring is not
Reinstall	Ea.		10.22		10.22	included.
Clean	Ea.	.03	2.65		2.68	
Minimum Charge	Job		192		192	
Run Wiring						
Install	Ea.	19.35	15.35		34.70	Includes labor and material to install a
Minimum Charge	Job		192		192	electric fixture utility box with
						non-metallic cable.
Fluorescent						
Good Quality						
Demolish	Ea.		19.35		19.35	Includes material and labor to install
Install	Ea.	48.50	54.50		103	standard quality 12" wide by 48" long
Demolish and Install	Ea.	48.50	73.85		122.35	decorative surface mounted fluorescent
Reinstall	Ea.		43.79		43.79	fixture w/acrylic diffuser.
Clean	Ea.	.03	2.65		2.68	
Minimum Charge	Job		192		192	
Better Quality						
Demolish	Ea.		19.35		19.35	Includes material and labor to install
Install	Ea.	99.50	62		161.50	an interior fluorescent lighting fixture.
Demolish and Install	Ea.	99.50	81.35		180.85	
Reinstall	Ea.		49.45		49.45	
Clean	Ea.	.03	2.65		2.68	
Minimum Charge	Job		192		192	
Premium Quality						
Demolish	Ea.		19.35		19.35	Includes material and labor to install
Install	Ea.	103	72.50		175.50	deluxe quality 24" wide by 48" long
Demolish and Install	Ea.	103	91.85		194.85	decorative surface mounted fluorescent
Reinstall	Ea.		57.84		57.84	fixture w/acrylic diffuser.
Clean	Ea.	.03	2.65		2.68	
Minimum Charge	Job		192		192	
Run Wiring						
Install	Ea.	19.35	15.35		34.70	Includes labor and material to install a
Minimum Charge	Job		192		192	electric fixture utility box with
						non-metallic cable.
Exterior						
Good Quality						
Demolish	Ea.		17.05		17.05	Includes material and labor to install
Install	Ea.	34.50	24		58.50	standard quality incandescent outdoor
Demolish and Install	Ea.	34.50	41.05		75.55	light fixture. Wiring is not included.
Reinstall	Ea.		19.16		19.16	
Clean	Ea.	.15	10.60		10.75	
Minimum Charge	Job		192		192	
Better Quality						
Demolish	Ea.		17.05		17.05	Includes material and labor to install
Install	Ea.	99	24		123	custom quality incandescent outdoor
Demolish and Install	Ea.	99	41.05		140.05	light fixture. Wiring is not included.
Reinstall	Ea.		19.16		19.16	
Clean	Ea.	.15	10.60		10.75	
Minimum Charge	Job		192		192	
Premium Quality						
Demolish	Ea.		17.05		17.05	Includes material and labor to install
Install	Ea.	165	96		261	deluxe quality incandescent outdoor
Demolish and Install	Ea.	165	113.05		278.05	light fixture. Wiring is not included.
Reinstall	Ea.		76.64		76.64	
Clean	Ea.	.15	10.60		10.75	
Minimum Charge	Job		192		192	

Finish Electrical

Residential Light Fixture	Unit	Material	Labor	Equip.	Total	Specification
Outdoor w / Pole						
Demolish	Ea.		17.05		17.05	Includes material and labor to install
Install	Ea.	121	96		217	post lantern type incandescent outdoor
Demolish and Install	Ea.	121	113.05		234.05	light fixture. Wiring is not included.
Reinstall	Ea.		76.64		76.64	
Clean	Ea.	.15	10.60		10.75	
Minimum Charge	Job		192		192	
Run Wiring						
Install	Ea.	19.35	15.35		34.70	Includes labor and material to install a
Minimum Charge	Job		192		192	electric fixture utility box with
						non-metallic cable.
Track Lighting						
Demolish	L.F.		5.90		5.90	Includes material and labor to install
Install	L.F.	13.95	15.95		29.90	track lighting.
Demolish and Install	L.F.	13.95	21.85		35.80	
Reinstall	L.F.		12.77		12.77	
Clean	L.F.	.15	5.10		5.25	
Minimum Charge	Job		192		192	

Commercial Light Fixture	Unit	Material	Labor	Equip.	Total	Specification
Fluorescent						
2 Bulb, 2'						
Demolish	Ea.		17.05		17.05	Includes material and labor to install a
Install	Ea.	34.50	48		82.50	2' long surface mounted fluorescent
Demolish and Install	Ea.	34.50	65.05		99.55	light fixture with two 40 watt bulbs.
Reinstall	Ea.		38.32		38.32	
Clean	Ea.	.78	18.15		18.93	
Minimum Charge	Job		192		192	
2 Bulb, 4'						
Demolish	Ea.		17.05		17.05	Includes material and labor to install a
Install	Ea.	65	54.50		119.50	4' long surface mounted fluorescent
Demolish and Install	Ea.	65	71.55		136.55	light fixture with two 40 watt bulbs.
Reinstall	Ea.		43.79		43.79	
Clean	Ea.	.78	18.15		18.93	
Minimum Charge	Job		192		192	
3 Bulb, 4'						
Demolish	Ea.		17.05		17.05	Includes material and labor to install a
Install	Ea.	103	54.50		157.50	4' long pendant mounted fluorescent
Demolish and Install	Ea.	103	71.55		174.55	light fixture with three 34 watt bulbs.
Reinstall	Ea.		43.79		43.79	
Clean	Ea.	.78	18.15		18.93	
Minimum Charge	Job		192		192	
4 Bulb, 4'						
Demolish	Ea.		17.05		17.05	Includes material and labor to install a
Install	Ea.	93.50	59		152.50	4' long pendant mounted fluorescent
Demolish and Install	Ea.	93.50	76.05		169.55	light fixture with four 34 watt bulbs.
Reinstall	Ea.		47.16		47.16	
Clean	Ea.	.78	18.15		18.93	
Minimum Charge	Job		192		192	
2 Bulb, 8'						
Demolish	Ea.		21.50		21.50	Includes material and labor to install a
Install	Ea.	99.50	72.50		172	8' long surface mounted fluorescent
Demolish and Install	Ea.	99.50	94		193.50	light fixture with two 110 watt bulbs.
Reinstall	Ea.		57.84		57.84	
Clean	Ea.	.78	18.15		18.93	
Minimum Charge	Job		192		192	

Finish Electrical

Commercial Light Fixture		Unit	Material	Labor	Equip.	Total	Specification
Run Wiring							
	Install	Ea.	19.35	15.35		34.70	Includes labor and material to install a
	Minimum Charge	Job		192		192	electric fixture utility box with
							non-metallic cable.
Ceiling, Recessed							
Eyeball Type							
	Demolish	Ea.		17.75		17.75	Includes material and labor to install
	Install	Ea.	60.50	13.70		74.20	custom quality recessed eyeball
	Demolish and Install	Ea.	60.50	31.45		91.95	spotlight with housing. Wiring is not
	Reinstall	Ea.		10.95		10.95	included.
	Clean	Ea.	.03	2.65		2.68	
	Minimum Charge	Job		192		192	
Shower Type							
	Demolish	Ea.		17.75		17.75	Includes material and labor to install
	Install	Ea.	50.50	12.75		63.25	deluxe quality recessed eyeball
	Demolish and Install	Ea.	50.50	30.50		81	spotlight with housing. Wiring is not
	Reinstall	Ea.		10.22		10.22	included.
	Clean	Ea.	.03	2.65		2.68	
	Minimum Charge	Job		192		192	
Run Wiring							
	Install	Ea.	19.35	15.35		34.70	Includes labor and material to install a
	Minimum Charge	Job		192		192	electric fixture utility box with
							non-metallic cable.
Ceiling Grid Type							
2 Tube							
	Demolish	Ea.		25		25	Cost includes material and labor to
	Install	Ea.	59.50	72.50		132	install 2 lamp fluorescent light fixture
	Demolish and Install	Ea.	59.50	97.50		157	in 2' x 4' ceiling grid system including
	Reinstall	Ea.		57.84		57.84	lens and tubes.
	Clean	Ea.	.78	18.15		18.93	
	Minimum Charge	Job		192		192	
4 Tube							
	Demolish	Ea.		28.50		28.50	Cost includes material and labor to
	Install	Ea.	67	81.50		148.50	install 4 lamp fluorescent light fixture
	Demolish and Install	Ea.	67	110		177	in 2' x 4' ceiling grid system including
	Reinstall	Ea.		65.23		65.23	lens and tubes.
	Clean	Ea.	.78	18.15		18.93	
	Minimum Charge	Job		192		192	
Acrylic Diffuser							
	Demolish	Ea.		4.44		4.44	Includes material and labor to install
	Install	Ea.	49.50	11.60		61.10	fluorescent light fixture w/acrylic
	Demolish and Install	Ea.	49.50	16.04		65.54	diffuser.
	Reinstall	Ea.		9.29		9.29	
	Clean	Ea.	.50	3.18		3.68	
	Minimum Charge	Job		192		192	
Louver Diffuser							
	Demolish	Ea.		5.30		5.30	Includes material and labor to install
	Install	Ea.	55.50	11.60		67.10	fluorescent light fixture w/louver
	Demolish and Install	Ea.	55.50	16.90		72.40	diffuser.
	Reinstall	Ea.		9.29		9.29	
	Clean	Ea.	.50	3.18		3.68	
	Minimum Charge	Job		192		192	
Run Wiring							
	Install	Ea.	19.35	15.35		34.70	Includes labor and material to install a
	Minimum Charge	Job		192		192	electric fixture utility box with
							non-metallic cable.

Finish Electrical

Commercial Light Fixture	Unit	Material	Labor	Equip.	Total	Specification
Track Lighting						
Demolish	L.F.		5.90		5.90	Includes material and labor to install
Install	L.F.	13.95	15.95		29.90	track lighting.
Demolish and Install	L.F.	13.95	21.85		35.80	
Reinstall	L.F.		12.77		12.77	
Clean	L.F.	.15	5.10		5.25	
Minimum Charge	Job		192		192	
Emergency						
Demolish	Ea.		30.50		30.50	Includes material and labor to install
Install	Ea.	134	96		230	battery-powered emergency lighting
Demolish and Install	Ea.	134	126.50		260.50	unit with two (2) lights, including
Reinstall	Ea.		76.64		76.64	wiring.
Clean	Ea.	.39	18.15		18.54	
Minimum Charge	Job		192		192	
Run Wiring						
Install	Ea.	19.35	15.35		34.70	Includes labor and material to install a
Minimum Charge	Job		192		192	electric fixture utility box with
						non-metallic cable.

Fan	Unit	Material	Labor	Equip.	Total	Specification
Ceiling Paddle						
Good Quality						
Demolish	Ea.		23.50		23.50	Includes material and labor to install
Install	Ea.	86.50	38.50		125	standard ceiling fan.
Demolish and Install	Ea.	86.50	62		148.50	
Reinstall	Ea.		30.66		30.66	
Clean	Ea.	.31	15.90		16.21	
Minimum Charge	Job		192		192	
Better Quality						
Demolish	Ea.		23.50		23.50	Includes material and labor to install
Install	Ea.	144	38.50		182.50	better ceiling fan.
Demolish and Install	Ea.	144	62		206	
Reinstall	Ea.		30.66		30.66	
Clean	Ea.	.31	15.90		16.21	
Minimum Charge	Job		192		192	
Premium Quality						
Demolish	Ea.		23.50		23.50	Includes material and labor to install
Install	Ea.	340	48		388	deluxe ceiling fan.
Demolish and Install	Ea.	340	71.50		411.50	
Reinstall	Ea.		38.32		38.32	
Clean	Ea.	.31	15.90		16.21	
Minimum Charge	Job		192		192	
Light Kit						
Install	Ea.	31.50	48		79.50	Includes labor and material to install a
Minimum Charge	Job		192		192	ceiling fan light kit.
Bathroom Exhaust						
Ceiling Mounted						
Demolish	Ea.		19.40		19.40	Cost includes material and labor to
Install	Ea.	46	38		84	install 60 CFM ceiling mounted
Demolish and Install	Ea.	46	57.40		103.40	bathroom exhaust fan, up to 10' of
Reinstall	Ea.		30.56		30.56	wiring with one wall switch.
Clean	Ea.	.15	6.35		6.50	
Minimum Charge	Job		192		192	

Finish Electrical

Fan		Unit	Material	Labor	Equip.	Total	Specification
Wall Mounted							
	Demolish	Ea.		19.40		19.40	Cost includes material and labor to
	Install	Ea.	62	38		100	install 110 CFM wall mounted
	Demolish and Install	Ea.	62	57.40		119.40	bathroom exhaust fan, up to 10' of
	Reinstall	Ea.		30.56		30.56	wiring with one wall switch.
	Clean	Ea.	.15	6.35		6.50	
	Minimum Charge	Job		192		192	
Lighted							
	Demolish	Ea.		19.40		19.40	Cost includes material and labor to
	Install	Ea.	44.50	38		82.50	install 50 CFM ceiling mounted lighted
	Demolish and Install	Ea.	44.50	57.40		101.90	exhaust fan, up to 10' of wiring with
	Reinstall	Ea.		30.56		30.56	one wall switch..
	Clean	Ea.	.15	6.35		6.50	
	Minimum Charge	Job		192		192	
Blower Heater w / Light							
	Demolish	Ea.		19.40		19.40	Includes material and labor to install
	Install	Ea.	91	63.50		154.50	ceiling blower type ventilator with
	Demolish and Install	Ea.	91	82.90		173.90	switch, light and snap-on grill. Includes
	Reinstall	Ea.		50.94		50.94	1300 watt/120 volt heater.
	Clean	Ea.	.15	6.35		6.50	
	Minimum Charge	Job		192		192	
Fan Heater w / Light							
	Demolish	Ea.		19.40		19.40	Includes material and labor to install
	Install	Ea.	117	63.50		180.50	circular ceiling fan heater unit, 1500
	Demolish and Install	Ea.	117	82.90		199.90	watt, surface mounted.
	Reinstall	Ea.		50.94		50.94	
	Clean	Ea.	.15	6.35		6.50	
	Minimum Charge	Job		192		192	
Whole House Exhaust							
36" Ceiling Mounted							
	Install	Ea.	620	95.50		715.50	Includes material and labor to install a
	Clean	Ea.	.04	4.56		4.60	whole house exhaust fan.
	Minimum Charge	Job		192		192	
Kitchen Exhaust							
Standard Model							
	Install	Ea.	66.50	25.50		92	Includes labor and materials to install
	Clean	Ea.	.04	4.56		4.60	a bathroom or kitchen vent fan.
	Minimum Charge	Job		192		192	
Low Noise Model							
	Install	Ea.	87	25.50		112.50	Includes labor and materials to install
	Clean	Ea.	.04	4.56		4.60	a bathroom or kitchen vent fan, low
	Minimum Charge	Job		192		192	noise model.

Low Voltage Systems		Unit	Material	Labor	Equip.	Total	Specification
Smoke Detector							
Pre-wired							
	Demolish	Ea.		8.85		8.85	Includes material and labor to install
	Install	Ea.	19.65	72		91.65	AC type smoke detector including
	Demolish and Install	Ea.	19.65	80.85		100.50	wiring.
	Reinstall	Ea.		57.52		57.52	
	Clean	Ea.	.07	7.95		8.02	
	Minimum Charge	Job		192		192	

Finish Electrical

Low Voltage Systems		Unit	Material	Labor	Equip.	Total	Specification
Battery-powered							
	Demolish	Ea.		8.85		8.85	Includes material and labor to install
	Install	Ea.	24	24		48	battery-powered type smoke detector.
	Demolish and Install	Ea.	24	32.85		56.85	
	Reinstall	Ea.		19.16		19.16	
	Clean	Ea.	.07	7.95		8.02	
	Minimum Charge	Job		192		192	
Run Wiring							
	Install	Ea.	19.35	15.35		34.70	Includes labor and material to install a
	Minimum Charge	Job		192		192	electric fixture utility box with
							non-metallic cable.
## Door Bell / Chimes							
Good Quality							
	Demolish	Ea.		26.50		26.50	Includes material and labor to install
	Install	Ea.	52.50	96		148.50	surface mounted 2 note door bell.
	Demolish and Install	Ea.	52.50	122.50		175	Wiring is not included.
	Reinstall	Ea.		76.64		76.64	
	Clean	Ea.	.15	7.95		8.10	
	Minimum Charge	Job		192		192	
Better Quality							
	Demolish	Ea.		26.50		26.50	Includes material and labor to install
	Install	Ea.	67.50	96		163.50	standard quality surface mounted 2
	Demolish and Install	Ea.	67.50	122.50		190	note door bell. Wiring is not included.
	Reinstall	Ea.		76.64		76.64	
	Clean	Ea.	.15	7.95		8.10	
	Minimum Charge	Job		192		192	
Premium Quality							
	Demolish	Ea.		26.50		26.50	Includes material and labor to install
	Install	Ea.	76.50	96		172.50	deluxe quality surface mounted 2 note
	Demolish and Install	Ea.	76.50	122.50		199	door bell. Wiring is not included.
	Reinstall	Ea.		76.64		76.64	
	Clean	Ea.	.15	7.95		8.10	
	Minimum Charge	Job		192		192	
Run Wiring							
	Install	Ea.	19.35	15.35		34.70	Includes labor and material to install a
	Minimum Charge	Job		192		192	electric fixture utility box with
							non-metallic cable.
## Intercom System							
Intercom (master)							
	Demolish	Ea.		71		71	Includes material and labor to install
	Install	Ea.	535	192		727	master intercom station with AM/FM
	Demolish and Install	Ea.	535	263		798	music and room monitoring. Wiring is
	Reinstall	Ea.		153.28		153.28	not included.
	Clean	Ea.	.15	21		21.15	
	Minimum Charge	Job		192		192	
Intercom (remote)							
	Demolish	Ea.		42.50		42.50	Includes material and labor to install
	Install	Ea.	48.50	48		96.50	remote station for master intercom
	Demolish and Install	Ea.	48.50	90.50		139	system. Wiring is not included.
	Reinstall	Ea.		38.32		38.32	
	Clean	Ea.	.15	7.95		8.10	
	Minimum Charge	Job		192		192	

Finish Electrical

Low Voltage Systems

Low Voltage Systems	Unit	Material	Labor	Equip.	Total	Specification
Security System						
Alarm Panel						
Demolish	Ea.		85		85	Includes material and labor to install a
Install	Ea.	127	192		319	6 zone burglar alarm panel.
Demolish and Install	Ea.	127	277		404	
Reinstall	Ea.		153.28		153.28	
Clean	Ea.	.15	7.95		8.10	
Minimum Charge	Job		192		192	
Motion Detector						
Demolish	Ea.		61		61	Includes material and labor to install a
Install	Ea.	231	167		398	ultrasonic motion detector for a
Demolish and Install	Ea.	231	228		459	security system.
Reinstall	Ea.		133.29		133.29	
Clean	Ea.	.15	7.95		8.10	
Minimum Charge	Job		192		192	
Door Switch						
Demolish	Ea.		26.50		26.50	Includes material and labor to install a
Install	Ea.	35	72.50		107.50	door / window contact for a security
Demolish and Install	Ea.	35	99		134	system.
Reinstall	Ea.		57.84		57.84	
Clean	Ea.	.15	2.54		2.69	
Minimum Charge	Job		192		192	
Window Switch						
Demolish	Ea.		26.50		26.50	Includes material and labor to install a
Install	Ea.	35	72.50		107.50	door / window contact for a security
Demolish and Install	Ea.	35	99		134	system.
Reinstall	Ea.		57.84		57.84	
Clean	Ea.	.15	2.54		2.69	
Minimum Charge	Job		192		192	
Satellite System						
Dish						
Demolish	Ea.		118		118	Includes material, labor and
Install	Ea.	760	320		1080	equipment to install 10' mesh dish.
Demolish and Install	Ea.	760	438		1198	
Reinstall	Ea.		255.47		255.47	
Clean	Ea.	3.11	32		35.11	
Minimum Charge	Job		192		192	
Motor						
Demolish	Ea.		59		59	Includes material and labor to install a
Install	Ea.	340	160		500	television satellite dish motor unit.
Demolish and Install	Ea.	340	219		559	
Reinstall	Ea.		127.73		127.73	
Clean	Ea.	.15	7.95		8.10	
Minimum Charge	Job		192		192	
Television Antenna						
40' Pole						
Demolish	Ea.		66.50		66.50	Includes material and labor to install
Install	Ea.	23	120		143	stand-alone 1"-2" metal pole, 10' long.
Demolish and Install	Ea.	23	186.50		209.50	Cost does not include antenna.
Reinstall	Ea.		95.80		95.80	
Clean	Ea.	.63	32		32.63	
Paint	Ea.	7.10	38		45.10	
Minimum Charge	Job		192		192	
Rotor Unit						
Demolish	Ea.		53		53	Includes material and labor to install a
Install	Ea.	107	48		155	television antenna rotor.
Demolish and Install	Ea.	107	101		208	
Reinstall	Ea.		38.32		38.32	
Clean	Ea.	.63	32		32.63	
Minimum Charge	Job		192		192	

Finish Electrical

Low Voltage Systems

	Unit	Material	Labor	Equip.	Total	Specification
Single Booster						Includes material and labor to install
Demolish	Ea.		53		53	standard grade VHF television signal
Install	Ea.	36.50	48		84.50	booster unit and all hardware required
Demolish and Install	Ea.	36.50	101		137.50	for connection to TV cable.
Reinstall	Ea.		38.32		38.32	
Clean	Ea.	.03	3.97		4	
Minimum Charge	Job		192		192	

| Gazebo | Gutter | Shutter |

Window Treatment	Unit	Material	Labor	Equip.	Total	Specification
Drapery						
Good Grade						
Install	L.F.	36.50	5.10		41.60	Includes material and labor to install
Reinstall	L.F.		4.06		4.06	good quality drapery.
Clean	L.F.	2.66	.13		2.79	
Minimum Charge	Job		127		127	
Premium Grade						
Install	L.F.	64	5.10		69.10	Includes material and labor to install
Reinstall	L.F.		4.06		4.06	premium quality drapery.
Clean	L.F.	2.66	.14		2.80	
Minimum Charge	Job		127		127	
Lining						
Install	L.F.	15.85	5.10		20.95	Includes material and labor to install
Reinstall	L.F.		4.06		4.06	drapery lining.
Clean	L.F.	1.33	.07		1.40	
Minimum Charge	Job		127		127	
Valance Board						
Demolish	L.F.		1.41		1.41	Includes material and labor to install a
Install	L.F.	4.65	10.15		14.80	drapery valance board.
Demolish and Install	L.F.	4.65	11.56		16.21	
Reinstall	L.F.		8.13		8.13	
Clean	L.F.	1.33	.25		1.58	
Minimum Charge	Job		127		127	
Rod						
Demolish	L.F.		3.53		3.53	Includes material and labor to install
Install	L.F.	2.93	5.90		8.83	standard traverse drapery hardware.
Demolish and Install	L.F.	2.93	9.43		12.36	
Reinstall	L.F.		4.73		4.73	
Clean	L.F.	.06	.23		.29	
Paint	L.F.	.06	.59		.65	
Minimum Charge	Job		127		127	
Blinds						
Mini-type						
Demolish	S.F.		.76		.76	Includes material and labor to install
Install	S.F.	3.41	.59		4	horizontal 1″ slat blinds.
Demolish and Install	S.F.	3.41	1.35		4.76	
Reinstall	S.F.		.47		.47	
Clean	S.F.	.51	.03		.54	
Minimum Charge	Job		127		127	

Window Treatment	Unit	Material	Labor	Equip.	Total	Specification
Vertical Type						
Demolish	S.F.		.76		.76	Includes material and labor to install
Install	S.F.	7.90	.76		8.66	vertical 3″ - 5″ cloth or PVC blinds.
Demolish and Install	S.F.	7.90	1.52		9.42	
Reinstall	S.F.		.61		.61	
Clean	S.F.	.26	.03		.29	
Minimum Charge	Job		127		127	
Vertical Type per L.F.						
Demolish	L.F.		1.18		1.18	Includes material and labor to install
Install	L.F.	47.50	3.39		50.89	vertical 3″ - 5″ cloth or PVC blinds x
Demolish and Install	L.F.	47.50	4.57		52.07	72″ long.
Reinstall	L.F.		2.71		2.71	
Clean	L.F.	1.88	.18		2.06	
Minimum Charge	Job		127		127	
Venetian Type						
Demolish	S.F.		.76		.76	Includes material and labor to install
Install	S.F.	4.95	.59		5.54	horizontal 2″ slat blinds.
Demolish and Install	S.F.	4.95	1.35		6.30	
Reinstall	S.F.		.47		.47	
Clean	S.F.	.26	.03		.29	
Minimum Charge	Job		127		127	
Wood Shutter						
27″ x 36″, 4 Panel						
Demolish	Ea.		24		24	Includes material and labor to install
Install	Pr.	91.50	20.50		112	moveable louver, up to 33″ x 36″, 4
Demolish and Install	Pr.	91.50	44.50		136	panel shutter.
Reinstall	Pr.		16.41		16.41	
Clean	Ea.	.06	7.60		7.66	
Paint	Ea.	1.13	22		23.13	
Minimum Charge	Job		174		174	
33″ x 36″, 4 Panel						
Demolish	Ea.		24		24	Includes material and labor to install
Install	Pr.	91.50	20.50		112	moveable louver, up to 33″ x 36″, 4
Demolish and Install	Pr.	91.50	44.50		136	panel shutter.
Reinstall	Pr.		16.41		16.41	
Clean	Ea.	.06	7.60		7.66	
Paint	Ea.	1.19	23.50		24.69	
Minimum Charge	Job		174		174	
39″ x 36″, 4 Panel						
Demolish	Ea.		24		24	Includes material and labor to install
Install	Pr.	103	20.50		123.50	moveable louver, up to 47″ x 36″, 4
Demolish and Install	Pr.	103	44.50		147.50	panel shutter.
Reinstall	Pr.		16.41		16.41	
Clean	Ea.	.06	7.60		7.66	
Paint	Ea.	1.19	23.50		24.69	
Minimum Charge	Job		174		174	
47″ x 36″, 4 Panel						
Demolish	Ea.		24		24	Includes material and labor to install
Install	Pr.	103	20.50		123.50	moveable louver, up to 47″ x 36″, 4
Demolish and Install	Pr.	103	44.50		147.50	panel shutter.
Reinstall	Pr.		16.41		16.41	
Clean	Ea.	.06	7.60		7.66	
Paint	Ea.	1.23	25.50		26.73	
Minimum Charge	Job		174		174	

Stairs

Stairs	Unit	Material	Labor	Equip.	Total	Specification
Concrete Steps						
Precast						
Demolish	Ea.		48.50	31.50	80	Includes material and labor to install
Install	Flight	335	120	47	502	precast concrete front entrance stairs,
Demolish and Install	Flight	335	168.50	78.50	582	4' wide, with 48" platform and 2
Clean	Ea.		3.53		3.53	risers.
Paint	Ea.	3.55	38		41.55	
Minimum Charge	Job		310		310	
Metal						
Concrete Filled						
Demolish	Ea.		14.10		14.10	Includes material and labor to install
Install	Riser	490	58.50	4.22	552.72	steel, cement filled metal pan stairs
Demolish and Install	Riser	490	72.60	4.22	566.82	with picket rail.
Clean	Ea.	.10	2.21		2.31	
Paint	Ea.	.41	2.26		2.67	
Minimum Charge	Job		216		216	
Landing						
Demolish	S.F.		3.53		3.53	Includes material and labor to install
Install	S.F.	101	6.90	.50	108.40	pre-erected conventional steel pan
Demolish and Install	S.F.	101	10.43	.50	111.93	landing.
Clean	S.F.	.10	.45		.55	
Paint	S.F.	.14	.65		.79	
Minimum Charge	Job		216		216	
Railing						
Demolish	L.F.		1.95		1.95	Includes material and labor to install
Install	L.F.	13.35	10	.72	24.07	primed steel pipe.
Demolish and Install	L.F.	13.35	11.95	.72	26.02	
Clean	L.F.	.63	1.59		2.22	
Paint	L.F.	.68	5.10		5.78	
Minimum Charge	Job		216		216	

Elevator

Elevator	Unit	Material	Labor	Equip.	Total	Specification
2 Stop Residential						
Install	Ea.	9825	4675		14500	Includes labor, material and
Clean	Ea.	.72	15.90		16.62	equipment to install 2 stop elevator for
Minimum Charge	Job		470		470	residential use.
2 Stop Commercial						
Install	Ea.	37300	9350		46650	Includes labor, material and
Clean	Ea.	.72	15.90		16.62	equipment to install 2 stop elevator for
Minimum Charge	Job		470		470	apartment or commercial building with
						2,000 lb/13 passenger capacity.
3 Stop Commercial						
Install	Ea.	46800	14400		61200	Includes labor, material and
Clean	Ea.	.72	15.90		16.62	equipment to install 3 stop elevator for
Minimum Charge	Job		470		470	apartment or commercial building with
						2,000 lb/13 passenger capacity.
4 Stop Commercial						
Install	Ea.	56000	19500		75500	Includes labor, material and
Clean	Ea.	.72	15.90		16.62	equipment to install 4 stop elevator for
Minimum Charge	Job		470		470	apartment or commercial building with
						2,000 lb/13 passenger capacity.

Improvements / Appliances / Treatments

Elevator		Unit	Material	Labor	Equip.	Total	Specification
5 Stop Commercial							Includes labor, material and
	Install	Ea.	65500	24600		90100	equipment to install 5 stop elevator for
	Clean	Ea.	.72	15.90		16.62	apartment or commercial building with
	Minimum Charge	Job		470		470	2,000 lb/13 passenger capacity.

Rangetop		Unit	Material	Labor	Equip.	Total	Specification
4-burner							
	Demolish	Ea.		25.50		25.50	Cost includes material and labor to
	Install	Ea.	237	64		301	install 4-burner electric surface unit.
	Demolish and Install	Ea.	237	89.50		326.50	
	Reinstall	Ea.		51.09		51.09	
	Clean	Ea.	.04	23		23.04	
	Minimum Charge	Job		63.50		63.50	
6-burner							
	Demolish	Ea.		25.50		25.50	Includes material and labor to install a
	Install	Ea.	935	76.50		1011.50	6-burner cooktop unit.
	Demolish and Install	Ea.	935	102		1037	
	Reinstall	Ea.		61.31		61.31	
	Clean	Ea.	.04	23		23.04	
	Minimum Charge	Job		63.50		63.50	
Premium Brand w / Grill							
	Demolish	Ea.		25.50		25.50	Includes material and labor to install
	Install	Ea.	955	76.50		1031.50	electric downdraft cooktop with grill.
	Demolish and Install	Ea.	955	102		1057	
	Reinstall	Ea.		61.31		61.31	
	Clean	Ea.	.04	23		23.04	
	Minimum Charge	Job		63.50		63.50	

Rangehood		Unit	Material	Labor	Equip.	Total	Specification
Vented							
30"							
	Demolish	Ea.		23.50		23.50	Includes material and labor to install a
	Install	Ea.	44.50	76.50		121	vented range hood.
	Demolish and Install	Ea.	44.50	100		144.50	
	Reinstall	Ea.		61.31		61.31	
	Clean	Ea.	.04	6.50		6.54	
	Minimum Charge	Job		63.50		63.50	
36"							
	Demolish	Ea.		23.50		23.50	Includes material and labor to install a
	Install	Ea.	123	76.50		199.50	vented range hood.
	Demolish and Install	Ea.	123	100		223	
	Reinstall	Ea.		61.31		61.31	
	Clean	Ea.	.04	6.50		6.54	
	Minimum Charge	Job		63.50		63.50	
42"							
	Demolish	Ea.		23.50		23.50	Includes material and labor to install a
	Install	Ea.	266	89		355	vented range hood.
	Demolish and Install	Ea.	266	112.50		378.50	
	Reinstall	Ea.		71.14		71.14	
	Clean	Ea.	.04	6.50		6.54	
	Minimum Charge	Job		63.50		63.50	

Improvements / Appliances / Treatments

Rangehood	Unit	Material	Labor	Equip.	Total	Specification
Ventless						
30″						
Demolish	Ea.		23.50		23.50	Includes material and labor to install a
Install	Ea.	47.50	48		95.50	ventless range hood.
Demolish and Install	Ea.	47.50	71.50		119	
Reinstall	Ea.		38.32		38.32	
Clean	Ea.	.04	6.50		6.54	
Minimum Charge	Job		63.50		63.50	
36″						
Demolish	Ea.		23.50		23.50	Includes material and labor to install a
Install	Ea.	51.50	76.50		128	ventless range hood.
Demolish and Install	Ea.	51.50	100		151.50	
Reinstall	Ea.		61.31		61.31	
Clean	Ea.	.04	6.50		6.54	
Minimum Charge	Job		63.50		63.50	
42″						
Demolish	Ea.		23.50		23.50	Includes material and labor to install a
Install	Ea.	238	64		302	ventless range hood.
Demolish and Install	Ea.	238	87.50		325.50	
Reinstall	Ea.		51.09		51.09	
Clean	Ea.	.04	6.50		6.54	
Minimum Charge	Job		63.50		63.50	

Oven	Unit	Material	Labor	Equip.	Total	Specification
Free Standing						
Demolish	Ea.		25.50		25.50	Cost includes material and labor to
Install	Ea.	445	51		496	install 30″ free standing electric range
Demolish and Install	Ea.	445	76.50		521.50	with self-cleaning oven.
Reinstall	Ea.		40.64		40.64	
Clean	Ea.	.04	23		23.04	
Minimum Charge	Job		63.50		63.50	
Single Wall Oven						
Demolish	Ea.		25.50		25.50	Includes material and labor to install a
Install	Ea.	470	96		566	built-in oven.
Demolish and Install	Ea.	470	121.50		591.50	
Reinstall	Ea.		76.64		76.64	
Clean	Ea.	.04	23		23.04	
Minimum Charge	Job		63.50		63.50	
Double Wall Oven						
Demolish	Ea.		25.50		25.50	Includes material and labor to install a
Install	Ea.	1425	174		1599	conventional double oven.
Demolish and Install	Ea.	1425	199.50		1624.50	
Reinstall	Ea.		139.52		139.52	
Clean	Ea.	.04	23		23.04	
Minimum Charge	Job		63.50		63.50	
Drop-in Type w / Range						
Demolish	Ea.		25.50		25.50	Includes material and labor to install a
Install	Ea.	515	64		579	built-in cooking range with oven.
Demolish and Install	Ea.	515	89.50		604.50	
Reinstall	Ea.		51.09		51.09	
Clean	Ea.	.04	23		23.04	
Minimum Charge	Job		63.50		63.50	

Improvements / Appliances / Treatments

Oven	Unit	Material	Labor	Equip.	Total	Specification
Hi/Lo w / Range, Microwave						
Demolish	Ea.		25.50		25.50	Includes material and labor to install a
Install	Ea.	1825	350		2175	double oven, one conventional, one
Demolish and Install	Ea.	1825	375.50		2200.50	microwave.
Reinstall	Ea.		279.04		279.04	
Clean	Ea.	.04	23		23.04	
Minimum Charge	Job		63.50		63.50	
Hi/Lo w/ Range / Double Oven						
Demolish	Ea.		25.50		25.50	Includes material and labor to install a
Install	Ea.	1425	174		1599	conventional double oven.
Demolish and Install	Ea.	1425	199.50		1624.50	
Reinstall	Ea.		139.52		139.52	
Clean	Ea.	.04	23		23.04	
Minimum Charge	Job		63.50		63.50	
Countertop Microwave						
Demolish	Ea.		8.80		8.80	Includes material and labor to install a
Install	Ea.	113	32		145	1.0 cubic foot microwave oven.
Demolish and Install	Ea.	113	40.80		153.80	
Reinstall	Ea.		25.40		25.40	
Clean	Ea.	.04	11.40		11.44	
Minimum Charge	Job		63.50		63.50	
Cabinet / Wall Mounted Microwave						
Demolish	Ea.		17.65		17.65	Includes material and labor to install a
Install	Ea.	590	96		686	space saver microwave oven.
Demolish and Install	Ea.	590	113.65		703.65	
Reinstall	Ea.		76.64		76.64	
Clean	Ea.	.04	11.40		11.44	
Minimum Charge	Job		63.50		63.50	

Dryer	Unit	Material	Labor	Equip.	Total	Specification
Basic Grade						
Demolish	Ea.		11.75		11.75	Includes material and labor to install
Install	Ea.	296	201		497	electric or gas dryer. Vent kit not
Demolish and Install	Ea.	296	212.75		508.75	included.
Reinstall	Ea.		160.85		160.85	
Clean	Ea.	.04	23		23.04	
Minimum Charge	Job		63.50		63.50	
Good Grade						
Demolish	Ea.		11.75		11.75	Includes material and labor to install
Install	Ea.	370	201		571	electric or gas dryer. Vent kit not
Demolish and Install	Ea.	370	212.75		582.75	included.
Reinstall	Ea.		160.85		160.85	
Clean	Ea.	.04	23		23.04	
Minimum Charge	Job		63.50		63.50	
Premium Grade						
Demolish	Ea.		11.75		11.75	Includes material and labor to install
Install	Ea.	730	300		1030	electric or gas dryer. Vent kit not
Demolish and Install	Ea.	730	311.75		1041.75	included.
Reinstall	Ea.		241.28		241.28	
Clean	Ea.	.04	23		23.04	
Minimum Charge	Job		63.50		63.50	

Improvements / Appliances / Treatments

Dryer

	Unit	Material	Labor	Equip.	Total	Specification
Washer / Dryer Combination Unit						
Demolish	Ea.		11.75		11.75	Includes material and labor to install
Install	Ea.	975	51		1026	electric washer/dryer combination
Demolish and Install	Ea.	975	62.75		1037.75	unit, 27" wide, stackable or unitized.
Reinstall	Ea.		40.64		40.64	
Clean	Ea.	.04	23		23.04	
Minimum Charge	Job		63.50		63.50	

Washer

	Unit	Material	Labor	Equip.	Total	Specification
Basic Grade						
Demolish	Ea.		11.75		11.75	Includes material and labor to install
Install	Ea.	340	130		470	washing machine. New hoses not
Demolish and Install	Ea.	340	141.75		481.75	included.
Reinstall	Ea.		103.68		103.68	
Clean	Ea.	.04	23		23.04	
Minimum Charge	Job		63.50		63.50	
Good Grade						
Demolish	Ea.		11.75		11.75	Includes material and labor to install
Install	Ea.	560	194		754	washing machine. New hoses not
Demolish and Install	Ea.	560	205.75		765.75	included.
Reinstall	Ea.		155.52		155.52	
Clean	Ea.	.04	23		23.04	
Minimum Charge	Job		63.50		63.50	
Premium Grade						
Demolish	Ea.		11.75		11.75	Includes material and labor to install
Install	Ea.	1200	390		1590	washing machine with digital
Demolish and Install	Ea.	1200	401.75		1601.75	computer readouts and large capacity.
Reinstall	Ea.		311.04		311.04	New hoses not included.
Clean	Ea.	.04	23		23.04	
Minimum Charge	Job		63.50		63.50	
Washer / Dryer Combination Unit						
Demolish	Ea.		11.75		11.75	Includes material and labor to install
Install	Ea.	975	51		1026	electric washer/dryer combination
Demolish and Install	Ea.	975	62.75		1037.75	unit, 27" wide, stackable or unitized.
Reinstall	Ea.		40.64		40.64	
Clean	Ea.	.04	23		23.04	
Minimum Charge	Job		63.50		63.50	

Refrigerator

	Unit	Material	Labor	Equip.	Total	Specification
12 Cubic Foot						
Demolish	Ea.		35.50		35.50	Includes material and labor to install a
Install	Ea.	450	51		501	refrigerator.
Demolish and Install	Ea.	450	86.50		536.50	
Reinstall	Ea.		40.64		40.64	
Clean	Ea.	.04	23		23.04	
Minimum Charge	Job		63.50		63.50	
16 Cubic Foot						
Demolish	Ea.		35.50		35.50	Includes material and labor to install a
Install	Ea.	490	56.50		546.50	refrigerator.
Demolish and Install	Ea.	490	92		582	
Reinstall	Ea.		45.16		45.16	
Clean	Ea.	.04	23		23.04	
Minimum Charge	Job		63.50		63.50	

Improvements / Appliances / Treatments

Refrigerator

	Unit	Material	Labor	Equip.	Total	Specification
18 Cubic Foot						
Demolish	Ea.		35.50		35.50	Includes material and labor to install a
Install	Ea.	605	63.50		668.50	refrigerator.
Demolish and Install	Ea.	605	99		704	
Reinstall	Ea.		50.80		50.80	
Clean	Ea.	.04	23		23.04	
Minimum Charge	Job		63.50		63.50	
21 Cubic Foot						
Demolish	Ea.		35.50		35.50	Includes material and labor to install a
Install	Ea.	775	72.50		847.50	refrigerator.
Demolish and Install	Ea.	775	108		883	
Reinstall	Ea.		58.06		58.06	
Clean	Ea.	.04	23		23.04	
Minimum Charge	Job		63.50		63.50	
24 Cubic Foot						
Demolish	Ea.		35.50		35.50	Includes material and labor to install a
Install	Ea.	1025	72.50		1097.50	refrigerator.
Demolish and Install	Ea.	1025	108		1133	
Reinstall	Ea.		58.06		58.06	
Clean	Ea.	.04	23		23.04	
Minimum Charge	Job		63.50		63.50	
27 Cubic Foot						
Demolish	Ea.		35.50		35.50	Includes material and labor to install a
Install	Ea.	1875	51		1926	refrigerator.
Demolish and Install	Ea.	1875	86.50		1961.50	
Reinstall	Ea.		40.64		40.64	
Clean	Ea.	.04	23		23.04	
Minimum Charge	Job		63.50		63.50	
Ice Maker						
Install	Ea.	73	48.50		121.50	Includes labor and material to install
Minimum Charge	Job		63.50		63.50	an automatic ice maker in a refrigerator.

Freezer

	Unit	Material	Labor	Equip.	Total	Specification
15 Cubic Foot						
Demolish	Ea.		35.50		35.50	Cost includes material and labor to
Install	Ea.	525	51		576	install 15 cubic foot freezer.
Demolish and Install	Ea.	525	86.50		611.50	
Reinstall	Ea.		40.64		40.64	
Clean	Ea.	.04	23		23.04	
Minimum Charge	Job		63.50		63.50	
18 Cubic Foot						
Demolish	Ea.		35.50		35.50	Cost includes material and labor to
Install	Ea.	700	51		751	install 18 cubic foot freezer.
Demolish and Install	Ea.	700	86.50		786.50	
Reinstall	Ea.		40.64		40.64	
Clean	Ea.	.04	23		23.04	
Minimum Charge	Job		63.50		63.50	
21 Cubic Foot						
Demolish	Ea.		35.50		35.50	Includes material and labor to install a
Install	Ea.	595	51		646	15 to 23 cubic foot deep freezer.
Demolish and Install	Ea.	595	86.50		681.50	
Reinstall	Ea.		40.64		40.64	
Clean	Ea.	.04	23		23.04	
Minimum Charge	Job		63.50		63.50	

Improvements / Appliances / Treatments

Freezer

Freezer	Unit	Material	Labor	Equip.	Total	Specification
24 Cubic Foot						
Demolish	Ea.		35.50		35.50	Includes material and labor to install a
Install	Ea.	660	102		762	24 cubic foot deep freezer.
Demolish and Install	Ea.	660	137.50		797.50	
Reinstall	Ea.		81.28		81.28	
Clean	Ea.	.04	23		23.04	
Minimum Charge	Job		63.50		63.50	
Chest-type						
Demolish	Ea.		35.50		35.50	Cost includes material and labor to
Install	Ea.	525	51		576	install 15 cubic foot freezer.
Demolish and Install	Ea.	525	86.50		611.50	
Clean	Ea.	.04	23		23.04	
Minimum Charge	Job		63.50		63.50	

Dishwasher

Dishwasher	Unit	Material	Labor	Equip.	Total	Specification
4 Cycle						
Demolish	Ea.		28		28	Includes material and labor to install
Install	Ea.	315	121		436	good grade automatic dishwasher.
Demolish and Install	Ea.	315	149		464	
Reinstall	Ea.		96.92		96.92	
Clean	Ea.	.04	23		23.04	
Minimum Charge	Job		63.50		63.50	

Garbage Disposal

Garbage Disposal	Unit	Material	Labor	Equip.	Total	Specification
In Sink Unit						
Demolish	Ea.		27		27	Cost includes material and labor to
Install	Ea.	76.50	48.50		125	install 1/2 HP custom disposal unit.
Demolish and Install	Ea.	76.50	75.50		152	
Reinstall	Ea.		38.77		38.77	
Clean	Ea.	.02	6.50		6.52	
Minimum Charge	Job		63.50		63.50	

Trash Compactor

Trash Compactor	Unit	Material	Labor	Equip.	Total	Specification
4 to 1 Compaction						
Demolish	Ea.		26.50		26.50	Includes material and labor to install a
Install	Ea.	645	70		715	trash compactor.
Demolish and Install	Ea.	645	96.50		741.50	
Reinstall	Ea.		55.81		55.81	
Clean	Ea.	.04	15.90		15.94	
Minimum Charge	Job		63.50		63.50	

Kitchenette Unit

Kitchenette Unit	Unit	Material	Labor	Equip.	Total	Specification
Range, Refrigerator and Sink						
Demolish	Ea.		106		106	Includes material and labor to install
Install	Ea.	3400	405		3805	range/refrigerator/sink unit, 72"
Demolish and Install	Ea.	3400	511		3911	wide, base unit only.
Reinstall	Ea.		323.07		323.07	
Clean	Ea.	.78	11.40		12.18	
Minimum Charge	Job		63.50		63.50	

Improvements / Appliances / Treatments

Bath Accessories

Bath Accessories		Unit	Material	Labor	Equip.	Total	Specification
Towel Bar							
	Demolish	Ea.		5.90		5.90	Includes material and labor to install
	Install	Ea.	41.50	14.55		56.05	towel bar up to 24″ long.
	Demolish and Install	Ea.	41.50	20.45		61.95	
	Reinstall	Ea.		11.63		11.63	
	Clean	Ea.		3.18		3.18	
	Minimum Charge	Job		174		174	
Toothbrush Holder							
	Demolish	Ea.		5.90		5.90	Includes material and labor to install a
	Install	Ea.	26	17.45		43.45	surface mounted tumbler and
	Demolish and Install	Ea.	26	23.35		49.35	toothbrush holder.
	Reinstall	Ea.		13.95		13.95	
	Clean	Ea.		3.18		3.18	
	Minimum Charge	Job		174		174	
Grab Rail							
	Demolish	Ea.		8.80		8.80	Includes material and labor to install a
	Install	Ea.	26.50	15.15		41.65	24″ long grab bar. Blocking is not
	Demolish and Install	Ea.	26.50	23.95		50.45	included.
	Reinstall	Ea.		12.13		12.13	
	Clean	Ea.	.31	5.30		5.61	
	Minimum Charge	Job		174		174	
Soap Dish							
	Demolish	Ea.		5.90		5.90	Includes material and labor to install a
	Install	Ea.	23	17.45		40.45	surface mounted soap dish.
	Demolish and Install	Ea.	23	23.35		46.35	
	Reinstall	Ea.		13.95		13.95	
	Clean	Ea.		3.18		3.18	
	Minimum Charge	Job		174		174	
Toilet Paper Roller							
	Demolish	Ea.		5.90		5.90	Includes material and labor to install a
	Install	Ea.	14.85	11.65		26.50	surface mounted toilet tissue dispenser.
	Demolish and Install	Ea.	14.85	17.55		32.40	
	Reinstall	Ea.		9.30		9.30	
	Clean	Ea.	.31	7.95		8.26	
	Minimum Charge	Job		174		174	
Dispenser							
Soap							
	Demolish	Ea.		7.85		7.85	Includes material and labor to install a
	Install	Ea.	56.50	17.45		73.95	surface mounted soap dispenser.
	Demolish and Install	Ea.	56.50	25.30		81.80	
	Reinstall	Ea.		13.95		13.95	
	Clean	Ea.	.31	7.95		8.26	
	Minimum Charge	Job		174		174	
Towel							
	Demolish	Ea.		8.80		8.80	Includes material and labor to install a
	Install	Ea.	43	22		65	surface mounted towel dispenser.
	Demolish and Install	Ea.	43	30.80		73.80	
	Reinstall	Ea.		17.44		17.44	
	Clean	Ea.	.31	10.60		10.91	
	Minimum Charge	Job		174		174	
Seat Cover							
	Demolish	Ea.		5.90		5.90	Includes material and labor to install a
	Install	Ea.	32.50	23.50		56	surface mounted toilet seat cover
	Demolish and Install	Ea.	32.50	29.40		61.90	dispenser.
	Reinstall	Ea.		18.60		18.60	
	Clean	Ea.	.31	7.95		8.26	
	Minimum Charge	Job		174		174	

Improvements / Appliances / Treatments

Bath Accessories		Unit	Material	Labor	Equip.	Total	Specification
Sanitary Napkin							
	Demolish	Ea.		14.85		14.85	Includes material and labor to install a
	Install	Ea.	26	36.50		62.50	surface mounted sanitary napkin
	Demolish and Install	Ea.	26	51.35		77.35	dispenser.
	Reinstall	Ea.		29.37		29.37	
	Clean	Ea.	.15	10.60		10.75	
	Minimum Charge	Job		174		174	
Partition							
Urinal							
	Demolish	Ea.		17.65		17.65	Includes material and labor to install a
	Install	Ea.	218	87		305	urinal screen.
	Demolish and Install	Ea.	218	104.65		322.65	
	Reinstall	Ea.		69.76		69.76	
	Clean	Ea.	.46	7.05		7.51	
	Minimum Charge	Job		174		174	
Toilet							
	Demolish	Ea.		52.50		52.50	Includes material and labor to install
	Install	Ea.	515	140		655	baked enamel standard size partition,
	Demolish and Install	Ea.	515	192.50		707.50	floor or ceiling mounted with one wall
	Reinstall	Ea.		111.62		111.62	and one door.
	Clean	Ea.	3.11	15.90		19.01	
	Minimum Charge	Job		174		174	
Handicap							
	Demolish	Ea.		52.50		52.50	Includes material and labor to install
	Install	Ea.	845	140		985	baked enamel handicap partition,
	Demolish and Install	Ea.	845	192.50		1037.50	floor or ceiling mounted with one wall
	Reinstall	Ea.		111.62		111.62	and one door.
	Clean	Ea.	3.11	15.90		19.01	
	Minimum Charge	Job		174		174	

Fire Prevention / Protection		Unit	Material	Labor	Equip.	Total	Specification
Fire Extinguisher							
5# Carbon Dioxide							
	Install	Ea.	143			143	Includes labor and material to install one wall mounted 5 lb. carbon dioxide factory charged unit, complete with hose, horn and wall mounting bracket.
10# Carbon Dioxide							
	Install	Ea.	176			176	Includes labor and material to install one wall mounted 10 lb. carbon dioxide factory charged unit, complete with hose, horn and wall mounting bracket.
15# Carbon Dioxide							
	Install	Ea.	209			209	Includes labor and material to install one wall mounted 15 lb. carbon dioxide factory charged unit, complete with hose, horn and wall mounting bracket.

Improvements / Appliances / Treatments

Fire Prevention / Protection		Unit	Material	Labor	Equip.	Total	Specification
5# Dry Chemical							Includes labor and material to install one wall mounted 5 lb. dry chemical factory charged unit, complete with hose, horn and wall mounting bracket.
	Install	Ea.	66			66	
20# Dry Chemical							Includes labor and material to install one wall mounted 20 lb. dry chemical factory charged unit, complete with hose, horn and wall mounting bracket.
	Install	Ea.	121			121	

Gutters		Unit	Material	Labor	Equip.	Total	Specification
Galvanized							Cost includes material and labor to install 5" galvanized steel gutter with enamel finish, fittings, hangers, and corners.
	Demolish	L.F.		1.18		1.18	
	Install	L.F.	1.49	3.22		4.71	
	Demolish and Install	L.F.	1.49	4.40		5.89	
	Reinstall	L.F.		2.58		2.58	
	Clean	L.F.		.34		.34	
	Paint	L.F.	.19	.94		1.13	
	Minimum Charge	Job		193		193	
Aluminum							Cost includes material and labor to install 5" aluminum gutter with enamel finish, fittings, hangers, and corners.
	Demolish	L.F.		1.18		1.18	
	Install	L.F.	1.94	3.22		5.16	
	Demolish and Install	L.F.	1.94	4.40		6.34	
	Reinstall	L.F.		2.58		2.58	
	Clean	L.F.		.34		.34	
	Paint	L.F.	.19	.94		1.13	
	Minimum Charge	Job		193		193	
Copper							Cost includes material and labor to install 5" half-round copper gutter with fittings, hangers, and corners.
	Demolish	L.F.		1.18		1.18	
	Install	L.F.	8.05	3.22		11.27	
	Demolish and Install	L.F.	8.05	4.40		12.45	
	Reinstall	L.F.		2.58		2.58	
	Clean	L.F.		.34		.34	
	Minimum Charge	Job		193		193	
Plastic							Cost includes material and labor to install 4" white vinyl gutter fittings, hangers, and corners.
	Demolish	L.F.		1.18		1.18	
	Install	L.F.	1.01	3.17		4.18	
	Demolish and Install	L.F.	1.01	4.35		5.36	
	Reinstall	L.F.		2.54		2.54	
	Clean	L.F.		.34		.34	
	Paint	L.F.	.19	.94		1.13	
	Minimum Charge	Job		193		193	

Downspouts		Unit	Material	Labor	Equip.	Total	Specification
Galvanized							Cost includes material and labor to install 4" diameter downspout with enamel finish.
	Demolish	L.F.		.81		.81	
	Install	L.F.	2.37	2.67		5.04	
	Demolish and Install	L.F.	2.37	3.48		5.85	
	Reinstall	L.F.		2.13		2.13	
	Clean	L.F.		.34		.34	
	Paint	L.F.	.19	.94		1.13	
	Minimum Charge	Job		193		193	

Improvements / Appliances / Treatments

Downspouts

Downspouts	Unit	Material	Labor	Equip.	Total	Specification
Aluminum						
Demolish	L.F.		.81		.81	Includes material and labor to install
Install	L.F.	3.84	2.76		6.60	aluminum enameled downspout.
Demolish and Install	L.F.	3.84	3.57		7.41	
Reinstall	L.F.		2.21		2.21	
Clean	L.F.		.34		.34	
Paint	L.F.	.19	.94		1.13	
Minimum Charge	Job		193		193	
Copper						
Demolish	L.F.		.81		.81	Includes material and labor to install
Install	L.F.	8.10	2.04		10.14	round copper downspout.
Demolish and Install	L.F.	8.10	2.85		10.95	
Reinstall	L.F.		1.63		1.63	
Clean	L.F.		.34		.34	
Paint	L.F.	.19	.94		1.13	
Minimum Charge	Job		193		193	
Plastic						
Demolish	L.F.		.81		.81	Cost includes material and labor to
Install	L.F.	.83	1.84		2.67	install 2" x 3" vinyl rectangular
Demolish and Install	L.F.	.83	2.65		3.48	downspouts.
Reinstall	L.F.		1.47		1.47	
Clean	L.F.		.34		.34	
Paint	L.F.	.19	.94		1.13	
Minimum Charge	Job		193		193	

Shutters

Shutters	Unit	Material	Labor	Equip.	Total	Specification
Exterior						
25" High						
Demolish	Ea.		8.80		8.80	Includes material and labor to install
Install	Pr.	102	35		137	pair of exterior shutters 25" x 16",
Demolish and Install	Pr.	102	43.80		145.80	unpainted, installed as fixed shutters
Reinstall	Pr.		34.88		34.88	on concrete or wood frame.
Clean	Ea.	.06	7.60		7.66	
Paint	Ea.	1.51	16.40		17.91	
Minimum Charge	Job		174		174	
39" High						
Demolish	Ea.		8.80		8.80	Includes material and labor to install
Install	Pr.	110	35		145	pair of shutters 39" x 16", unpainted,
Demolish and Install	Pr.	110	43.80		153.80	installed as fixed shutters on concrete
Reinstall	Pr.		34.88		34.88	or wood frame.
Clean	Ea.	.06	7.60		7.66	
Paint	Ea.	2.34	22		24.34	
Minimum Charge	Job		174		174	
51" High						
Demolish	Ea.		8.80		8.80	Includes material and labor to install
Install	Pr.	122	35		157	pair of shutters 51" x 16", unpainted,
Demolish and Install	Pr.	122	43.80		165.80	installed as fixed shutters on concrete
Reinstall	Pr.		34.88		34.88	or wood frame.
Clean	Ea.	.06	7.60		7.66	
Paint	Ea.	3.05	23		26.05	
Minimum Charge	Job		174		174	

Shutters	Unit	Material	Labor	Equip.	Total	Specification
59″ High						
Demolish	Ea.		8.80		8.80	Includes material and labor to install
Install	Pr.	173	35		208	pair of shutters 59″ x 16″, unpainted,
Demolish and Install	Pr.	173	43.80		216.80	installed as fixed shutters on concrete
Reinstall	Pr.		34.88		34.88	or wood frame.
Clean	Ea.	.06	7.60		7.66	
Paint	Ea.	3.52	25		28.52	
Minimum Charge	Job		174		174	
67″ High						
Demolish	Ea.		8.80		8.80	Includes material and labor to install
Install	Pr.	174	39		213	pair of shutters 67″ x 16″, unpainted,
Demolish and Install	Pr.	174	47.80		221.80	installed as fixed shutters on concrete
Reinstall	Pr.		38.76		38.76	or wood frame.
Clean	Ea.	.06	7.60		7.66	
Paint	Ea.	4.02	28.50		32.52	
Minimum Charge	Job		174		174	
Interior Wood						
23″ x 24″, 2 Panel						
Demolish	Ea.		8.80		8.80	Cost includes material and labor to
Install	Ea.	92.50	20.50		113	install 23″ x 24″, 4 panel interior
Demolish and Install	Ea.	92.50	29.30		121.80	moveable pine shutter, 1-1/4″ louver,
Reinstall	Ea.		16.41		16.41	knobs and hooks and hinges.
Clean	Ea.	.06	7.60		7.66	
Paint	Ea.	1.13	22		23.13	
Minimum Charge	Job		174		174	
27″ x 24″, 4 Panel						
Demolish	Ea.		8.80		8.80	Cost includes material and labor to
Install	Ea.	107	20.50		127.50	install 27″ x 24″, 4 panel interior
Demolish and Install	Ea.	107	29.30		136.30	moveable pine shutter, 1-1/4″ louver,
Reinstall	Ea.		16.41		16.41	knobs and hooks and hinges.
Clean	Ea.	.06	7.60		7.66	
Paint	Ea.	1.13	22		23.13	
Minimum Charge	Job		174		174	
31″ x 24″, 4 Panel						
Demolish	Ea.		8.80		8.80	Cost includes material and labor to
Install	Ea.	123	20.50		143.50	install 31″ x 24″, 4 panel interior
Demolish and Install	Ea.	123	29.30		152.30	moveable pine shutter, 1-1/4″ louver,
Reinstall	Ea.		16.41		16.41	knobs and hooks and hinges.
Clean	Ea.	.06	7.60		7.66	
Paint	Ea.	1.13	22		23.13	
Minimum Charge	Job		174		174	
35″ x 24″, 4 Panel						
Demolish	Ea.		8.80		8.80	Cost includes material and labor to
Install	Ea.	123	20.50		143.50	install 35″ x 24″, 4 panel interior
Demolish and Install	Ea.	123	29.30		152.30	moveable pine shutter, 1-1/4″ louver,
Reinstall	Ea.		16.41		16.41	knobs and hooks and hinges.
Clean	Ea.	.06	7.60		7.66	
Paint	Ea.	1.13	22		23.13	
Minimum Charge	Job		174		174	
39″ x 24″, 4 Panel						
Demolish	Ea.		8.80		8.80	Cost includes material and labor to
Install	Ea.	143	20.50		163.50	install 39″ x 24″, 4 panel interior
Demolish and Install	Ea.	143	29.30		172.30	moveable pine shutter, 1-1/4″ louver,
Reinstall	Ea.		16.41		16.41	knobs and hooks and hinges.
Clean	Ea.	.06	7.60		7.66	
Paint	Ea.	1.13	22		23.13	
Minimum Charge	Job		174		174	

Improvements / Appliances / Treatments

Shutters	Unit	Material	Labor	Equip.	Total	Specification
47″ x 24″, 4 Panel						
Demolish	Ea.		8.80		8.80	Cost includes material and labor to
Install	Ea.	157	20.50		177.50	install 47″ x 24″, 4 panel interior
Demolish and Install	Ea.	157	29.30		186.30	moveable pine shutter, 1-1/4″ louver,
Reinstall	Ea.		16.41		16.41	knobs and hooks and hinges.
Clean	Ea.	.06	7.60		7.66	
Paint	Ea.	1.19	23.50		24.69	
Minimum Charge	Job		174		174	

Awning	Unit	Material	Labor	Equip.	Total	Specification
Standard Aluminum						
36″ Wide x 30″ Deep						
Demolish	Ea.		8.80		8.80	Includes material and labor to install
Install	Ea.	208	29		237	one aluminum awning with
Demolish and Install	Ea.	208	37.80		245.80	weather-resistant finish.
Reinstall	Ea.		23.25		23.25	
Clean	Ea.		10.60		10.60	
Minimum Charge	Job		174		174	
48″ Wide x 30″ Deep						
Demolish	Ea.		8.80		8.80	Includes material and labor to install
Install	Ea.	244	35		279	one aluminum awning with
Demolish and Install	Ea.	244	43.80		287.80	weather-resistant finish.
Reinstall	Ea.		27.90		27.90	
Clean	Ea.		10.60		10.60	
Minimum Charge	Job		174		174	
60″ Wide x 30″ Deep						
Demolish	Ea.		17.65		17.65	Includes material and labor to install
Install	Ea.	268	39		307	one aluminum awning with
Demolish and Install	Ea.	268	56.65		324.65	weather-resistant finish.
Reinstall	Ea.		31		31	
Clean	Ea.		10.60		10.60	
Minimum Charge	Job		174		174	
72″ Wide x 30″ Deep						
Demolish	Ea.		17.65		17.65	Includes material and labor to install
Install	Ea.	294	43.50		337.50	one aluminum awning with
Demolish and Install	Ea.	294	61.15		355.15	weather-resistant finish.
Reinstall	Ea.		34.88		34.88	
Clean	Ea.		10.60		10.60	
Minimum Charge	Job		174		174	
Roll-up Type						
Demolish	S.F.		.64		.64	Includes material and labor to install
Install	S.F.	8.45	3.49		11.94	an aluminum roll-up awning.
Demolish and Install	S.F.	8.45	4.13		12.58	
Reinstall	S.F.		2.79		2.79	
Clean	S.F.	.03	.53		.56	
Minimum Charge	Job		174		174	
Canvas						
30″ Wide						
Demolish	Ea.		5.30		5.30	Includes material and labor to install a
Install	Ea.	201	29		230	canvas window awning.
Demolish and Install	Ea.	201	34.30		235.30	
Reinstall	Ea.		23.25		23.25	
Clean	Ea.		3.18		3.18	
Minimum Charge	Job		174		174	

Improvements / Appliances / Treatments

Awning		Unit	Material	Labor	Equip.	Total	Specification
36" Wide							
	Demolish	Ea.		5.30		5.30	Includes material and labor to install a
	Install	Ea.	213	35		248	canvas window awning.
	Demolish and Install	Ea.	213	40.30		253.30	
	Reinstall	Ea.		27.90		27.90	
	Clean	Ea.		3.48		3.48	
	Minimum Charge	Job		174		174	
42" Wide							
	Demolish	Ea.		5.30		5.30	Includes material and labor to install a
	Install	Ea.	227	43.50		270.50	canvas window awning.
	Demolish and Install	Ea.	227	48.80		275.80	
	Reinstall	Ea.		34.88		34.88	
	Clean	Ea.		4.13		4.13	
	Minimum Charge	Job		174		174	
48" Wide							
	Demolish	Ea.		5.30		5.30	Includes material and labor to install a
	Install	Ea.	239	50		289	canvas window awning.
	Demolish and Install	Ea.	239	55.30		294.30	
	Reinstall	Ea.		39.86		39.86	
	Clean	Ea.		4.79		4.79	
	Minimum Charge	Job		174		174	
Cloth Canopy Cover							
8' x 10'							
	Demolish	Ea.		295		295	Includes material and labor to install
	Install	Ea.	375	174		549	one 8' x 10' cloth canopy patio cover
	Demolish and Install	Ea.	375	469		844	with front bar and tension support
	Reinstall	Ea.		139.52		139.52	rafters and 9" valance.
	Clean	Ea.	2.49	42.50		44.99	
	Minimum Charge	Job		350		350	
8' x 15'							
	Demolish	Ea.		295		295	Includes material and labor to install
	Install	Ea.	505	174		679	one 8' x 15' cloth canopy patio cover
	Demolish and Install	Ea.	505	469		974	with front bar and tension support
	Reinstall	Ea.		139.52		139.52	rafters and 9" valance.
	Clean	Ea.	2.49	63.50		65.99	
	Minimum Charge	Job		350		350	

Vent		Unit	Material	Labor	Equip.	Total	Specification
Small Roof Turbine							
	Demolish	Ea.		8.80		8.80	Includes material and labor to install
	Install	Ea.	52.50	35		87.50	spinner ventilator, wind driven,
	Demolish and Install	Ea.	52.50	43.80		96.30	galvanized, 4" neck diam, 180 CFM.
	Reinstall	Ea.		27.86		27.86	
	Clean	Ea.		5.65		5.65	
	Minimum Charge	Job		194		194	
Large Roof Turbine							
	Demolish	Ea.		7.05		7.05	Includes material and labor to install
	Install	Ea.	60.50	50		110.50	spinner ventilator, wind driven,
	Demolish and Install	Ea.	60.50	57.05		117.55	galvanized, 8" neck diam, 360 CFM.
	Reinstall	Ea.		39.80		39.80	
	Clean	Ea.		6.35		6.35	
	Minimum Charge	Job		194		194	

Improvements / Appliances / Treatments

Vent		Unit	Material	Labor	Equip.	Total	Specification
Roof Ventilator							
	Demolish	Ea.		8.80		8.80	Includes material and labor to install a
	Install	Ea.	171	43.50		214.50	roof ventilator.
	Demolish and Install	Ea.	171	52.30		223.30	
	Reinstall	Ea.		34.82		34.82	
	Clean	Ea.	.63	6.35		6.98	
	Paint	Ea.	1.42	7.65		9.07	
	Minimum Charge	Job		194		194	
Gable Louvered							
	Demolish	Ea.		17.65		17.65	Includes material and labor to install a
	Install	Ea.	38.50	11.65		50.15	vinyl gable end vent.
	Demolish and Install	Ea.	38.50	29.30		67.80	
	Reinstall	Ea.		9.30		9.30	
	Clean	Ea.		3.39		3.39	
	Paint	Ea.	.95	4.70		5.65	
	Minimum Charge	Job		174		174	
Ridge Vent							
	Demolish	L.F.		.91		.91	Includes material and labor to install a
	Install	L.F.	2.95	2.50		5.45	mill finish aluminum ridge vent strip.
	Demolish and Install	L.F.	2.95	3.41		6.36	
	Minimum Charge	Job		174		174	
Foundation							
	Demolish	Ea.		6.15		6.15	Includes material and labor to install
	Install	Ea.	23.50	11.80		35.30	galvanized foundation block vent.
	Demolish and Install	Ea.	23.50	17.95		41.45	
	Reinstall	Ea.		9.45		9.45	
	Clean	Ea.	.15	6.35		6.50	
	Minimum Charge	Job		174		174	
Frieze							
	Demolish	L.F.		.63		.63	Includes material and labor to install
	Install	L.F.	.97	.79		1.76	pine frieze board.
	Demolish and Install	L.F.	.97	1.42		2.39	
	Reinstall	L.F.		.63		.63	
	Clean	L.F.	.03	.21		.24	
	Minimum Charge	Job		174		174	
Water Heater Cap							
	Demolish	Ea.		13.50		13.50	Includes material and labor to install a
	Install	Ea.	25.50	48.50		74	galvanized steel water heater cap.
	Demolish and Install	Ea.	25.50	62		87.50	
	Reinstall	Ea.		38.68		38.68	
	Clean	Ea.	1.09	15.90		16.99	

Cupolas		Unit	Material	Labor	Equip.	Total	Specification
22″ x 22″							
	Demolish	Ea.		168		168	Includes material and labor to install a
	Install	Ea.	340	87		427	cedar cupola with aluminum roof
	Demolish and Install	Ea.	340	255		595	covering.
	Reinstall	Ea.		69.76		69.76	
	Minimum Charge	Job		174		174	
29″ x 29″							
	Demolish	Ea.		168		168	Includes material and labor to install a
	Install	Ea.	450	87		537	cedar cupola with aluminum roof
	Demolish and Install	Ea.	450	255		705	covering.
	Reinstall	Ea.		69.76		69.76	
	Minimum Charge	Job		174		174	

Improvements / Appliances / Treatments

Cupolas	Unit	Material	Labor	Equip.	Total	Specification
35" x 35"						
Demolish	Ea.		168		168	Includes material and labor to install a
Install	Ea.	705	116		821	cedar cupola with aluminum roof
Demolish and Install	Ea.	705	284		989	covering.
Reinstall	Ea.		93.01		93.01	
Minimum Charge	Job		174		174	
47" x 47"						
Demolish	Ea.		168		168	Includes material and labor to install a
Install	Ea.	1250	174		1424	cedar cupola with aluminum roof
Demolish and Install	Ea.	1250	342		1592	covering.
Reinstall	Ea.		139.52		139.52	
Minimum Charge	Job		174		174	
Weather Vane						
Install	Ea.	143	43.50		186.50	Includes minimum labor and
Minimum Charge	Job		174		174	equipment to install residential type weathervane.

Aluminum Carport	Unit	Material	Labor	Equip.	Total	Specification
Natural Finish						
Demolish	S.F.		1.76		1.76	Includes material, labor and
Install	S.F.	23.50	5.85	1.02	30.37	equipment to install an aluminum
Demolish and Install	S.F.	23.50	7.61	1.02	32.13	carport. Foundations are not included.
Reinstall	S.F.		4.68	.82	5.50	
Clean	S.F.	.03	.19		.22	
Minimum Charge	Job		350		350	
Enamel Finish						
Demolish	S.F.		1.76		1.76	Includes material, labor and
Install	S.F.	23.50	5.85	1.02	30.37	equipment to install an aluminum
Demolish and Install	S.F.	23.50	7.61	1.02	32.13	carport. Foundations are not included.
Reinstall	S.F.		4.68	.82	5.50	
Clean	S.F.	.03	.19		.22	
Minimum Charge	Job		350		350	
20' x 20' Complete						
Demolish	Ea.		176		176	Includes material, labor and
Install	Ea.	7475	585	102	8162	equipment to install an aluminum
Demolish and Install	Ea.	7475	761	102	8338	carport. Foundations are not included.
Reinstall	Ea.		467.52	81.50	549.02	
Clean	Ea.	.17	63.50		63.67	
Minimum Charge	Job		350		350	
24' x 24' Complete						
Demolish	Ea.		176		176	Includes material, labor and
Install	Ea.	10200	780	136	11116	equipment to install an aluminum
Demolish and Install	Ea.	10200	956	136	11292	carport. Foundations are not included.
Reinstall	Ea.		623.36	108.67	732.03	
Clean	Ea.	.17	127		127.17	
Minimum Charge	Job		350		350	
Metal Support Posts						
Demolish	Ea.		2.35		2.35	Includes material and labor to install
Install	Ea.	16.60	43.50		60.10	metal support posts for a carport.
Demolish and Install	Ea.	16.60	45.85		62.45	
Reinstall	Ea.		34.88		34.88	
Clean	Ea.	.17	2.65		2.82	
Paint	Ea.	.13	6.10		6.23	
Minimum Charge	Job		350		350	

Improvements / Appliances / Treatments

Storage Shed	Unit	Material	Labor	Equip.	Total	Specification
Aluminum						
Pre-fab (8' x 10')						
Demolish	Ea.		51.50		51.50	Includes material and labor to install
Install	Ea.	1850	505		2355	an aluminum storage shed.
Demolish and Install	Ea.	1850	556.50		2406.50	
Clean	Ea.	1.03	32		33.03	
Minimum Charge	Ea.		430		430	
Pre-fab (10' x 12')						
Demolish	Ea.		98.50		98.50	Includes material and labor to install
Install	Ea.	2275	505		2780	an aluminum storage shed.
Demolish and Install	Ea.	2275	603.50		2878.50	
Clean	Ea.	1.03	35.50		36.53	
Minimum Charge	Ea.		430		430	
Wood						
Pre-fab (8' x 10')						
Demolish	Ea.		51.50		51.50	Includes material and labor to install a
Install	Ea.	1350	505		1855	pre-fabricated wood storage shed.
Demolish and Install	Ea.	1350	556.50		1906.50	
Clean	Ea.	1.03	32		33.03	
Minimum Charge	Ea.		430		430	
Pre-fab (10' x 12')						
Demolish	Ea.		98.50		98.50	Includes material and labor to install a
Install	Ea.	1725	505		2230	pre-fabricated wood storage shed.
Demolish and Install	Ea.	1725	603.50		2328.50	
Clean	Ea.	1.03	35.50		36.53	
Minimum Charge	Ea.		430		430	

Gazebo	Unit	Material	Labor	Equip.	Total	Specification
Good Grade						
Demolish	S.F.		2.82		2.82	Includes material and labor to install a
Install	S.F.	22.50	26		48.50	gazebo.
Demolish and Install	S.F.	22.50	28.82		51.32	
Reinstall	S.F.		20.93		20.93	
Clean	S.F.	.12	1.69		1.81	
Minimum Charge	Job		174		174	
Custom Grade						
Demolish	S.F.		2.82		2.82	Includes material and labor to install a
Install	S.F.	33.50	26		59.50	gazebo.
Demolish and Install	S.F.	33.50	28.82		62.32	
Reinstall	S.F.		20.93		20.93	
Clean	S.F.	.12	1.69		1.81	
Minimum Charge	Job		174		174	
Bench Seating						
Demolish	L.F.		4.70		4.70	Includes material and labor to install
Install	L.F.	2.41	17.45		19.86	pressure treated bench seating.
Demolish and Install	L.F.	2.41	22.15		24.56	
Reinstall	L.F.		13.95		13.95	
Clean	L.F.	.12	1.27		1.39	
Paint	L.F.	.56	2.44		3	
Minimum Charge	Job		174		174	
Re-screen						
Install	S.F.	1.16	1.40		2.56	Includes labor and material to
Minimum Charge	Job		174		174	re-screen wood frame.

Pool	Unit	Material	Labor	Equip.	Total	Specification
Screen Enclosure						
Re-screen						
Install	SF Wall		4.32		4.32	Includes material and labor to install a
Minimum Charge	Job		174		174	screen enclosure for a pool.
Screen Door						
Demolish	Ea.		6.90		6.90	Includes material and labor to install
Install	Ea.	144	58		202	wood screen door with aluminum cloth
Demolish and Install	Ea.	144	64.90		208.90	screen. Frame and hardware not
Reinstall	Ea.		46.51		46.51	included.
Clean	Ea.	.12	9.10		9.22	
Paint	Ea.	3.10	12.20		15.30	
Minimum Charge	Job		174		174	
Heater / Motor						
Demolish	Ea.		34		34	Includes material and labor to install
Install	Ea.	2425	435		2860	swimming pool heater, not incl. base
Demolish and Install	Ea.	2425	469		2894	or pad, elec., 12 kW, 4,800 gal
Reinstall	Ea.		348.83		348.83	pool, incl. pump.
Clean	Ea.	3.11	32		35.11	
Minimum Charge	Job		196		196	
Filter System						
Demolish	Ea.		56.50		56.50	Includes material, labor and
Install	Total	720	310		1030	equipment to install sand filter system
Demolish and Install	Total	720	366.50		1086.50	tank including hook-up to existing
Reinstall	Total		248.83		248.83	equipment. Pump and motor not
Minimum Charge	Job		194		194	included.
Pump / Motor						
Demolish	Ea.		34		34	Includes material and labor to install a
Install	Ea.	400	78		478	22 GPM pump.
Demolish and Install	Ea.	400	112		512	
Reinstall	Ea.		62.21		62.21	
Clean	Ea.	3.11	32		35.11	
Minimum Charge	Job		196		196	
Pump Water						
Install	Day		350	77	427	Includes labor and equipment to pump
Minimum Charge	Job		160		160	water from a swimming pool.
Replaster						
Install	SF Surf	1.12	8.30		9.42	Includes labor and material to
Minimum Charge	Job		160		160	replaster gunite pool with a surface area of 500 to 600 S.F.
Solar Panels						
Demolish	S.F.		1.44		1.44	Includes material and labor to install
Install	S.F.	23	2.56		25.56	solar panel.
Demolish and Install	S.F.	23	4		27	
Reinstall	S.F.		2.05		2.05	
Minimum Charge	Ea.		430		430	

Construction Clean-up	Unit	Material	Labor	Equip.	Total	Specification
Final						
Clean	S.F.		.22		.22	Includes labor for final site/job
Minimum Charge	Job		127		127	clean-up including detail cleaning.

Location Factors

Costs shown in *RSMeans cost data publications* are based on National Averages for materials and installation. To adjust these costs to a specific location, simply multiply the base cost by the factor for that city. The data is arranged alphabetically by state and postal zip code numbers. For a city not listed, use the factor for a nearby city with similar economic characteristics.

STATE	CITY	Residential
ALABAMA		
350-352	Birmingham	.86
354	Tuscaloosa	.73
355	Jasper	.71
356	Decatur	.76
357-358	Huntsville	.84
359	Gadsden	.73
360-361	Montgomery	.75
362	Anniston	.68
363	Dothan	.74
364	Evergreen	.70
365-366	Mobile	.79
367	Selma	.72
368	Phenix City	.73
369	Butler	.71
ALASKA		
995-996	Anchorage	1.27
997	Fairbanks	1.29
998	Juneau	1.27
999	Ketchikan	1.29
ARIZONA		
850,853	Phoenix	.86
852	Mesa/Tempe	.83
855	Globe	.79
856-857	Tucson	.84
859	Show Low	.81
860	Flagstaff	.86
863	Prescott	.81
864	Kingman	.83
865	Chambers	.80
ARKANSAS		
716	Pine Bluff	.81
717	Camden	.70
718	Texarkana	.75
719	Hot Springs	.70
720-722	Little Rock	.87
723	West Memphis	.81
724	Jonesboro	.79
725	Batesville	.76
726	Harrison	.78
727	Fayetteville	.72
728	Russellville	.77
729	Fort Smith	.79
CALIFORNIA		
900-902	Los Angeles	1.06
903-905	Inglewood	1.05
906-908	Long Beach	1.04
910-912	Pasadena	1.05
913-916	Van Nuys	1.08
917-918	Alhambra	1.09
919-921	San Diego	1.04
922	Palm Springs	1.04
923-924	San Bernardino	1.05
925	Riverside	1.05
926-927	Santa Ana	1.06
928	Anaheim	1.05
930	Oxnard	1.07
931	Santa Barbara	1.06
932-933	Bakersfield	1.03
934	San Luis Obispo	1.08
935	Mojave	1.06
936-938	Fresno	1.09
939	Salinas	1.12
940-941	San Francisco	1.23
942,956-958	Sacramento	1.11
943	Palo Alto	1.18
944	San Mateo	1.22
945	Vallejo	1.15
946	Oakland	1.21
947	Berkeley	1.24
948	Richmond	1.24
949	San Rafael	1.22
950	Santa Cruz	1.15
951	San Jose	1.19
952	Stockton	1.09
953	Modesto	1.08
954	Santa Rosa	1.16

STATE	CITY	Residential
955	Eureka	1.12
959	Marysville	1.10
960	Redding	1.10
961	Susanville	1.09
COLORADO		
800-802	Denver	.94
803	Boulder	.93
804	Golden	.91
805	Fort Collins	.90
806	Greeley	.80
807	Fort Morgan	.93
808-809	Colorado Springs	.90
810	Pueblo	.91
811	Alamosa	.88
812	Salida	.90
813	Durango	.91
814	Montrose	.87
815	Grand Junction	.92
816	Glenwood Springs	.90
CONNECTICUT		
060	New Britain	1.08
061	Hartford	1.08
062	Willimantic	1.08
063	New London	1.08
064	Meriden	1.08
065	New Haven	1.08
066	Bridgeport	1.09
067	Waterbury	1.09
068	Norwalk	1.09
069	Stamford	1.10
D.C.		
200-205	Washington	.95
DELAWARE		
197	Newark	.99
198	Wilmington	1.00
199	Dover	.99
FLORIDA		
320,322	Jacksonville	.77
321	Daytona Beach	.84
323	Tallahassee	.73
324	Panama City	.67
325	Pensacola	.78
326,344	Gainesville	.77
327-328,347	Orlando	.84
329	Melbourne	.86
330-332,340	Miami	.85
333	Fort Lauderdale	.84
334,349	West Palm Beach	.84
335-336,346	Tampa	.86
337	St. Petersburg	.76
338	Lakeland	.83
339,341	Fort Myers	.80
342	Sarasota	.84
GEORGIA		
300-303,399	Atlanta	.90
304	Statesboro	.71
305	Gainesville	.79
306	Athens	.79
307	Dalton	.75
308-309	Augusta	.81
310-312	Macon	.82
313-314	Savannah	.82
315	Waycross	.76
316	Valdosta	.73
317,398	Albany	.79
318-319	Columbus	.83
HAWAII		
967	Hilo	1.21
968	Honolulu	1.23
STATES & POSS.		
969	Guam	.89

STATE	CITY	Residential
IDAHO		
832	Pocatello	.86
833	Twin Falls	.74
834	Idaho Falls	.75
835	Lewiston	.96
836-837	Boise	.87
838	Coeur d'Alene	.94
ILLINOIS		
600-603	North Suburban	1.10
604	Joliet	1.10
605	South Suburban	1.10
606-608	Chicago	1.16
609	Kankakee	1.00
610-611	Rockford	1.04
612	Rock Island	.96
613	La Salle	1.02
614	Galesburg	.99
615-616	Peoria	.98
617	Bloomington	.98
618-619	Champaign	.99
620-622	East St. Louis	1.00
623	Quincy	.98
624	Effingham	.96
625	Decatur	.97
626-627	Springfield	.97
628	Centralia	1.00
629	Carbondale	.95
INDIANA		
460	Anderson	.91
461-462	Indianapolis	.95
463-464	Gary	1.01
465-466	South Bend	.91
467-468	Fort Wayne	.91
469	Kokomo	.92
470	Lawrenceburg	.87
471	New Albany	.86
472	Columbus	.92
473	Muncie	.91
474	Bloomington	.94
475	Washington	.91
476-477	Evansville	.90
478	Terre Haute	.90
479	Lafayette	.91
IOWA		
500-503,509	Des Moines	.91
504	Mason City	.77
505	Fort Dodge	.76
506-507	Waterloo	.79
508	Creston	.81
510-511	Sioux City	.87
512	Sibley	.73
513	Spencer	.74
514	Carroll	.74
515	Council Bluffs	.81
516	Shenandoah	.75
520	Dubuque	.86
521	Decorah	.76
522-524	Cedar Rapids	.94
525	Ottumwa	.84
526	Burlington	.87
527-528	Davenport	.97
KANSAS		
660-662	Kansas City	.99
664-666	Topeka	.80
667	Fort Scott	.85
668	Emporia	.72
669	Belleville	.78
670-672	Wichita	.80
673	Independence	.85
674	Salina	.76
675	Hutchinson	.77
676	Hays	.82
677	Colby	.83
678	Dodge City	.82
679	Liberal	.79
KENTUCKY		
400-402	Louisville	.92
403-405	Lexington	.89
406	Frankfort	.89
407-409	Corbin	.78
410	Covington	1.00
411-412	Ashland	.98

STATE	CITY	Residential
413-414	Campton	.79
415-416	Pikeville	.86
417-418	Hazard	.73
420	Paducah	.90
421-422	Bowling Green	.90
423	Owensboro	.89
424	Henderson	.91
425-426	Somerset	.78
427	Elizabethtown	.87
LOUISIANA		
700-701	New Orleans	.86
703	Thibodaux	.84
704	Hammond	.79
705	Lafayette	.82
706	Lake Charles	.83
707-708	Baton Rouge	.82
710-711	Shreveport	.79
712	Monroe	.74
713-714	Alexandria	.74
MAINE		
039	Kittery	.79
040-041	Portland	.90
042	Lewiston	.89
043	Augusta	.82
044	Bangor	.88
045	Bath	.80
046	Machias	.81
047	Houlton	.85
048	Rockland	.81
049	Waterville	.80
MARYLAND		
206	Waldorf	.85
207-208	College Park	.87
209	Silver Spring	.86
210-212	Baltimore	.90
214	Annapolis	.85
215	Cumberland	.86
216	Easton	.68
217	Hagerstown	.86
218	Salisbury	.75
219	Elkton	.81
MASSACHUSETTS		
010-011	Springfield	1.04
012	Pittsfield	1.01
013	Greenfield	1.00
014	Fitchburg	1.12
015-016	Worcester	1.14
017	Framingham	1.12
018	Lowell	1.13
019	Lawrence	1.13
020-022, 024	Boston	1.19
023	Brockton	1.12
025	Buzzards Bay	1.10
026	Hyannis	1.09
027	New Bedford	1.12
MICHIGAN		
480,483	Royal Oak	1.03
481	Ann Arbor	1.05
482	Detroit	1.07
484-485	Flint	.97
486	Saginaw	.94
487	Bay City	.95
488-489	Lansing	.97
490	Battle Creek	.93
491	Kalamazoo	.92
492	Jackson	.95
493,495	Grand Rapids	.82
494	Muskegon	.89
496	Traverse City	.80
497	Gaylord	.83
498-499	Iron Mountain	.90
MINNESOTA		
550-551	Saint Paul	1.13
553-555	Minneapolis	1.17
556-558	Duluth	1.09
559	Rochester	1.05
560	Mankato	1.03
561	Windom	.83
562	Willmar	.85
563	St. Cloud	1.07
564	Brainerd	.98

Location Factors

STATE	CITY	Residential
565	Detroit Lakes	.96
566	Bemidji	.96
567	Thief River Falls	.95
MISSISSIPPI		
386	Clarksdale	.62
387	Greenville	.69
388	Tupelo	.64
389	Greenwood	.65
390-392	Jackson	.73
393	Meridian	.66
394	Laurel	.63
395	Biloxi	.75
396	McComb	.74
397	Columbus	.65
MISSOURI		
630-631	St. Louis	1.03
633	Bowling Green	.94
634	Hannibal	.87
635	Kirksville	.80
636	Flat River	.94
637	Cape Girardeau	.87
638	Sikeston	.82
639	Poplar Bluff	.82
640-641	Kansas City	1.03
644-645	St. Joseph	.95
646	Chillicothe	.84
647	Harrisonville	.94
648	Joplin	.85
650-651	Jefferson City	.88
652	Columbia	.88
653	Sedalia	.85
654-655	Rolla	.88
656-658	Springfield	.86
MONTANA		
590-591	Billings	.87
592	Wolf Point	.83
593	Miles City	.85
594	Great Falls	.88
595	Havre	.80
596	Helena	.87
597	Butte	.83
598	Missoula	.83
599	Kalispell	.81
NEBRASKA		
680-681	Omaha	.89
683-685	Lincoln	.78
686	Columbus	.70
687	Norfolk	.77
688	Grand Island	.77
689	Hastings	.76
690	Mccook	.70
691	North Platte	.75
692	Valentine	.66
693	Alliance	.66
NEVADA		
889-891	Las Vegas	1.00
893	Ely	.87
894-895	Reno	.94
897	Carson City	.95
898	Elko	.93
NEW HAMPSHIRE		
030	Nashua	.95
031	Manchester	.95
032-033	Concord	.92
034	Keene	.72
035	Littleton	.80
036	Charleston	.70
037	Claremont	.71
038	Portsmouth	.89
NEW JERSEY		
070-071	Newark	1.13
072	Elizabeth	1.15
073	Jersey City	1.11
074-075	Paterson	1.12
076	Hackensack	1.11
077	Long Branch	1.13
078	Dover	1.12
079	Summit	1.12
080,083	Vineland	1.09
081	Camden	1.10

STATE	CITY	Residential
082,084	Atlantic City	1.13
085-086	Trenton	1.11
087	Point Pleasant	1.10
088-089	New Brunswick	1.12
NEW MEXICO		
870-872	Albuquerque	.85
873	Gallup	.85
874	Farmington	.85
875	Santa Fe	.85
877	Las Vegas	.85
878	Socorro	.85
879	Truth/Consequences	.84
880	Las Cruces	.82
881	Clovis	.84
882	Roswell	.85
883	Carrizozo	.85
884	Tucumcari	.85
NEW YORK		
100-102	New York	1.34
103	Staten Island	1.25
104	Bronx	1.27
105	Mount Vernon	1.15
106	White Plains	1.18
107	Yonkers	1.19
108	New Rochelle	1.19
109	Suffern	1.12
110	Queens	1.26
111	Long Island City	1.29
112	Brooklyn	1.31
113	Flushing	1.28
114	Jamaica	1.28
115,117,118	Hicksville	1.19
116	Far Rockaway	1.27
119	Riverhead	1.21
120-122	Albany	.95
123	Schenectady	.96
124	Kingston	1.03
125-126	Poughkeepsie	1.04
127	Monticello	1.05
128	Glens Falls	.89
129	Plattsburgh	.93
130-132	Syracuse	.96
133-135	Utica	.93
136	Watertown	.90
137-139	Binghamton	.95
140-142	Buffalo	1.05
143	Niagara Falls	1.01
144-146	Rochester	.99
147	Jamestown	.89
148-149	Elmira	.86
NORTH CAROLINA		
270,272-274	Greensboro	.85
271	Winston-Salem	.85
275-276	Raleigh	.87
277	Durham	.85
278	Rocky Mount	.75
279	Elizabeth City	.75
280	Gastonia	.86
281-282	Charlotte	.88
283	Fayetteville	.84
284	Wilmington	.83
285	Kinston	.75
286	Hickory	.80
287-288	Asheville	.83
289	Murphy	.74
NORTH DAKOTA		
580-581	Fargo	.79
582	Grand Forks	.76
583	Devils Lake	.79
584	Jamestown	.74
585	Bismarck	.79
586	Dickinson	.77
587	Minot	.80
588	Williston	.77
OHIO		
430-432	Columbus	.94
433	Marion	.91
434-436	Toledo	.99
437-438	Zanesville	.90
439	Steubenville	.95
440	Lorain	.99
441	Cleveland	1.02

Location Factors

STATE	CITY	Residential
442-443	Akron	.98
444-445	Youngstown	.96
446-447	Canton	.94
448-449	Mansfield	.95
450	Hamilton	.93
451-452	Cincinnati	.93
453-454	Dayton	.91
455	Springfield	.93
456	Chillicothe	.96
457	Athens	.88
458	Lima	.90
OKLAHOMA		
730-731	Oklahoma City	.80
734	Ardmore	.79
735	Lawton	.82
736	Clinton	.78
737	Enid	.78
738	Woodward	.77
739	Guymon	.68
740-741	Tulsa	.79
743	Miami	.82
744	Muskogee	.73
745	Mcalester	.74
746	Ponca City	.78
747	Durant	.77
748	Shawnee	.76
749	Poteau	.78
OREGON		
970-972	Portland	1.02
973	Salem	1.01
974	Eugene	1.01
975	Medford	.99
976	Klamath Falls	1.00
977	Bend	1.02
978	Pendleton	.99
979	Vale	.98
PENNSYLVANIA		
150-152	Pittsburgh	.99
153	Washington	.94
154	Uniontown	.90
155	Bedford	.88
156	Greensburg	.94
157	Indiana	.91
158	Dubois	.90
159	Johnstown	.89
160	Butler	.92
161	New Castle	.92
162	Kittanning	.93
163	Oil City	.90
164-165	Erie	.95
166	Altoona	.88
167	Bradford	.90
168	State College	.91
169	Wellsboro	.90
170-171	Harrisburg	.94
172	Chambersburg	.89
173-174	York	.91
175-176	Lancaster	.91
177	Williamsport	.84
178	Sunbury	.91
179	Pottsville	.91
180	Lehigh Valley	1.01
181	Allentown	1.04
182	Hazleton	.91
183	Stroudsburg	.92
184-185	Scranton	.96
186-187	Wilkes-Barre	.93
188	Montrose	.90
189	Doylestown	1.04
190-191	Philadelphia	1.15
193	Westchester	1.09
194	Norristown	1.08
195-196	Reading	.97
PUERTO RICO		
009	San Juan	.74
RHODE ISLAND		
028	Newport	1.06
029	Providence	1.07
SOUTH CAROLINA		
290-292	Columbia	.82
293	Spartanburg	.81

STATE	CITY	Residential
294	Charleston	.82
295	Florence	.76
296	Greenville	.80
297	Rock Hill	.72
298	Aiken	.97
299	Beaufort	.75
SOUTH DAKOTA		
570-571	Sioux Falls	.76
572	Watertown	.72
573	Mitchell	.74
574	Aberdeen	.75
575	Pierre	.75
576	Mobridge	.73
577	Rapid City	.75
TENNESSEE		
370-372	Nashville	.85
373-374	Chattanooga	.77
375,380-381	Memphis	.83
376	Johnson City	.72
377-379	Knoxville	.75
382	Mckenzie	.73
383	Jackson	.71
384	Columbia	.73
385	Cookeville	.72
TEXAS		
750	Mckinney	.75
751	Waxahackie	.75
752-753	Dallas	.82
754	Greenville	.69
755	Texarkana	.73
756	Longview	.67
757	Tyler	.74
758	Palestine	.66
759	Lufkin	.71
760-761	Fort Worth	.82
762	Denton	.77
763	Wichita Falls	.79
764	Eastland	.73
765	Temple	.75
766-767	Waco	.78
768	Brownwood	.69
769	San Angelo	.72
770-772	Houston	.86
773	Huntsville	.70
774	Wharton	.71
775	Galveston	.84
776-777	Beaumont	.83
778	Bryan	.74
779	Victoria	.75
780	Laredo	.73
781-782	San Antonio	.81
783-784	Corpus Christi	.78
785	Mc Allen	.76
786-787	Austin	.80
788	Del Rio	.67
789	Giddings	.70
790-791	Amarillo	.78
792	Childress	.76
793-794	Lubbock	.76
795-796	Abilene	.75
797	Midland	.76
798-799,885	El Paso	.75
UTAH		
840-841	Salt Lake City	.81
842,844	Ogden	.79
843	Logan	.80
845	Price	.72
846-847	Provo	.81
VERMONT		
050	White River Jct.	.73
051	Bellows Falls	.75
052	Bennington	.82
053	Brattleboro	.78
054	Burlington	.80
056	Montpelier	.81
057	Rutland	.80
058	St. Johnsbury	.80
059	Guildhall	.79
VIRGINIA		
220-221	Fairfax	1.02
222	Arlington	1.04

STATE	CITY	Residential
223	Alexandria	1.06
224-225	Fredericksburg	.95
226	Winchester	.93
227	Culpeper	1.00
228	Harrisonburg	.90
229	Charlottesville	.92
230-232	Richmond	1.01
233-235	Norfolk	1.02
236	Newport News	1.01
237	Portsmouth	.92
238	Petersburg	.99
239	Farmville	.91
240-241	Roanoke	.99
242	Bristol	.86
243	Pulaski	.84
244	Staunton	.93
245	Lynchburg	.97
246	Grundy	.85
WASHINGTON		
980-981,987	Seattle	1.02
982	Everett	1.05
983-984	Tacoma	1.01
985	Olympia	1.01
986	Vancouver	.98
988	Wenatchee	.93
989	Yakima	.97
990-992	Spokane	.99
993	Richland	.97
994	Clarkston	.97
WEST VIRGINIA		
247-248	Bluefield	.88
249	Lewisburg	.89
250-253	Charleston	.97
254	Martinsburg	.86
255-257	Huntington	1.01
258-259	Beckley	.90
260	Wheeling	.93
261	Parkersburg	.92
262	Buckhannon	.92
263-264	Clarksburg	.92
265	Morgantown	.93
266	Gassaway	.92
267	Romney	.88
268	Petersburg	.90
WISCONSIN		
530,532	Milwaukee	1.07
531	Kenosha	1.04
534	Racine	1.02
535	Beloit	1.00
537	Madison	.99
538	Lancaster	.97
539	Portage	.96
540	New Richmond	1.00
541-543	Green Bay	1.01
544	Wausau	.95
545	Rhinelander	.95
546	La Crosse	.94
547	Eau Claire	.98
548	Superior	.99
549	Oshkosh	.95
WYOMING		
820	Cheyenne	.84
821	Yellowstone Nat. Pk.	.75
822	Wheatland	.75
823	Rawlins	.76
824	Worland	.75
825	Riverton	.74
826	Casper	.78
827	Newcastle	.74
828	Sheridan	.80
829-831	Rock Springs	.79

STATE	CITY	Residential
CANADIAN FACTORS (reflect Canadian Currency)		
ALBERTA		
	Calgary	1.14
	Edmonton	1.13
	Fort McMurray	1.09
	Lethbridge	1.10
	Lloydminster	1.09
	Medicine Hat	1.10
	Red Deer	1.10
BRITISH COLUMBIA		
	Kamloops	1.08
	Prince George	1.08
	Vancouver	1.09
	Victoria	1.03
MANITOBA		
	Brandon	1.06
	Portage la Prairie	1.06
	Winnipeg	1.05
NEW BRUNSWICK		
	Bathurst	.97
	Dalhousie	.97
	Fredericton	1.05
	Moncton	.98
	Newcastle	.97
	Saint John	1.05
NEWFOUNDLAND		
	Corner Brook	.99
	St. John's	1.01
NORTHWEST TERRITORIES		
	Yellowknife	1.10
NOVA SCOTIA		
	Dartmouth	1.00
	Halifax	1.02
	New Glasgow	1.00
	Sydney	.99
	Yarmouth	1.00
ONTARIO		
	Barrie	1.17
	Brantford	1.19
	Cornwall	1.19
	Hamilton	1.19
	Kingston	1.19
	Kitchener	1.11
	London	1.17
	North Bay	1.15
	Oshawa	1.17
	Ottawa	1.19
	Owen Sound	1.15
	Peterborough	1.16
	Sarnia	1.19
	Sudbury	1.09
	Thunder Bay	1.15
	Toronto	1.20
	Windsor	1.14
PRINCE EDWARD ISLAND		
	Charlottetown	.95
	Summerside	.94
QUEBEC		
	Cap-de-la-Madeleine	1.18
	Charlesbourg	1.18
	Chicoutimi	1.20
	Gatineau	1.16
	Laval	1.17
	Montreal	1.21
	Quebec	1.22
	Sherbrooke	1.17
	Trois Rivieres	1.18
SASKATCHEWAN		
	Moose Jaw	.97
	Prince Albert	.96
	Regina	.99
	Saskatoon	.97
YUKON		
	Whitehorse	.96

Abbreviations

A	Area
ASTM	American Society for Testing and Materials
B.F.	Board feet
Carp.	Carpenter
C.F.	Cubic feet
CWJ	Composite wood joist
C.Y.	Cubic yard
Ea.	Each
Equip.	Equipment
Exp.	Exposure
Ext.	Exterior
F	Fahrenheit
Ft.	Foot, feet
Gal.	Gallon
Hr.	Hour
in.	Inch, inches
Inst.	Installation
Int.	Interior
Lb.	Pound
L.F.	Linear feet
LVL	Laminated veneer lumber
Mat.	Material
Max.	Maximum
MBF	Thousand board feet
MBM	Thousand feet board measure
MSF	Thousand square feet
Min.	Minimum
O.C.	On center
O&P	Overhead and profit
OWJ	Open web wood joist
Oz.	Ounce
Pr.	Pair
Quan.	Quantity
S.F.	Square foot
Sq.	Square, 100 square feet
S.Y.	Square yard
V.L.F.	Vertical linear feet
'	Foot, feet
"	Inch, inches
°	Degrees

Contractor's Pricing Guides

For more information
visit RSMeans Web site
at www.rsmeans.com

Contractor's Pricing Guide:
Residential Detailed
Costs 2007

Every aspect of residential construction, from overhead costs to residential lighting and wiring, is in here. All the detail you need to accurately estimate the costs of your work with or without markups—labor-hours, typical crews, and equipment are included as well. When you need a detailed estimate, this publication has all the costs to help you come up with a complete, on the money, price you can rely on to win profitable work.

Unit Prices Now Updated to MasterFormat 2004!

$39.95 per copy
Available Nov. 2006
Catalog No. 60337

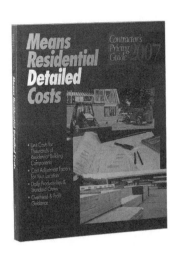

Contractor's Pricing Guide:
Residential Repair &
Remodeling Costs 2007

This book provides total unit price costs for every aspect of the most common repair and remodeling projects. Organized in the order of construction by component and activity, it includes demolition and installation, cleaning, painting, and more.

With simplified estimating methods, clear, concise descriptions, and technical specifications for each component, the book is a valuable tool for contractors who want to speed up their estimating time, while making sure their costs are on target.

$39.95 per copy
Available Nov. 2006
Catalog No. 60347

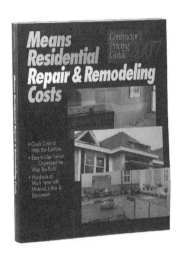

Contractor's Pricing Guide:
Residential Square Foot
Costs 2007

Now available in one concise volume, all you need to know to plan and budget the cost of new homes. If you are looking for a quick reference, the model home section contains costs for over 250 different sizes and types of residences, with hundreds of easily applied modifications. If you need even more detail, the Assemblies Section lets you build your own costs or modify the model costs further. Hundreds of graphics are provided, along with forms and procedures to help you get it right.

$39.95 per copy
Available Nov. 2006
Catalog No. 60327

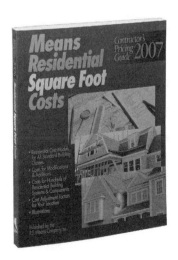

Annual Cost Guides

For more information
visit RSMeans Web site
at www.rsmeans.com

RSMeans Building Construction Cost Data 2007

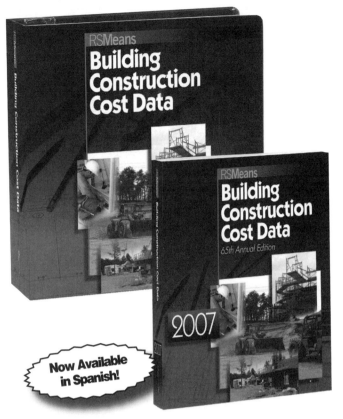

Available in Both Softbound and Looseleaf Editions

Many customers enjoy the convenience and flexibility of the looseleaf binder, which increases the usefulness of *RSMeans Building Construction Cost Data 2007* by making it easy to add and remove pages. You can insert your own cost information pages, so everything is in one place. Copying pages for faxing is easier also. Whichever edition you prefer, softbound or the convenient looseleaf edition, you'll be eligible to receive *RSMeans Quarterly Update Service* FREE. Current subscribers can receive *RSMeans Quarterly Update Service* via e-mail.

Now Available in Spanish!

Unit Prices Now Updated to MasterFormat 2004!

$169.95 per copy, Looseleaf
Available Oct. 2006
Catalog No. 61017

RSMeans Building Construction Cost Data 2007

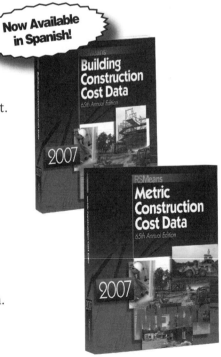

Now Available in Spanish!

Offers you unchallenged unit price reliability in an easy-to-use arrangement. Whether used for complete, finished estimates or for periodic checks, it supplies more cost facts better and faster than any comparable source. Over 20,000 unit prices for 2007. The City Cost Indexes and Location Factors cover over 930 areas, for indexing to any project location in North America. Order and get *RSMeans Quarterly Update Service* FREE. You'll have year-long access to the RSMeans Estimating **HOTLINE** FREE with your subscription. Expert assistance when using RSMeans data is just a phone call away.

$136.95 per copy (English or Spanish)
Catalog No. 60017 (English) Available Oct. 2006
Catalog No. 60717 (Spanish) Available Jan. 2007

Unit Prices Now Updated to MasterFormat 2004!

RSMeans Metric Construction Cost Data 2007

A massive compendium of all the data from both the 2007 *RSMeans Building Construction Cost Data* AND *Heavy Construction Cost Data*, in **metric** format! Access all of this vital information from one complete source. It contains more than 600 pages of unit costs and 40 pages of assemblies costs. The Reference Section contains over 200 pages of tables, charts and other estimating aids. A great way to stay in step with today's construction trends and rapidly changing costs.

$162.95 per copy
Available Dec. 2006
Catalog No. 63017

For more information
visit RSMeans Web site
at www.rsmeans.com

Annual Cost Guides

RSMeans Mechanical Cost Data 2007

- **HVAC**
- **Controls**

Total unit and systems price guidance for mechanical construction... materials, parts, fittings, and complete labor cost information. Includes prices for piping, heating, air conditioning, ventilation, and all related construction.

Plus new 2007 unit costs for:

- Over 2500 installed HVAC/controls assemblies
- "On Site" Location Factors for over 930 cities and towns in the U.S. and Canada
- Crews, labor, and equipment

$136.95 per copy
Available Oct. 2006
Catalog No. 60027

Unit Prices Now Updated to MasterFormat 2004!

RSMeans Plumbing Cost Data 2007

Comprehensive unit prices and assemblies for plumbing, irrigation systems, commercial and residential fire protection, point-of-use water heaters, and the latest approved materials. This publication and its companion, *RSMeans Mechanical Cost Data*, provide full-range cost estimating coverage for all the mechanical trades.

New for '07: More lines of no-hub CI soil pipe fittings, more flange-type escutcheons, fiberglass pipe insulation in a full range of sizes for 2-1/2" and 3" wall thicknesses, 220 lines of grease duct, and much more.

$136.95 per copy
Available Oct. 2006
Catalog No. 60217

RSMeans Electrical Cost Data 2007

Pricing information for every part of electrical cost planning. More than 13,000 unit and systems costs with design tables; clear specifications and drawings; engineering guides; illustrated estimating procedures; complete labor-hour and materials costs for better scheduling and procurement; and the latest electrical products and construction methods.

- A variety of special electrical systems including cathodic protection
- Costs for maintenance, demolition, HVAC/ mechanical, specialties, equipment, and more

$136.95 per copy
Available Oct. 2006
Catalog No. 60037

Unit Prices Now Updated to MasterFormat 2004!

RSMeans Electrical Change Order Cost Data 2007

RSMeans Electrical Change Order Cost Data 2007 provides you with electrical unit prices exclusively for pricing change orders—based on the recent, direct experience of contractors and suppliers. Analyze and check your own change order estimates against the experience others have had doing the same work. It also covers productivity analysis and change order cost justifications. With useful information for calculating the effects of change orders and dealing with their administration.

$136.95 per copy
Available Nov. 2006
Catalog No. 60237

RSMeans Square Foot Costs 2007

It's Accurate and Easy To Use!

- **Updated 2007 price information**, based on nationwide figures from suppliers, estimators, labor experts, and contractors
- "How-to-Use" sections, with **clear examples** of commercial, residential, industrial, and institutional structures
- Realistic graphics, offering true-to-life illustrations of building projects
- Extensive information on using square foot cost data, including sample estimates and alternate pricing methods

$148.95 per copy
Over 450 pages, illustrated, available Nov. 2006
Catalog No. 60057

RSMeans Repair & Remodeling Cost Data 2007

Commercial/Residential

Use this valuable tool to estimate commercial and residential renovation and remodeling.

Includes: **New costs for hundreds of unique methods, materials, and conditions that only come up in repair and remodeling, PLUS:**

- Unit costs for over 15,000 construction components
- Installed costs for over 90 assemblies
- Over 930 "On-Site" localization factors for the U.S. and Canada.

Unit Prices Now Updated to MasterFormat 2004!
$116.95 per copy
Available Nov. 2006
Catalog No. 60047

Annual Cost Guides

For more information
visit RSMeans Web site
at www.rsmeans.com

RSMeans Facilities Construction Cost Data 2007

For the maintenance and construction of commercial, industrial, municipal, and institutional properties. Costs are shown for new and remodeling construction and are broken down into materials, labor, equipment, overhead, and profit. Special emphasis is given to sections on mechanical, electrical, furnishings, site work, building maintenance, finish work, and demolition.

More than 43,000 unit costs, plus assemblies costs and a comprehensive Reference Section are included.

$323.95 per copy
Available Dec. 2006
Catalog No. 60207

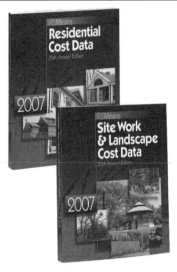

**Unit Prices Now Updated to
MasterFormat 2004!**

RSMeans Light Commercial Cost Data 2007

Specifically addresses the light commercial market, which is a specialized niche in the construction industry. Aids you, the owner/designer/contractor, in preparing all types of estimates—from budgets to detailed bids. Includes new advances in methods and materials.

Assemblies Section allows you to evaluate alternatives in the early stages of design/planning.

Over 11,000 unit costs for 2007 ensure you have the prices you need... when you need them.

$116.95 per copy
Available Dec. 2006
Catalog No. 60187

RSMeans Residential Cost Data 2007

Contains square foot costs for 30 basic home models with the look of today, plus hundreds of custom additions and modifications you can quote right off the page. With costs for the 100 residential systems you're most likely to use in the year ahead. Complete with blank estimating forms, sample estimates, and step-by-step instructions.

Now contains line items for cultured stone and brick, PVC trim lumber, and TPO roofing.

$116.95 per copy
Available Oct. 2006
Catalog No. 60177

**Unit Prices Now Updated to
MasterFormat 2004!**

RSMeans Site Work & Landscape Cost Data 2007

Includes unit and assemblies costs for earthwork, sewerage, piped utilities, site improvements, drainage, paving, trees & shrubs, street openings/repairs, underground tanks, and more. Contains 57 tables of Assemblies Costs for accurate conceptual estimates.

2007 update includes:
- Estimating for infrastructure improvements
- Environmentally-oriented construction
- ADA-mandated handicapped access
- Hazardous waste line items

$136.95 per copy
Available Nov. 2006
Catalog No. 60287

RSMeans Assemblies Cost Data 2007

RSMeans Assemblies Cost Data 2007 takes the guesswork out of preliminary or conceptual estimates. Now you don't have to try to calculate the assembled cost by working up individual component costs. We've done all the work for you.

Presents detailed illustrations, descriptions, specifications, and costs for every conceivable building assembly—240 types in all—arranged in the easy-to-use UNIFORMAT II system. Each illustrated "assembled" cost includes a complete grouping of materials and associated installation costs, including the installing contractor's overhead and profit.

$223.95 per copy
Available Oct. 2006
Catalog No. 60067

RSMeans Open Shop Building Construction Cost Data 2007

The latest costs for accurate budgeting and estimating of new commercial and residential construction... renovation work... change orders... cost engineering.

RSMeans Open Shop "BCCD" will assist you to:
- Develop benchmark prices for change orders
- Plug gaps in preliminary estimates and budgets
- Estimate complex projects
- Substantiate invoices on contracts
- Price ADA-related renovations

Unit Prices Now Updated to MasterFormat 2004!

$136.95 per copy
Available Dec. 2006
Catalog No. 60157

or more information
isit RSMeans Web site
t www.rsmeans.com

Annual Cost Guides

RSMeans Building Construction Cost Data 2007

Western Edition

This regional edition provides more precise cost information for western North America. Labor rates are based on union rates from 13 western states and western Canada. Included are western practices and materials not found in our national edition: tilt-up concrete walls, glu-lam structural systems, specialized timber construction, seismic restraints, and landscape and irrigation systems.

136.95 per copy
Available Dec. 2006
Catalog No. 60227

Unit Prices Now Updated to MasterFormat 2004!

RSMeans Heavy Construction Cost Data 2007

A comprehensive guide to heavy construction costs. Includes costs for highly specialized projects such as tunnels, dams, highways, airports, and waterways. Information on labor rates, equipment, and material costs is included. Features unit price costs, systems costs, and numerous reference tables for costs and design.

$136.95 per copy
Available Dec. 2006
Catalog No. 60167

RSMeans Construction Cost Indexes 2007

Who knows what 2007 holds? What materials and labor costs will change unexpectedly? By how much?

- Breakdowns for 316 major cities
- National averages for 30 key cities
- Expanded five major city indexes
- Historical construction cost indexes

294.00 per year (subscription)
73.50 individual quarters
Catalog No. 60147 A,B,C,D

RSMeans Interior Cost Data 2007

Provides you with prices and guidance needed to make accurate interior work estimates. Contains costs on materials, equipment, hardware, custom installations, furnishings, and labor costs... for new and remodel commercial and industrial interior construction, including updated information on office furnishings, and reference information.

Unit Prices Now Updated to MasterFormat 2004!

$136.95 per copy
Available Nov. 2006
Catalog No. 60097

RSMeans Concrete & Masonry Cost Data 2007

Provides you with cost facts for virtually all concrete/masonry estimating needs, from complicated formwork to various sizes and face finishes of brick and block—all in great detail. The comprehensive unit cost section contains more than 7,500 selected entries. Also contains an Assemblies Cost section, and a detailed Reference section that supplements the cost data.

$124.95 per copy
Available Dec. 2006
Catalog No. 60117

Unit Prices Now Updated to MasterFormat 2004!

RSMeans Labor Rates for the Construction Industry 2007

Complete information for estimating labor costs, making comparisons, and negotiating wage rates by trade for over 300 U.S. and Canadian cities. With 46 construction trades listed by local union number in each city, and historical wage rates included for comparison. Each city chart lists the county and is alphabetically arranged with handy visual flip tabs for quick reference.

$296.95 per copy
Available Dec. 2006
Catalog No. 60127

RSMeans Facilities Maintenance & Repair Cost Data 2007

RSMeans Facilities Maintenance & Repair Cost Data gives you a complete system to manage and plan your facility repair and maintenance costs and budget efficiently. Guidelines for auditing a facility and developing an annual maintenance plan. Budgeting is included, along with reference tables on cost and management, and information on frequency and productivity of maintenance operations.

The only nationally recognized source of maintenance and repair costs. Developed in cooperation with the Civil Engineering Research Laboratory (CERL) of the Army Corps of Engineers.

$296.95 per copy
Available Dec. 2006
Catalog No. 60307

Reference Books

For more information
visit RSMeans Web site
at www.rsmeans.com

Home Addition & Renovation Project Costs

This essential home remodeling reference gives you 35 project estimates, and guidance for some of the most popular home renovation and addition projects... from opening up a simple interior wall to adding an entire second story. Each estimate includes a floor plan, color photos, and detailed costs. Use the project estimates as backup for pricing, to check your own estimates, or as a cost reference for preliminary discussion with homeowners.

Includes:

- Case studies—with creative solutions and design ideas.
- Alternate materials costs—so you can match the estimates to the particulars of your projects.
- Location Factors—easy multipliers to adjust the book's costs to your own location.

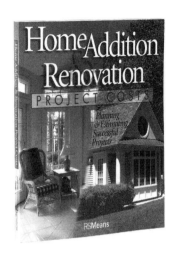

$29.95 per copy
Over 200 pages, illustrated, Softcover
Catalog No. 67349

Kitchen & Bath Project Costs:
Planning & Estimating
Successful Projects

Project estimates for 35 of the most popular kitchen and bath renovations... from replacing a single fixture to whole-room remodels. Each estimate includes:

- All materials needed for the project
- Labor-hours to install (and demolish/remove) each item
- Subcontractor costs for certain trades and services
- An allocation for overhead and profit

PLUS! Takeoff and pricing worksheets—forms you can photocopy or access electronically from the book's Web site; alternate materials—unit costs for different finishes and fixtures; location factors—easy multipliers to adjust the costs to your location; and expert guidance on estimating methods, project design, contracts, marketing, working with homeowners, and tips for each of the estimated projects.

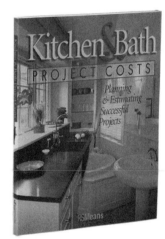

$29.95 per copy
Over 175 pages
Catalog No. 67347

Residential & Light Commercial Construction Standards 2nd Edition
By RSMeans and Contributing Authors

For contractors, subcontractors, owners, developers, architects, engineers, attorneys, and insurance personnel, this book provides authoritative requirements and recommendations compiled from the nation's leading professional associations, industry publications, and building code organizations.

It's an all-in-one reference for establishing a standard for workmanship, quickly resolving disputes, and avoiding defect claims. Includes practical guidance from professionals who are well-known in their respective fields for quality design and construction.

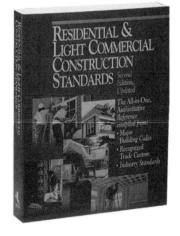

$59.95 per copy
600 pages, illustrated, Softcover
Catalog No. 67322A

For more information
visit RSMeans Web site
at www.rsmeans.com

Reference Books

Value Engineering: Practical Applications

By Alphonse Dell'Isola, PE

For Design, Construction, Maintenance & Operations

A tool for immediate application—for engineers, architects, facility managers, owners, and contractors. Includes: making the case for VE—the management briefing, integrating VE into planning, budgeting, and design, conducting life cycle costing, using VE methodology in design review and consultant selection, case studies, VE workbook, and a life cycle costing program on disk.

$79.95 per copy
Over 450 pages, illustrated, Softcover
Catalog No. 67319

The Building Professional's Guide to Contract Documents

3rd Edition
By Waller S. Poage, AIA, CSI, CVS

A comprehensive reference for owners, design professionals, contractors, and students.

* Structure your documents for maximum efficiency.
* Effectively communicate construction requirements.
* Understand the roles and responsibilities of construction professionals.
* Improve methods of project delivery.

$32.48 per copy, 400 pages
Diagrams and construction forms, Hardcover
Catalog No. 67261A

Cost Planning & Estimating for Facilities Maintenance

In this unique book, a team of facilities management authorities shares their expertise on:

* Evaluating and budgeting maintenance operations
* Maintaining and repairing key building components
* Applying *RSMeans Facilities Maintenance & Repair Cost Data* to your estimating

Covers special maintenance requirements of the ten major building types.

$89.95 per copy
Over 475 pages, Hardcover
Catalog No. 67314

Planning & Managing Interior Projects 2nd Edition

By Carol E. Farren, CFM

Expert guidance on managing renovation & relocation projects.

This book guides you through every step in relocating to a new space or renovating an old one. From initial meeting through design and construction, to post-project administration, it helps you get the most for your company or client. Includes sample forms, spec lists, agreements, drawings, and much more!

$69.95 per copy
200 pages, Softcover
Catalog No. 67245A

Facilities Operations & Engineering Reference

By the Association for Facilities Engineering and RSMeans

An all-in-one technical reference for planning and managing facility projects and solving day-to-day operations problems. Selected as the official Certified Plant Engineer reference, this handbook covers financial analysis, maintenance, HVAC and energy efficiency, and more.

$54.98 per copy
Over 700 pages, illustrated, Hardcover
Catalog No. 67318

Building Security: Strategies & Costs

By David Owen

This comprehensive resource will help you evaluate your facility's security needs, and design and budget for the materials and devices needed to fulfill them.

Includes over 130 pages of RSMeans cost data for installation of security systems and materials, plus a review of more than 50 security devices and construction solutions.

$44.98 per copy
350 pages, illustrated, Hardcover
Catalog No. 67339

Life Cycle Costing for Facilities

By Alphonse Dell'Isola and Dr. Steven Kirk

Guidance for achieving higher quality design and construction projects at lower costs! Cost-cutting efforts often sacrifice quality to yield the cheapest product. Life cycle costing enables building designers and owners to achieve both. The authors of this book show how LCC can work for a variety of projects — from roads to HVAC upgrades to different types of buildings.

$99.95 per copy
450 pages, Hardcover
Catalog No. 67341

Builder's Essentials: Best Business Practices for Builders & Remodelers

An Easy-to-Use Checklist System
By Thomas N. Frisby

A comprehensive guide covering all aspects of running a construction business, with more than 40 user-friendly checklists. Provides expert guidance on: increasing your revenue and keeping more of your profit, planning for long-term growth, keeping good employees, and managing subcontractors.

$29.95 per copy
Over 220 pages, Softcover
Catalog No. 67329

Reference Books

For more information
visit RSMeans Web site
at www.rsmeans.com

Interior Home Improvement
Costs New 9th Edition

Updated estimates for the most popular remodeling and repair projects—from small, do-it-yourself jobs to major renovations and new construction. Includes: Kitchens & Baths; New Living Space from your Attic, Basement, or Garage; New Floors, Paint, and Wallpaper; Tearing Out or Building New Walls; Closets, Stairs, and Fireplaces; New Energy-Saving Improvements, Home Theaters, and More!

$24.95 per copy
250 pages, illustrated, Softcover
Catalog No. 67308E

Exterior Home Improvement
Costs New 9th Edition

Updated estimates for the most popular remodeling and repair projects—from small, do-it-yourself jobs, to major renovations and new construction. Includes: Curb Appeal Projects—Landscaping, Patios, Porches, Driveways, and Walkways; New Windows and Doors; Decks, Greenhouses, and Sunrooms; Room Additions and Garages; Roofing, Siding, and Painting; "Green" Improvements to Save Energy & Water.

$24.95 per copy
Over 275 pages, illustrated, Softcover
Catalog No. 67309E

Builder's Essentials: Plan Reading & Material Takeoff
By Wayne J. DelPico

For Residential and Light Commercial Construction

A valuable tool for understanding plans and specs, and accurately calculating material quantities. Step-by-step instructions and takeoff procedures based on a full set of working drawings.

$35.95 per copy
Over 420 pages, Softcover
Catalog No. 67307

Builder's Essentials: Framing & Rough Carpentry 2nd Edition
By Scot Simpson

Develop and improve your skills with easy-to-follow instructions and illustrations. Learn proven techniques for framing walls, floors, roofs, stairs, doors, and windows. Updated guidance on standards, building codes, safety requirements, and more. Also available in Spanish!

$24.95 per copy
Over 150 pages, Softcover
Catalog No. 67298A
Spanish Catalog No. 67298AS

Concrete Repair and Maintenance Illustrated
By Peter Emmons

Hundreds of illustrations show users how to analyze, repair, clean, and maintain concrete structures for optimal performance and cost effectiveness. From parking garages to roads and bridges to structural concrete, this comprehensive book describes the causes, effects, and remedies for concrete wear and failure. Invaluable for planning jobs, selecting materials, and training employees, this book is a must-have for concrete specialists, general contractors, facility managers, civil and structural engineers, and architects.

$34.98 per copy
300 pages, illustrated, Softcover
Catalog No. 67146

Means Unit Price Estimating Methods
New 3rd Edition

This new edition includes up-to-date cost data and estimating examples, updated to reflect changes to the CSI numbering system and new features of RSMeans cost data. It describes the most productive, universally accepted ways to estimate, and uses checklists and forms to illustrate shortcuts and timesavers. A model estimate demonstrates procedures. A new chapter explores computer estimating alternatives.

$29.98 per copy
Over 350 pages, illustrated, Hardcover
Catalog No. 67303A

Total Productive Facilities Management
By Richard W. Sievert, Jr.

Today, facilities are viewed as strategic resources... elevating the facility manager to the role of asset manager supporting the organization's overall business goals. Now, Richard Sievert Jr., in this well-articulated guidebook, sets forth a new operational standard for the facility manager's emerging role... a comprehensive program for managing facilities as a true profit center.

$29.98 per copy
275 pages, Softcover
Catalog No. 67321

Means Environmental Remediation Estimating Methods 2nd Edition
By Richard R. Rast

Guidelines for estimating 50 standard remediation technologies. Use it to prepare preliminary budgets, develop estimates, compare costs and solutions, estimate liability, review quotes, negotiate settlements.

$49.98 per copy
Over 750 pages, illustrated, Hardcover
Catalog No. 64777A

For more information
visit RSMeans Web site
at www.rsmeans.com

Reference Books

Means Illustrated Construction Dictionary Condensed, 2nd Edition

Recognized in the industry as the best resource of its kind.

This essential tool has been further enhanced with updates to existing terms and the addition of hundreds of new terms and illustrations—in keeping with recent developments. For contractors, architects, insurance and real estate personnel, homeowners, and anyone who needs quick, clear definitions for construction terms.

$59.95 per copy
Over 500 pages, Softcover
Catalog No. 67282A

Means Repair & Remodeling Estimating New 4th Edition

By Edward B. Wetherill & RSMeans

This important reference focuses on the unique problems of estimating renovations of existing structures, and helps you determine the true costs of remodeling through careful evaluation of architectural details and a site visit.

New section on disaster restoration costs.

$69.95 per copy
Over 450 pages, illustrated, Hardcover
Catalog No. 67265B

Facilities Planning & Relocation

New, lower price and user-friendly format.

By David D. Owen

A complete system for planning space needs and managing relocations. Includes step-by-step manual, over 50 forms, and extensive reference section on materials and furnishings.

$89.95 per copy
Over 450 pages, Softcover
Catalog No. 67301

Means Square Foot & Assemblies Estimating Methods 3rd Edition

Develop realistic square foot and assemblies costs for budgeting and construction funding. The new edition features updated guidance on square foot and assemblies estimating using UNIFORMAT II. An essential reference for anyone who performs conceptual estimates.

$34.98 per copy
Over 300 pages, illustrated, Hardcover
Catalog No. 67145B

Means Electrical Estimating Methods 3rd Edition

Expanded edition includes sample estimates and cost information in keeping with the latest version of the CSI MasterFormat and UNIFORMAT II. Complete coverage of fiber optic and uninterruptible power supply electrical systems, broken down by components, and explained in detail. Includes a new chapter on computerized estimating methods. A practical companion to *RSMeans Electrical Cost Data*.

$64.95 per copy
Over 325 pages, Hardcover
Catalog No. 67230B

Means Mechanical Estimating Methods 3rd Edition

This guide assists you in making a review of plans, specs, and bid packages, with suggestions for takeoff procedures, listings, substitutions, and pre-bid scheduling. Includes suggestions for budgeting labor and equipment usage. Compares materials and construction methods to allow you to select the best option.

$64.95 per copy
Over 350 pages, illustrated, Hardcover
Catalog No. 67294A

Means ADA Compliance Pricing Guide New Second Edition

By Adaptive Environments and RSMeans

Completely updated and revised to the new 2004 *Americans with Disabilities Act Accessibility Guidelines*, this book features more than 70 of the most commonly needed modifications for ADA compliance. Projects range from installing ramps and walkways, widening doorways and entryways, and installing and refitting elevators, to relocating light switches and signage.

$79.95 per copy
Over 350 pages, illustrated, Softcover
Catalog No. 67310A

Project Scheduling & Management for Construction New 3rd Edition

By David R. Pierce, Jr.

A comprehensive yet easy-to-follow guide to construction project scheduling and control—from vital project management principles through the latest scheduling, tracking, and controlling techniques. The author is a leading authority on scheduling, with years of field and teaching experience at leading academic institutions. Spend a few hours with this book and come away with a solid understanding of this essential management topic.

$64.95 per copy
Over 300 pages, illustrated, Hardcover
Catalog No. 67247B

Reference Books

For more information
visit RSMeans Web site
at www.rsmeans.com

The Practice of Cost Segregation Analysis

by Bruce A. Desrosiers and Wayne J. DelPico

This expert guide walks you through the practice of cost segregation analysis, which enables property owners to defer taxes and benefit from "accelerated cost recovery" through depreciation deductions on assets that are properly identified and classified.

With a glossary of terms, sample cost segregation estimates for various building types, key information resources, and updates via a dedicated Web site, this book is a critical resource for anyone involved in cost segregation analysis.

$99.95 per copy
Over 225 pages
Catalog No. 67345

Preventive Maintenance for Multi-Family Housing

by John C. Maciha

Prepared by one of the nation's leading experts on multi-family housing.

This complete PM system for apartment and condominium communities features expert guidance, checklists for buildings and grounds maintenance tasks and their frequencies, a reusable wall chart to track maintenance, and a dedicated Web site featuring customizable electronic forms. A must-have for anyone involved with multi-family housing maintenance and upkeep.

$89.95 per copy
225 pages
Catalog No. 67346

Means Landscape Estimating Methods

4th Edition

By Sylvia H. Fee

This revised edition offers expert guidance for preparing accurate estimates for new landscape construction and grounds maintenance. Includes a complete project estimate featuring the latest equipment and methods, and chapters on Life Cycle Costing and Landscape Maintenance Estimating.

$62.95 per copy
Over 300 pages, illustrated, Hardcover
Catalog No. 67295B

Job Order Contracting
Expediting Construction Project Delivery

by Allen Henderson

Expert guidance to help you implement JOC—fast becoming the preferred project delivery method for repair and renovation, minor new construction, and maintenance projects in the public sector and in many states and municipalities. The author, a leading JOC expert and practitioner, shows how to:

• Establish a JOC program

• Evaluate proposals and award contracts

• Handle general requirements and estimating

• Partner for maximum benefits

$89.95 per copy
192 pages, illustrated, Hardcover
Catalog No. 67348

Builder's Essentials: Estimating Building Costs

For the Residential & Light Commercial Contractor

By Wayne J. DelPico

Step-by-step estimating methods for residential and light commercial contractors. Includes a detailed look at every construction specialty—explaining all the components, takeoff units, and labor needed for well-organized, complete estimates. Covers correctly interpreting plans and specifications, and developing accurate and complete labor and material costs.

$29.95 per copy
Over 400 pages, illustrated, Softcover
Catalog No. 67343

Building & Renovating Schools

This all-inclusive guide covers every step of the school construction process—from initial planning, needs assessment, and design, right through moving into the new facility. A must-have resource for anyone concerned with new school construction or renovation. With square foot cost models for elementary, middle, and high school facilities, and real-life case studies of recently completed school projects.

The contributors to this book—architects, construction project managers, contractors, and estimators who specialize in school construction—provide start-to-finish, expert guidance on the process.

$99.95 per copy
Over 425 pages, Hardcover
Catalog No. 67342

r more information
sit RSMeans Web site
www.rsmeans.com

Reference Books

Historic Preservation: Project Planning & Estimating

by Swanke Hayden Connell Architects

xpert guidance on managing historic restoration, habilitation, and preservation building projects and termining and controlling their costs. Includes:

How to determine whether a structure qualifies as historic
Where to obtain funding and other assistance
How to evaluate and repair more than 75 historic building materials

9.98 per copy
ver 675 pages, Hardcover
atalog No. 67323

Designing & Building with the IBC 2nd Edition

by Rolf Jensen & Associates, Inc.

is updated, comprehensive guide helps building ofessionals make the transition to the 2003 International uilding Code®. Includes a side-by-side code comparison the IBC 2003 to the IBC 2000 and the three primary odel codes, a quick-find index, and professional code mmentary. With illustrations, abbreviations key, and an tensive Resource section.

9.95 per copy
ver 875 pages, Softcover
atalog No. 67328A

Builder's Essentials: Advanced Framing Methods

by Scot Simpson

highly illustrated, "framer-friendly" approach to advanced aming elements. Provides expert, but easy to interpret, struction for laying out and framing complex walls, roofs, d stairs, and special requirements for earthquake and rricane protection. Also helps bring framers up to date on e latest building code changes, and provides tips on the d framer's role and responsibilities, how to prepare for a , and how to get the crew started.

4.95 per copy
0 pages, illustrated, Softcover
atalog No. 67330

Preventive Maintenance Guidelines for School Facilities

by John C. Maciha

complete PM program for K-12 schools that ensures stained security, safety, property integrity, user tisfaction, and reasonable ongoing expenditures.

cludes schedules for weekly, monthly, semiannual, and nual maintenance in hard copy and electronic format.

149.95 per copy
ver 225 pages, Hardcover
atalog No. 67326

Means Illustrated Construction Dictionary

Unabridged 3rd Edition, with CD-ROM

Long regarded as the industry's finest, *Means Illustrated Construction Dictionary* is now even better. With the addition of over 1,000 new terms and hundreds of new illustrations, it is the clear choice for the most comprehensive and current information. The companion CD-ROM that comes with this new edition adds many extra features: larger graphics, expanded definitions, and links to both CSI MasterFormat numbers and product information.

$99.95 per copy
Over 790 pages, illustrated, Hardcover
Catalog No. 67292A

Means Plumbing Estimating Methods 3rd Edition

By Joseph Galeno and Sheldon Greene

Updated and revised! This practical guide walks you through a plumbing estimate, from basic materials and installation methods through change order analysis. *Plumbing Estimating Methods* covers residential, commercial, industrial, and medical systems, and features sample takeoff and estimate forms and detailed illustrations of systems and components.

$29.98 per copy
330+ pages, Softcover
Catalog No. 67283B

Means Estimating Handbook

2nd Edition

Updated Second Edition answers virtually any estimating technical question—all organized by CSI MasterFormat. This comprehensive reference covers the full spectrum of technical data required to estimate construction costs. The book includes information on sizing, productivity, equipment requirements, code-mandated specifications, design standards, and engineering factors.

$99.95 per copy
Over 900 pages, Hardcover
Catalog No. 67276A

Preventive Maintenance for Higher Education Facilities

By Applied Management Engineering, Inc.

An easy-to-use system to help facilities professionals establish the value of PM, and to develop and budget for an appropriate PM program for their college or university. Features interactive campus building models typical of those found in different-sized higher education facilities, and PM checklists linked to each piece of equipment or system in hard copy and electronic format.

$149.95 per copy
150 pages, Hardcover
Catalog No. 67337

For more information
visit RSMeans Web site
at www.rsmeans.com

Seminars

Means CostWorks® Training

This one-day seminar has been designed with the intention of assisting both new and existing users to become more familiar with the *Means CostWorks* program. The class is broken into two unique sections: (1) A one-half day presentation on the function of each icon; and each student will be shown how to use the software to develop a cost estimate. (2) Hands-on estimating exercises that will ensure that each student thoroughly understands how to use *CostWorks*. You must bring your own laptop computer to this course.

Means CostWorks Benefits/Features:
- Estimate in your own spreadsheet format
- Power of RSMeans National Database
- Database automatically regionalized
- Save time with keyword searches
- Save time by establishing common estimate items in "Bookmark" files
- Customize your spreadsheet template
- Hot Key to Product Manufacturers' listings and specs
- Merge capability for networking environments
- View crews and assembly components
- AutoSave capability
- Enhanced sorting capability

Unit Price Estimating

This interactive two-day seminar teaches attendees how to interpret project information and process it into final, detailed estimates with the greatest accuracy level.

The single most important credential an estimator can take to the job is the ability to visualize construction in the mind's eye, and thereby estimate accurately.

Some Of What You'll Learn:
- Interpreting the design in terms of cost
- The most detailed, time-tested methodology for accurate "pricing"
- Key cost drivers—material, labor, equipment, staging, and subcontracts
- Understanding direct and indirect costs for accurate job cost accounting and change order management

Who Should Attend: Corporate and government estimators and purchasers, architects, engineers... and others needing to produce accurate project estimates.

Square Foot and Assemblies Estimating

This two-day course teaches attendees how to quickly deliver accurate square foot estimates using limited budget and design information.

Some Of What You'll Learn:
- How square foot costing gets the estimate done faster
- Taking advantage of a "systems" or "assemblies" format
- The RSMeans "building assemblies/square foot cost approach"
- How to create a very reliable preliminary and systems estimate using bare-bones design information

Who Should Attend: Facilities managers, facilities engineers, estimators, planners, developers, construction finance professionals... and others needing to make quick, accurate construction cost estimates at commercial, government, educational, and medical facilities.

Repair and Remodeling Estimating

This two-day seminar emphasizes all the underlying considerations unique to repair/remodeling estimating and presents the correct methods for generating accurate, reliable R&R project costs using the unit price and assemblies methods.

Some Of What You'll Learn:
- Estimating considerations—like labor-hours, building code compliance, working within existing structures, purchasing materials in smaller quantities, unforeseen deficiencies
- Identifying problems and providing solutions to estimating building alterations
- Rules for factoring in minimum labor costs, accurate productivity estimates, and allowances for project contingencies
- R&R estimating examples calculated using unit price and assemblies data

Who Should Attend: Facilities managers, plant engineers, architects, contractors, estimators, builders... and others who are concerned with the proper preparation and/or evaluation of repair and remodeling estimates.

Mechanical and Electrical Estimating

This two-day course teaches attendees how to prepare more accurate and complete mechanical/electrical estimates, avoiding the pitfalls of omission and double-counting, while understanding the composition and rationale within the RSMeans Mechanical/Electrical database.

Some Of What You'll Learn:
- The unique way mechanical and electrical systems are interrelated
- M&E estimates–conceptual, planning, budgeting, and bidding stages
- Order of magnitude, square foot, assemblies, and unit price estimating
- Comparative cost analysis of equipment and design alternatives

Who Should Attend: Architects, engineers, facilities managers, mechanical and electrical contractors... and others needing a highly reliable method for developing, understanding, and evaluating mechanical and electrical contracts.

Plan Reading and Material Takeoff

This two-day program teaches attendees to read and understand construction documents and to use them in the preparation of material takeoffs.

Some of What You'll Learn:
- Skills necessary to read and understand typical contract documents—blueprints and specifications
- Details and symbols used by architects and engineers
- Construction specifications' importance in conjunction with blueprints
- Accurate takeoff of construction materials and industry-accepted takeoff methods

Who Should Attend: Facilities managers, construction supervisors, office managers... and others responsible for the execution and administration of a construction project, including government, medical, commercial, educational, or retail facilities.

Facilities Maintenance and Repair Estimating

This two-day course teaches attendees how to plan, budget, and estimate the cost of ongoing and preventive maintenance and repair for existing buildings and grounds.

Some Of What You'll Learn:
- The most financially favorable maintenance, repair, and replacement scheduling and estimating
- Auditing and value engineering facilities
- Preventive planning and facilities upgrading
- Determining both in-house and contract-out service costs
- Annual, asset-protecting M&R plan

Who Should Attend: Facility managers, maintenance supervisors, buildings and grounds superintendents, plant managers, planners, estimators... and others involved in facilities planning and budgeting.

Scheduling and Project Management

This two-day course teaches attendees the most current and proven scheduling and management techniques needed to bring projects in on time and on budget.

Some Of What You'll Learn:
- Crucial phases of planning and scheduling
- How to establish project priorities and develop realistic schedules and management techniques
- Critical Path and Precedence Methods
- Special emphasis on cost control

Who Should Attend: Construction project managers, supervisors, engineers, estimators, contractors... and others who want to improve their project planning, scheduling, and management skills.

Assessing Scope of Work for Facility Construction Estimating

This two-day course is a practical training program that addresses the vital importance of understanding the SCOPE of projects in order to produce accurate cost estimates in a facility repair and remodeling environment.

Some Of What You'll Learn:
- Discussions on site visits, plans/specs, record drawings of facilities, and site-specific lists
- Review of CSI divisions, including means, methods, materials, and the challenges of scoping each topic
- Exercises in SCOPE identification and SCOPE writing for accurate estimating of projects
- Hands-on exercises that require SCOPE, take-off, and pricing

Who Should Attend: Corporate and government estimators, planners, facility managers... and others needing to produce accurate project estimates.

Seminars

2007 RSMeans Seminar Schedule

Location	Dates
Las Vegas, NV	March
Washington, DC	April
Phoenix, AZ	April
Denver, CO	May
San Francisco, CA	June
Philadelphia, PA	June
Washington, DC	September
Dallas, TX	September
Las Vegas, NV	October
Orlando, FL	November
Atlantic City, NJ	November
San Diego, CA	December

Note: Call for exact dates and details.

Registration Information

Register Early... Save up to $100! Register 30 days before the start date of seminar and save $100 off your total fee. *Note: This discount can be applied only once per order. It cannot be applied to team discount registrations or any other special offer.*

How to Register Register by phone today! RSMeans' toll-free number for making reservations is: **1-800-334-3509.**

Individual Seminar Registration Fee $935. *Means CostWorks®* Training Registration Fee $375. To register by mail, complete the registration form and return with your full fee to: Seminar Division, Reed Construction Data, RSMeans Seminars, 63 Smiths Lane, Kingston, MA 02364.

Federal Government Pricing All federal government employees save 25% of regular seminar price. Other promotional discounts cannot be combined with federal Government discount.

Team Discount Program Two to four seminar registrations, call for pricing: 800-334-3509, Ext. 5115

Multiple Course Discounts When signing up for two or more courses, call for pricing.

Refund Policy Cancellations will be accepted up to ten days prior to the seminar start. There are no refunds for cancellations received later than ten working days prior to the first day of the seminar. A $150 processing fee will be applied for all cancellations. Written notice of cancellation is required. Substitutions can be made at any time before the session starts. **No-shows are subject to the full seminar fee.**

AACE Approved Courses Many seminars described and offered here have been approved for 14 hours (1.4 recertification credits) of credit by the AACE International Certification Board toward meeting the continuing education requirements for recertification as a Certified Cost Engineer/Certified Cost Consultant.

AIA Continuing Education We are registered with the AIA Continuing Education System (AIA/CES) and are committed to developing quality learning activities in accordance with the CES criteria. Many seminars meet the AIA/CES criteria for Quality Level 2. AIA members may receive (14) learning units (LUs) for each two-day RSMeans course.

NASBA CPE Sponsor Credits We are part of the National Registry of CPE Sponsors. Attendees may be eligible for (16) CPE credits.

Daily Course Schedule The first day of each seminar session begins at 8:30 A.M. and ends at 4:30 P.M. The second day is 8:00 A.M.–4:00 P.M. Participants are urged to bring a hand-held calculator since many actual problems will be worked out in each session.

Continental Breakfast Your registration includes the cost of a continental breakfast, a morning coffee break, and an afternoon break. These informal segments will allow you to discuss topics of mutual interest with other members of the seminar. (You are free to make your own lunch and dinner arrangements.)

Hotel/Transportation Arrangements RSMeans has arranged to hold a block of rooms at most host hotels. To take advantage of special group rates when making your reservation, be sure to mention that you are attending the RSMeans Seminar. You are, of course, free to stay at the lodging place of your choice. (**Hotel reservations and transportation arrangements should be made directly by seminar attendees.**)

Important Class sizes are limited, so please register as soon as possible.

Note: Pricing subject to change.

Registration Form   Call 1-800-334-3509 to register or FAX 1-800-632-6732. Visit our Web site: www.rsmeans.com

Please register the following people for the RSMeans Construction Seminars as shown here. We understand that we must make our own hotel reservations if overnight stays are necessary.

Full payment of $_____ enclosed.

☐ Bill me

Name of Registrant(s)
(to appear on certificate of completion) _____

_____

P.O. #: _____

GOVERNMENT AGENCIES MUST SUPPLY PURCHASE ORDER NUMBER OR TRAINING FORM.

Firm Name _____

Address _____

City/State/Zip _____

Telephone No. _____ Fax No. _____

E-mail Address _____

Charge our registration(s) to: ☐ MasterCard ☐ VISA ☐ American Express ☐ Discover

Account No. _____ Exp. Date _____

Cardholder's Signature _____

Seminar Name	City	Dates

Please mail check to: Seminar Division, Reed Construction Data, RSMeans Seminars, 63 Smiths Lane, P.O. Box 800, Kingston, MA 02364 USA

MeansData™

CONSTRUCTION COSTS FOR SOFTWARE APPLICATIONS
Your construction estimating software is only as good as your cost data.

A proven construction cost database is a mandatory part of any estimating package. The following list of software providers can offer you MeansData™ as an added feature for their estimating systems. See the table below for what types of products and services they offer (match their numbers). Visit online at **www.rsmeans.com/demosource/** for more information and free demos. Or call their numbers listed below.

1. **3D International**
   713-871-7000
   venegas@3di.com

2. **4Clicks-Solutions, LLC**
   719-574-7721
   mbrown@4clicks-solutions.com

3. **Aepco, Inc.**
   301-670-4642
   blueworks@aepco.com

4. **Applied Flow Technology**
   800-589-4943
   info@aft.com

5. **ArenaSoft Estimating**
   888-370-8806
   info@arenasoft.com

6. **Ares Corporation**
   925-299-6700
   sales@arescorporation.com

7. **Beck Technology**
   214-303-6293
   stewartcarroll@beckgroup.com

8. **BSD - Building Systems Design, Inc.**
   888-273-7638
   bsd@bsdsoftlink.com

9. **CMS - Computerized Micro Solutions**
   800-255-7407
   cms@proest.com

10. **Corecon Technologies, Inc.**
    714-895-7222
    sales@corecon.com

11. **CorVet Systems**
    301-622-9069
    sales@corvetsys.com

12. **Earth Tech**
    303-771-3103
    kyle.knudson@earthtech.com

13. **Estimating Systems, Inc.**
    800-967-8572
    esipulsar@adelphia.net

14. **HCSS**
    800-683-3196
    info@hcss.com

15. **MC² - Management Computer**
    800-225-5622
    vkeys@mc2-ice.com

16. **Maximus Asset Solutions**
    800-659-9001
    assetsolutions@maximus.com

17. **Sage Timberline Office**
    800-628-6583
    productinfo.timberline@sage.com

18. **Shaw Beneco Enterprises, Inc.**
    877-719-4748
    inquire@beneco.com

19. **US Cost, Inc.**
    800-372-4003
    sales@uscost.com

20. **Vanderweil Facility Advisors**
    617-451-5100
    info@VFA.com

21. **WinEstimator, Inc.**
    800-950-2374
    sales@winest.com

TYPE	1	2	3	4	5	6	7	8	9	10	11	12	13	14	15	16	17	18	19	20	21
BID					•			•	•		•			•	•		•				•
Estimating		•			•	•	•	•	•	•	•	•	•	•	•		•	•	•		•
DOC/JOC/SABER		•			•			•			•		•			•	•	•			•
ID/IQ		•									•		•			•		•			•
Asset Mgmt.																•				•	•
Facility Mgmt.	•		•													•				•	
Project Mgmt.	•	•					•		•	•				•			•	•			
TAKE-OFF					•				•	•	•	•		•	•		•		•		
EARTHWORK										•		•		•	•						
Pipe Flow				•										•							
HVAC/Plumbing					•				•		•										
Roofing					•				•		•										
Design	•				•		•									•	•				•
Other Offers/Links:																					
Accounting/HR		•			•										•		•	•			
Scheduling					•	•										•			•		•
CAD							•									•					
PDA																•			•		
Lt. Versions		•							•							•		•			
Consulting	•	•			•		•				•	•				•	•		•	•	•
Training		•			•		•	•	•	•	•	•	•	•	•	•	•	•	•	•	•

Qualified re-seller applications now being accepted. Call Carol Polio, Ext. 5107.

FOR MORE INFORMATION
CALL 1-800-448-8182, EXT. 5107 OR FAX 1-800-632-6732

For more information
visit RSMeans Web site
t www.rsmeans.com

New Titles

Understanding & Negotiating Construction Contracts

By Kit Werremeyer

Take advantage of the author's 30 years' experience in small-to-large (including international) construction projects. Learn how to identify, understand, and evaluate high risk terms and conditions typically found in all construction contracts—then negotiate to lower or eliminate the risk, improve terms of payment, and reduce exposure to claims and disputes. The author avoids "legalese" and gives real-life examples from actual projects.

$69.95 per copy
300 pages, Softcover
Catalog No. 67350

Green Building: Project Planning & Cost Estimating, 2nd Edition

This new edition has been completely updated with the latest in green building technologies, design concepts, standards, and costs. Now includes a 2007 Green Building *CostWorks* CD with more than 300 green building assemblies and over 5,000 unit price line items for sustainable building. The new edition is also full-color with all new case studies—plus a new chapter on deconstruction, a key aspect of green building.

$129.95 per copy
350 pages, Softcover
Catalog No. 67338A

How to Estimate with Means Data & CostWorks

New 3rd Edition

By RSMeans and Saleh A. Mubarak, Ph.D.

New 3rd Edition—fully updated with new chapters, plus new CD with updated *CostWorks* cost data and MasterFormat organization. Includes all major construction items—with more than 300 exercises and two sets of plans that show how to estimate for a broad range of construction items and systems—including general conditions and equipment costs.

$59.95 per copy
272 pages, Softcover
Includes CostWorks CD
Catalog No. 67324B

Means Spanish/English Construction Dictionary

2nd Edition

By RSMeans and the International Code Council

This expanded edition features thousands of the most common words and useful phrases in the construction industry with easy-to-follow pronunciations in both Spanish and English. Over 800 new terms, phrases, and illustrations have been added. It also features a new stand-alone "Safety & Emergencies" section, with colored pages for quick access. Unique to this dictionary are the systems illustrations showing the relationship of components in the most common building systems for all major trades.

$23.95 per copy
Over 400 pages
Catalog No. 67327A

Construction Business Management

By Nick Ganaway

Only 43% of construction firms stay in business after four years. Make sure your company thrives with valuable guidance from a pro with 25 years of success as a commercial contractor. Find out what it takes to build all aspects of a business that is profitable, enjoyable, and enduring. With a bonus chapter on retail construction.

$49.95 per copy
200 pages, Softcover
Catalog No. 67352

The Homeowner's Guide to Mold

Expert guidance to protect your health and your home.

Mold, whether caused by leaks, humidity or flooding, is a real health and financial issue—for homeowners and contractors. This full-color book explains:

- Construction and maintenance practices to prevent mold
- How to inspect for and remove mold
- Mold remediation procedures and costs
- What to do after a flood
- How to deal with insurance companies if you're thinking of submitting a mold damages claim

$21.95 per copy
144 pages, Softcover
Catalog No. 67344

2007 Order Form

Qty.	Book No.	COST ESTIMATING BOOKS	Unit Price	Total
	60067	Assemblies Cost Data 2007	$223.95	
	60017	Building Construction Cost Data 2007	136.95	
	61017	Building Const. Cost Data–Looseleaf Ed. 2007	169.95	
	60717	Building Const. Cost Data–Spanish 2007	136.95	
	60227	Building Const. Cost Data–Western Ed. 2007	136.95	
	60117	Concrete & Masonry Cost Data 2007	124.95	
	50147	Construction Cost Indexes 2007 (subscription)	294.00	
	60147A	Construction Cost Index–January 2007	73.50	
	60147B	Construction Cost Index–April 2007	73.50	
	60147C	Construction Cost Index–July 2007	73.50	
	60147D	Construction Cost Index–October 2007	73.50	
	60347	Contr. Pricing Guide: Resid. R & R Costs 2007	39.95	
	60337	Contr. Pricing Guide: Resid. Detailed 2007	39.95	
	60327	Contr. Pricing Guide: Resid. Sq. Ft. 2007	39.95	
	60237	Electrical Change Order Cost Data 2007	136.95	
	60037	Electrical Cost Data 2007	136.95	
	60207	Facilities Construction Cost Data 2007	323.95	
	60307	Facilities Maintenance & Repair Cost Data 2007	296.95	
	60167	Heavy Construction Cost Data 2007	136.95	
	60097	Interior Cost Data 2007	136.95	
	60127	Labor Rates for the Const. Industry 2007	296.95	
	60187	Light Commercial Cost Data 2007	116.95	
	60027	Mechanical Cost Data 2007	136.95	
	63017	Metric Construction Cost Data 2007	162.95	
	60157	Open Shop Building Const. Cost Data 2007	136.95	
	60217	Plumbing Cost Data 2007	136.95	
	60047	Repair and Remodeling Cost Data 2007	116.95	
	60177	Residential Cost Data 2007	116.95	
	60287	Site Work & Landscape Cost Data 2007	136.95	
	60057	Square Foot Costs 2007	148.95	
	62017	Yardsticks for Costing (2007)	136.95	
	62016	Yardsticks for Costing (2006)	126.95	
		REFERENCE BOOKS		
	67310A	ADA Compliance Pricing Guide, 2nd Ed.	79.95	
	67330	Bldrs Essentials: Adv. Framing Methods	24.95	
	67329	Bldrs Essentials: Best Bus. Practices for Bldrs	29.95	
	67298A	Bldrs Essentials: Framing/Carpentry 2nd Ed.	24.95	
	67298AS	Bldrs Essentials: Framing/Carpentry Spanish	24.95	
	67307	Bldrs Essentials: Plan Reading & Takeoff	35.95	
	67342	Building & Renovating Schools	99.95	
	67261A	Bldg. Prof. Guide to Contract Documents, 3rd Ed.	32.48	
	67339	Building Security: Strategies & Costs	44.98	
	67146	Concrete Repair & Maintenance Illustrated	34.98	
	67352	Construction Business Management	49.95	
	67314	Cost Planning & Est. for Facil. Maint.	89.95	
	67328A	Designing & Building with the IBC, 2nd Ed.	99.95	
	67230B	Electrical Estimating Methods, 3rd Ed.	64.95	
	64777A	Environmental Remediation Est. Methods, 2nd Ed.	49.98	
	67343	Estimating Bldg. Costs for Resi. & Lt. Comm.	29.95	

Qty.	Book No.	REFERENCE BOOKS (Cont.)	Unit Price	Total
	67276A	Estimating Handbook, 2nd Ed.	$99.95	
	67318	Facilities Operations & Engineering Reference	54.98	
	67301	Facilities Planning & Relocation	89.95	
	67338A	Green Building: Proj. Planning & Cost Est., 2nd Ed.	129.95	
	67323	Historic Preservation: Proj. Planning & Est.	49.98	
	67349	Home Addition & Renovation Project Costs	29.95	
	67308E	Home Improvement Costs–Int. Projects, 9th Ed.	24.95	
	67309E	Home Improvement Costs–Ext. Projects, 9th Ed.	24.95	
	67344	Homeowner's Guide to Mold	21.95	
	67324B	How to Est.w/Means Data & CostWorks, 3rd Ed.	59.95	
	67282A	Illustrated Const. Dictionary, Condensed, 2nd Ed.	59.95	
	67292A	Illustrated Const. Dictionary, w/CD-ROM, 3rd Ed.	99.95	
	67348	Job Order Contracting	89.95	
	67347	Kitchen & Bath Project Costs	29.95	
	67295B	Landscape Estimating Methods, 4th Ed.	62.95	
	67341	Life Cycle Costing for Facilities	99.95	
	67294A	Mechanical Estimating Methods, 3rd Ed.	64.95	
	67245A	Planning & Managing Interior Projects, 2nd Ed.	69.95	
	67283B	Plumbing Estimating Methods, 3rd Ed.	29.98	
	67345	Practice of Cost Segregation Analysis	99.95	
	67337	Preventive Maint. for Higher Education Facilities	149.95	
	67346	Preventive Maint. for Multi-Family Housing	89.95	
	67326	Preventive Maint. Guidelines for School Facil.	149.95	
	67247B	Project Scheduling & Management for Constr. 3rd Ed.	64.95	
	67265B	Repair & Remodeling Estimating Methods, 4th Ed.	69.95	
	67322A	Resi. & Light Commercial Const. Stds., 2nd Ed.	59.95	
	67327A	Spanish/English Construction Dictionary, 2nd Ed.	23.95	
	67145B	Sq. Ft. & Assem. Estimating Methods, 3rd Ed.	34.98	
	67321	Total Productive Facilities Management	29.98	
	67350	Understanding and Negotiating Const. Contracts	69.95	
	67303A	Unit Price Estimating Methods, 3rd Ed.	29.98	
	67319	Value Engineering: Practical Applications	79.95	

MA residents add 5% state sales tax	
Shipping & Handling**	
Total (U.S. Funds)*	

Prices are subject to change and are for U.S. delivery only. *Canadian customers may call for current prices. **Shipping & handling charges: Add 7% of total order for check and credit card payments. Add 9% of total order for invoiced orders.

Send Order To: **ADDV-1000**

Name (Please Print) _____

Company _____

☐ **Company**

☐ **Home   Address** _____

_____

City/State/Zip _____

Phone # _____ P.O. # _____

(Must accompany all orders being billed)

Mail To: RSMeans, P.O. Box 800, Kingston, MA 02364-0800

Reed Construction Data, Inc.

Reed Construction Data, Inc. is a leading worldwide provider of total construction information solutions. The company's portfolio of information products and services is designed specifically to help construction industry professionals advance their businesses with timely, accurate, and actionable project, product, and cost data. Each of these groups offers a variety of innovative products and services created for the full spectrum of design, construction, distribution, and manufacturing professionals. Reed Construction Data is a division of Reed Business Information, a member of the Reed Elsevier plc group of companies.

Cost Information

RSMeans, the undisputed market leader and authority on construction costs, publishes current cost and estimating information in annual cost books, on *Means CostWorks®* CD, and on the Web. RSMeans furnishes the construction industry with a rich library of corresponding reference books and a series of professional seminars that are designed to sharpen personal skills and maximize the effective use of cost estimating and management tools. RSMeans also provides construction cost consulting for owners, manufacturers, designers, and contractors.

Project Data

Reed Construction Data maintains a significant role in the overall construction process by facilitating the assembly of public and private project data and delivering this information to contractors, distributors, and building product manufacturers in a secure, accurate, and timely manner. Acting on behalf of architects, engineers, owner/developers, and government agencies, Reed Construction Data is the construction community's premier resource for project leads and bid documents. Our wide variety of products and services provides the most up-to-date, enhanced details on many of the country's largest public sector, commercial, industrial, and multi-family residential projects.

Reed Bulletin and Reed CONNECT™ provide project leads and project data through all stages of construction. Customers are supplied industry data through leads, project reports, contact lists, plans, specifications, and addenda—either online, via e-mail, or in paper format.

Building Product Information

The Reed First Source suite of products is the only integrated building product information system offered to the commercial construction industry for searching, selecting, and specifying building products. These online and print resources include the *First Source* catalog, SPEC-DATA™, MANU-SPEC ™, First Source CAD, and manufacturer's catalogs. Written by industry professionals and organized using CSI MasterFormat 2004 criteria, the design and construction community uses this information to make better, more informed design decisions.

Research & Analytics

Reed Construction Data's forecasting tools cover most aspects of the construction business. Our vast network of resources makes us uniquely qualified to give you the information you need to keep your business profitable.

Associated Construction Publications (ACP)

Reed Construction Data's regional construction magazines cover the nation through a network of 14 regional magazines. Serving the construction market for more than 100 years, our magazines are a trusted source of news and information in the local and national construction communities.

International

Reed Construction Data Canada is the "voice of construction" in Canada, and leading provider of project information and construction market intelligence. Products and services include: *Building Reports*, Canadian ICI project leads tracking construction projects through life cycles—from concept to construction. *Building Reports* can be customized and delivered through CONNECT™, RCD's web-based sales and contact management system, or by fax or email; *Daily Commercial News*, Ontario's leading construction industry daily newspaper, contains ICI project leads, Building Reports, Bidders' Register, news and trends, construction tenders, certificates of substantial performance, and CanaData Economic Snapshot; *Journal of Commerce*, Alberta and British Columbia's twice-weekly construction industry newspaper, is a one-stop source for private and public projects, tenders and expressions of interest, top industry news and construction trends, and more; *CanaData*, a variety of products tracking market performance, economic analysis, forecasts, and trends in the construction market; *Buildcore*, organized by MasterFormat™, is Canada's most comprehensive listing of commercial building products and manufacturers active and available in Canada—in print and online at www.buildcore.com.

Reed Business Information – Scandinavia, with offices in Denmark, Norway, Finland, and Sweden, is the construction industry's source for project information in the Nordic region.

Reed Construction Data—Australia (formerly Cordell Building Information Services) is Australia's leading provider and respected industry authority on building and construction information.

For more information, please visit our website at www. reedconstructiondata.com

Reed Construction Data, Inc.
30 Technology Parkway South
Norcross, GA 30092-2912
(800) 322-6996
(800) 895-8661 (fax)
Email: info@reedbusiness.com